Edward Klein

Elements of Histology

Edward Klein

Elements of Histology

ISBN/EAN: 9783743343962

Manufactured in Europe, USA, Canada, Australia, Japa

Cover: Foto ©ninafisch / pixelio.de

Manufactured and distributed by brebook publishing software (www.brebook.com)

Edward Klein

Elements of Histology

ELEMENTS

OF

HISTOLOGY.

E. KLEIN, M.D., F.R.S.,

JOINT-LECTURER ON GENERAL ANATOMY AND PHYSIOLOGY IN THE
MEDICAL SCHOOL OF ST. BARTHOLOMEW'S HOSPITAL,
LONDON.

ILLUSTRATED WITH 181 ENGRAVINGS.

SECOND EDITION.

CASSELL & COMPANY, LIMITED.
LONDON, PARIS & NEW YORK.

[ALL RIGHTS RESERVED.]

1883.

To

WILLIAM BOWMAN, LL.D., F.R.S., &c., &c.,

THIS BOOK IS DEDICATED,

IN ACKNOWLEDGMENT OF HIS MANY AND GREAT DISCOVERIES

IN ANATOMY AND PHYSIOLOGY.

PREFACE.

This Work is intended as a Manual for Medical Students. The original plan was to embody in it only such matter as was absolutely required for the first year's student in the Medical Schools; but this plan was not carried out, since the book would then have been, in some measure, incomplete as a Manual of Histology. The work contains a good deal that will be acceptable to the advanced student, as well as to the beginner.

My best thanks are due to Mr. Charles Berjeau, F.L.S., for the readiness and artistic skill with which he has executed the illustrations on wood. Some of the figures are copied from the "Atlas of Histology," others from the "Handbook for the Physiological Laboratory." The former are marked "Atlas," the latter "Handbook."

E. KLEIN.

May, 1883.

CONTENTS.

CHAPTER I.
Cells

CHAPTER II.
Blood

CHAPTER III.
Epithelium

CHAPTER IV.
Endothelium

CHAPTER V.
Fibrous Connective Tissues

CHAPTER VI.
Cartilage

CHAPTER VII.
Bone

CHAPTER VIII.
Non-striped Muscular Tissue

CHAPTER IX.
Striped Muscular Tissue

CHAPTER X.
The Heart and Blood-vessels

CHAPTER XI.
The Lymphatic Vessels

CHAPTER XII.
SIMPLE LYMPHATIC GLANDS 92

CHAPTER XIII.
COMPOUND LYMPHATIC GLANDS 98

CHAPTER XIV.
NERVE-FIBRES 103

CHAPTER XV.
PERIPHERAL NERVE-ENDINGS 115

CHAPTER XVI.
THE SPINAL CORD 127

CHAPTER XVII.
THE MEDULLA OBLONGATA 142

CHAPTER XVIII.
THE CEREBRUM AND CEREBELLUM 149

CHAPTER XIX.
THE CEREBRO-SPINAL GANGLIA 163

CHAPTER XX.
THE SYMPATHETIC SYSTEM 166

CHAPTER XXI.
TEETH 171

CHAPTER XXII.
THE SALIVARY GLANDS 178

CHAPTER XXIII.
THE MOUTH, PHARYNX, AND TONGUE 187

CHAPTER XXIV.
THE ŒSOPHAGUS AND STOMACH 194

CHAPTER XXV.
THE SMALL AND LARGE INTESTINE 201

CHAPTER XXVI.
The Glands of Brunner, and the Pancreas 208

CHAPTER XXVII.
The Liver 210

CHAPTER XXVIII.
The Organs of Respiration 215

CHAPTER XXIX.
The Spleen 225

CHAPTER XXX.
The Kidney, Ureter, and Bladder 229

CHAPTER XXXI.
The Male Genital Organs 244

CHAPTER XXXII.
The Female Genital Organs 257

CHAPTER XXXIII.
The Mammary Gland 270

CHAPTER XXXIV.
The Skin 274

CHAPTER XXXV.
The Conjunctiva and its Glands 291

CHAPTER XXXVI.
The Cornea, Sclerotic, Ligamentum Pectinatum, and Ciliary Muscle 295

CHAPTER XXXVII.
The Iris, Ciliary Processes, and Choroidea . . . 300

CHAPTER XXXVIII.
The Lens and Vitreous Body 305

CHAPTER XXXIX.
The Retina 308

CHAPTER XL.
THE OUTER AND MIDDLE EAR 318

CHAPTER XLI.
THE INTERNAL EAR 320

CHAPTER XLII.
THE NASAL MUCOUS MEMBRANE 333

CHAPTER XLIII.
THE DUCTLESS GLANDS 338

INDEX 343

ELEMENTS OF HISTOLOGY.

CHAPTER I.

CELLS.

1. THE ripe **ovum** (Fig. 1) of man and mammals is a minute spherical clump of a soft, gelatinous, transparent, granular-looking substance, containing numerous minute particles—yolk globules. It is invested by a vertically-striated delicate membrane, called the zona pellucida. Inside this clump, and situated more or less excentrically, is a vesicle—the *germinal vesicle*—and inside this, one or more solid spots — the *germinal spot or spots*. The gelatinous transparent substance of the ovum, containing a very large percentage of proteid material, is called *Protoplasm*. Before and immediately after fertilisation, the protoplasm of the ovum shows distinct movement, consisting in contraction and expansion. These movements are spontaneous—*i.e.*, not caused by any directly visible external influence.

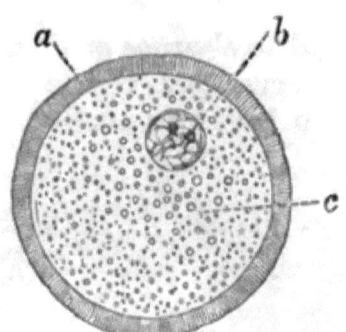

Fig. 1.—Ripe Ovum of Cat.
a, Zona pellucida; *b*, germinal vesicle; *c*, protoplasm.

The diameter of the ripe ovum in man and domestic animals varies between $\frac{1}{200}$ and $\frac{1}{120}$ of an inch. But before it ripens the ovum is considerably smaller—in fact, its size is in proportion to its state of development.

2. **Fertilisation** causes marked changes in the contractions of the protoplasm of the ovum; these

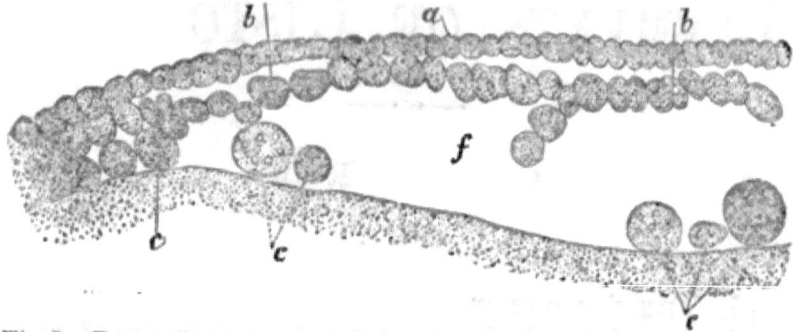

Fig. 2.—From a Section through the Blastoderm of Chick, unincubated. *a*, Cells forming the ectoderm; *b*, cells forming the endoderm; *c*, large formative cells; *f*, segmentation cavity. (Handbook.)

lead to *cleavage* or *division* of its body into two parts, the germinal vesicle having previously split up into two bodies or nuclei; so that we now find the ovum

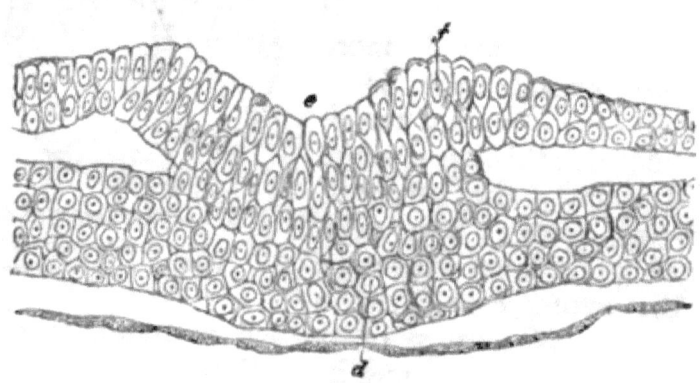

Fig. 3A.—From a Section through the Rudiment of the Embryo Chick. *e*, Primitive groove; *f*, dorsal laminæ of epiblast; *d*, mesoblast. The thin layer of spindle-shaped cells is the hypoblast. (Handbook.)

has originated two new elements, each of which consists of protoplasm, of the same substance as that of the original ovum, and each contains one nucleus or kernel. The investment of the ovum takes no part in this process of division. Not long after, each of the two daughter elements undergoes cleavage or division

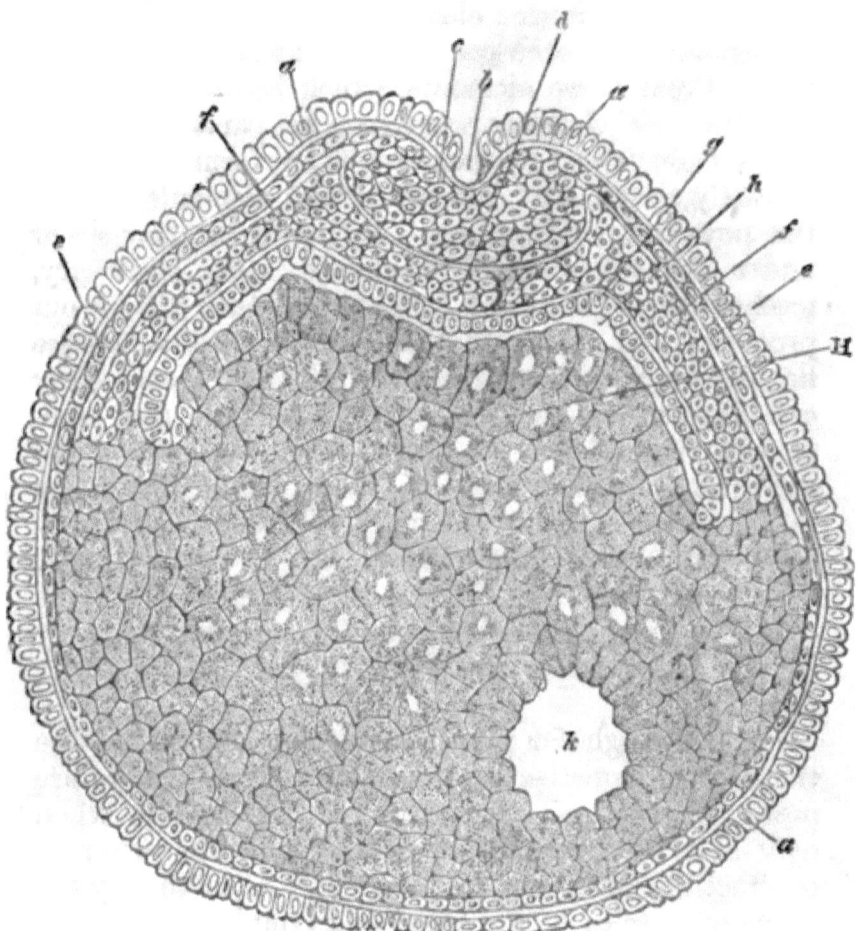

Fig. 3B.—Vertical Section through the Ovum of **Bufo** Cinereus, in the early stage of the Embryo Development.

a, Tegmental layer of epiblast; b, dorsal groove; c, rudiment of central nervous system; d, notochord; e, deep layer of epiblast; f, mesoblast; g, hypoblast; h, cavity of alimentary canal—Rusconi's cavity; H, central yolk; k, remainder of von Baer's or segmentation cavity. (Handbook.)

into two new elements, the nucleus having previously divided into two, so that each of the new offspring possesses its own nucleus. This process of division is continued in the same manner for many generations

(Figs. 2, 3A, 3B), so that after a few days we find within the original investment of the ovum a large number of minute elements, each consisting of protoplasm, and each containing a nucleus.

3. From these elements, which become smaller as the process of cleavage progresses, all parts and organs of the embryo and its membranes are formed. It can be easily shown that the individual elements possess the power of contractility. Either spontaneously or under the influence of moderate heat, electricity, mechanical or chemical stimulation, they throw out processes and withdraw them again, their substance flowing slowly but perceptibly along. Hence they can change their position. In this respect they completely resemble those lowest organisms which are known as amœbæ, each of these being likewise a nucleated mass of protoplasm. Wherefore this movement is termed *amœboid movement*. It can be further shown that they, like amœbæ, grow in size and divide —that is to say, the individuals of a generation grow in size before each gives rise to two new daughter individuals.

4. Although for some time during embryonal life the elements constituting the organs of the embryo are possessed of these characters, a time arrives when only a limited number of them retain the power of contractility in any marked degree. At birth only the white corpuscles of the blood and lymph, many of the elements of the lymphatic organs, and the muscular tissues, possess this power, while the others lose it, or at any rate do not show it except when dividing into two new elements. Some of these elements retain their protoplasmic basis; as a rule, each contains one nucleus (but some two or more) and is capable of giving origin by division to a new generation. Others, however, change their nature altogether, their protoplasm and nucleus disappear, and they

give origin to material other than protoplasm—*e.g.*, collagenous, osseous, elastic, and other substances.

5. Beginning with the ovum, and ending with the protoplasmic nucleated elements found in the organs and tissues of the embryo and adult, we have, then, one uninterrupted series of generations of elements, which with Schwann we call *cells* and with Brücke *elementary organisms*. Of these it can be said that not only is each of them derived from a cell (Virchow: *omnis cellula a cellula*), but each consists of the protoplasm of Max Schultze (Sarcode of Dujardin),

Fig. 4.—Amœboid movement of a White Blood Corpuscle of Man; various phases of movement. (Handbook.)

is without any investing membrane, and includes generally one nucleus, but may contain two or more. We can further say that each of these cells shows the phenomenon of growth, which presupposes nutrition, and reproduction. All of them in an early stage of their life history, and some of them throughout it, show the phenomenon of contractility, or amœboid movement (Fig. 4.)

Cells differ in shape according to kind, locality, and function, being spherical, irregular, polygonal, squamous, branched, spindle-shaped, cylindrical, prismatic, or conical. These various shapes will be more fully described when dealing in detail with the various kinds of cells. Cells in man and mammals differ in size within considerable limits: from the size of a small white blood corpuscle of about $\frac{1}{2500}$ of an inch to that of a large ganglion cell in the anterior horns of the spinal cord of about $\frac{1}{200}$ of an inch, or to that of a multinucleated cell of the bone marrow—myeloplax—some of which surpass in size even the ganglion cells. The same holds good of the nucleus. Between the nucleus of a ganglion cell of about $\frac{1}{800}$ to $\frac{1}{1200}$ of an inch in diameter and the nucleus of a white blood corpuscle of about $\frac{1}{8000}$ to $\frac{1}{10000}$ of an inch and less there are all intermediate sizes.

6. **Protoplasm** is a transparent homogeneous or granular-looking substance. On very careful examination with good and high powers, and especially when examined with certain reagents, in many instances it shows a more or less definite structure, composed of fibrils, more or less regular, and in some instances grouped into a honeycombed or fibrillar reticulum in the meshes of which is a homogeneous interstitial substance. The closer the meshes of the reticulum, the less there is of this interstitial substance, and the more regularly granular does it appear. In the meshes of the reticulum, however, may be included larger or smaller granules of fat, pigment, or other material. Water makes protoplasm swell up, and ultimately become disintegrated; so do dilute acids and alkalies. All substances that coagulate proteids have the same effect on protoplasm.

7. The **nucleus**, the size of which is generally in proportion to that of the cell, is usually spherical or oval. It is composed of a more or less distinct investing cuticle and the nuclear contents, which are in the ripe state arranged as an irregular or regular network, the parts of which may be uniform fibrils or septa, or irregularly-shaped trabeculæ. In the life history of each nucleus there may be stages in which one or more clumps or nucleoli are present in the nuclear network. The substance of the nucleus differs chemically from that of the cell, the former containing *nuclein*.

Immediately before division the nuclear membrane disappears, and also immediately after division the nuclear matter shows no definite boundary. The nuclear membrane when present is a condensed outer stratum of the nuclear matter.

In some instances it can be shown that the nuclear fibrils are in continuity with the fibrils of the cell substance. In the moving white blood corpuscles Stricker and Unger have seen the nucleus becoming one with the cell substance, and again afterwards differentiated by the appearance of a membrane.

8. During **division** of the cell the nucleus divides generally before the cell protoplasm. This division of the nucleus has previously been supposed to occur in the same manner as that of the cell protoplasm—*i.e.*, by simple cleavage. This mode is called the direct division, or Remak's mode of division. In this division the nucleus is supposed to become constricted, kidney-shaped and hour-glass shaped, and if the division is into more than two, lobed. Nuclei of these shapes are not uncommon; but they need not necessarily indicate direct division, because, being very soft structures, pressure exerted from outside, or the motion of the cell protoplasm, may produce these shapes; and, further, the contractility of the nucleus may and occasionally has been observed to cause these changes of shape. From the observations of recent investigators—Bütschli, Hertwig, Strassburger, **Mayzel, van** Beneden, **Balfour**, Eberth, Schleicher, Peremeschko, Flemming, Klein, Arnold, Pfützner, Retzius, Bizzozero, and many others—it is now known that in the embryo and adult, in plant and animal, vertebrates and invertebrates, all kinds of cells, before their protoplasm undergoes division, show complicated changes of their nucleus, leading to division. This manner of division is called the indirect division or karyokinesis. It has been observed by Mayzel, Schleicher, and Flemming, that the nuclear fibrils show movement, hence the name karyokinesis. This

process of the karyokinesis is represented in the adjoining figure, 5, and it consists of the following phases: (a) The nuclear network become very pronounced, while the nuclear membrane disappears, and the fibrils of the nuclear network becomes twisted and bent into a more or less dense *convolution;* at the same time the nucleus, as a whole, is considerably enlarged. (b) The fibrils unravel into loops, arranged

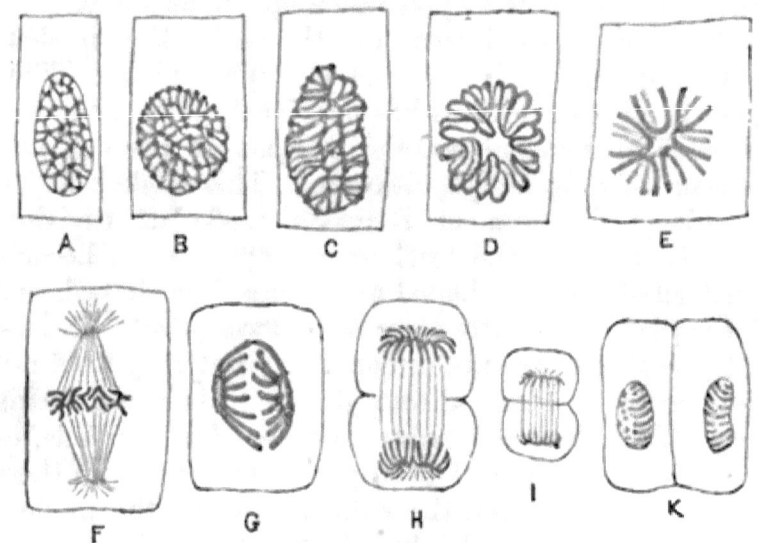

Fig. 5.—Karyokinesis.

A, Ordinary nucleus of a columnar epithelial cell; B, C, the same nucleus in the stage of convolution; D, the wreath, or rosette form; E, the aster, or single star; F, a nuclear spindle from the Descemet's endothelium of the frog's cornea; G, H, I, diaster; K, two daughter nuclei.

around the centre as a *wreath* or *rosette*. (c) The peripheral points of the loops become broken, and we obtain a star-shaped figure of single loops—the *aster*. (d) The loops separate into two groups or new centres: this is the *diaster*, or double star. (e) The two groups of threads become further apart, as if attracted by opposite poles; but the two groups remain still connected with one another by fine pale threads. These latter differ from the others in not staining with certain dyes, and representing the interstitial substance of the nuclear matter—*i.e.*, the pale substance contained in the original reticulum of the nucleus. Flemming calls this substance achromatin, whereas the threads forming the original network, the convolution, aster and diaster, he calls chromatin, on account of its readily staining with dyes.

In this stage the whole figure resembles a spindle, the nuclear spindle of Bütschli (Fig. 5, F).

(*f*) Further, all connection between the two sets of threads is broken—*i.e.*, between the stars of the diaster. (*g*) The threads of each set become greatly convoluted. (*h*) A membrane appears for each set. In this stage we speak of two new or *daughter nuclei*. The cell protoplasm may commence to divide at any stage between the one when the threads aggregate round two centres, and the one when two distinct daughter nuclei are present; or the division of the nucleus may not be followed by the division of the cell protoplasm, in which case we have a two-nucleated cell. In some instances, especially of invertebrates and lower vertebrates, a peculiar sun-like arrangement of fibrils of the cell protoplasm towards each of the two stars of the above nuclear fibrils has been observed. Martin has noticed, in pathological new growths, a simultaneous division into three and four daughter nuclei, after the mode of karyokinesis. Although this indirect mode of division of the nucleus has been observed in all kinds of cells in the embryo, and to a limited degree also in the adult, it is not proved to be the universal mode of nuclear division. On the contrary, there is strong evidence that in amœboid corpuscles division of the nucleus follows the direct mode, and it is also probable that other nuclei, under certain conditions, may undergo the direct mode of division.

CHAPTER II.

BLOOD.

9. UNDER the microscope blood appears as a transparent fluid, the *liquor sanguinis* or *plasma*, in which float vast numbers of formed bodies, the *blood corpuscles*. The great majority of these are coloured; a few of them are colourless. The latter are called *white* or *colourless blood corpuscles*, or leucocytes. The former are called *red* or *coloured blood corpuscles*, and appear red only when seen in a thick layer; when in a single layer they appear of a yellow-greenish colour, more yellow if of arterial, more green if of venous blood. The proportions of plasma and blood corpuscles are sixty-four of the former and thirty-six of the latter in one hundred volumes of blood. By measurement it has been found that there are a little over five millions of blood corpuscles in each cubic millimètre ($\frac{1}{625}$ of a cubic inch) of human blood. There appears to be in healthy human blood one white corpuscle for 600–1200 red ones. In man and mammals the relative number of blood corpuscles is greater than in birds, and in birds greater than in lower vertebrates.

10. The **red blood corpuscles** (Fig. 6) of man and mammals are homogeneous bi-concave discs (except in the camelidæ, where they are elliptical), and do not possess any nucleus. Being bi-concave in shape, they are thinner and more transparent in the centre than at the periphery. In other vertebrates they are oval and more or less flattened from side to side, and each possesses a central oval nucleus.

The diameter of the human red blood corpuscles is

about $\frac{1}{3000}$ of an inch on the broad, and $\frac{1}{12000}$ of an inch on the narrow side. But there are always corpuscles present which are much smaller by about one-third to one-half than the others. In normal blood these small red corpuscles are scarce; in certain abnormal conditions, especially anæmia, or scarcity of blood, they are more numerous.

Fig. 6.—Various kinds of Red Blood Corpuscles.

A, Two human, one seen from the broad, the other from the narrow side; B, a red corpuscle of the camel; C, two red corpuscles of the frog, one seen from the broad, the other from the narrow side.

According to Gulliver, Welcker, and others, the following are the measurements of the sizes of red blood corpuscles of various vertebrates:—man, $\frac{1}{3200}$; dog, $\frac{1}{3500}$; cat, $\frac{1}{4000}$; sheep, $\frac{1}{5000}$; elephant, $\frac{1}{2745}$; horse, $\frac{1}{4600}$; musk deer, $\frac{1}{12325}$; pigeon, $\frac{1}{2317}$; toad, $\frac{1}{1043}$; newt, $\frac{1}{814}$; proteus, $\frac{1}{400}$; pike, $\frac{1}{2000}$; shark, $\frac{1}{1142}$.

11. In a microscopic specimen of fresh unaltered blood (Fig. 7) the red blood corpuscles form peculiar shorter or longer rolls, like so many coins, becoming adherent to one another by their broad surfaces. Under various conditions—such as when isolated, or when blood is diluted with saline solution or solutions of other salts (sulphate of sodium or magnesium)—the corpuscles lose their smooth circular outline, shrinking and becoming *crenate* (Fig. 8, A). In a further stage of this process of shrinking they lose their discoid form, and

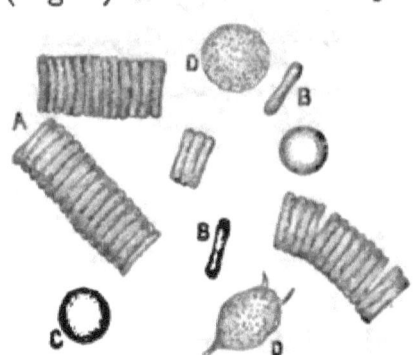

Fig. 7.—Human Blood, fresh.

A, Rouleaux of red corpuscles; B, isolated red corpuscle seen in profile; C, isolated red corpuscle seen from its broad surface; D, white corpuscles.

become smaller and spherical, but beset all over their surface with minute processes. This shape is called the horse-chestnut shape (Fig. 8, b, c). It is probably due to the corpuscles losing carbonic acid, as its addition brings back their discoid shape and smooth circular outline. On abstracting the carbonic acid they return to the horse-chestnut shape. Water, acid, alcohol, ether, the electric current, and many other reagents, produce discoloration of the red blood corpuscles; the colouring matter — generally the combination of the blood-colouring matter with globulin known as *hæmoglobin*—becoming dissolved in the plasma. What is left of the corpuscles is called the *stroma*. In newts' and frogs' blood a separation of

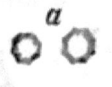

Fig. 8.—Human Red Blood Corpuscles.

a, Crenate; *b, c,* horse-chestnut shaped.

the stroma from the nucleus plus hæmoglobin can be effected by means of boracic acid (Fig. 9, B); the former is called by Brücke the Oikoid, the latter Zooid. This stroma contains amongst other things a good deal of paraglobulin. The stroma of the corpuscles of amphibians is seen, under certain reagents, to be of a reticulated

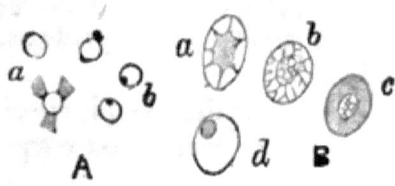

Fig. 9.—Red Blood Corpuscles of Man and Newt.

A, Human red corpuscles after the action of tannic acid: *a*, three red corpuscles, from which the hæmoglobin is passing out; *b*, Roberts's corpuscles.
B, Newt's red corpuscles after the action of boracic acid: *a*, a corpuscle showing Brücke's' zooid and oikoid; *b*, a corpuscle showing the reticulated stroma; *c*, a corpuscle showing the reticulum in the nucleus; *d*, the nucleus passing out.

structure, but in the fresh state appears homogeneous and pale. Discoloration of the blood corpuscles can be observed to take place also in blood without the addition of any or with that of perfectly harmless fluids, such as humor aqueus of the eye, hydrocele fluid, &c. The number of corpuscles undergoing discoloration is, however, small.

The elements of the blood described by Dr. William Norris, of Birmingham, as the invisible, pale, or third corpuscle, are simply red blood corpuscles that have become discoloured by the mode of preparation (Alice Hart).

12. The **hæmoglobin** of the red blood corpuscles forms crystals (Fig. 10), which differ in shape in the various mammals. They are always of microscopic size, and of a bright red colour.

In man and most mammals they are of the shape of prismatic needles, or rhombic plates; in the squirrel they are hexagonal plates, and in the guinea-pig they are tetrahedral or octahedral.

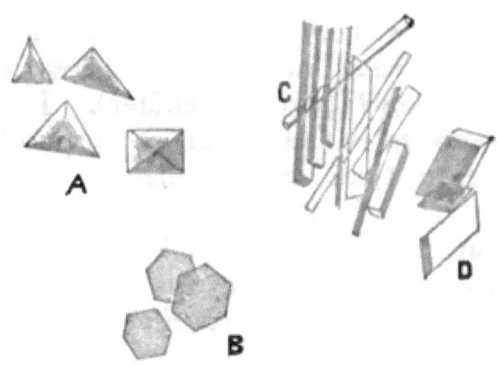

Fig. 10.—Hæmoglobin crystals. A, Of guinea-pig; B, of squirrel; C, D, human.

The blood pigment itself is an amorphous dark-brown or black powder—the *hæmatin*; but it can be obtained in a crystalline form as hydrochlorate of hæmatin (Fig. 11). These crystals also are of microscopic size, of a nut-brown colour, of the shape of narrow rhombic plates, and are called *hæmin crystals*, or *Teichmann's crystals*. In extravasated human blood crystals of a bright yellow or orange colour are occasionally met with; they are called by Virchow, their discoverer, *hæmatoidin*. They are supposed to be identical with bilirubin, obtainable from human bile.

Fig. 11.—Hæmin crystals.

13. The **white or colourless blood corpuscles** are in human blood of about $\frac{1}{2000}$ to $\frac{1}{2500}$ of an inch in diameter, and are spherical in the

circulating blood or in blood that has just been removed from the vessels. Their substance is transparent granular-looking protoplasm, containing larger or smaller bright granules. These granules are either of a fatty nature, or, as in some kinds of blood, notably horses', are of a reddish colour, and these corpuscles are supposed by some observers (Semmer and Alexander Schmidt) to be intermediate between red and white corpuscles. The protoplasm of the colourless corpuscles contains glycogen (Ranvier, Schäfer). In the blood of lower vertebrates the colourless corpuscles are considerably larger than in mammals. But in all cases they consist of protoplasm, include one, two, or more nuclei, and show amœboid movement. This may be observed in corpuscles without any addition to a fresh microscopic specimen of blood, but it always becomes much more pronounced on applying artificial heat of about the degree of mammals' blood. It is then seen that they throw out longer or shorter filamentous processes, which may gradually lengthen or be withdrawn, appearing again at another point of the periphery. The corpuscle changes its position either by a flowing movement of its protoplasm as a whole, thus rapidly creeping along the field of the microscope, or it may push out a filamentous process and shift the rest of its body into it. During this movement the corpuscle may take up granules from the surrounding fluid.

14. The white corpuscles of the same sample of blood differ in size and aspect within considerable limits, some being half the size of others, some much paler than others. The smaller examples generally possess one nucleus occupying the greater part of the corpuscle, the larger ones usually include two, three, or even more nuclei, and show more decided amœboid movement than the others. Division by

cleavage of the white corpuscles of the blood of lower vertebrates has been directly observed by Klein and Ranvier.

15. In every microscopic specimen of the blood of man and mammals are found a variable number of large granules, more or less angular, singly or in groups, which have been specially studied by Osler. According to Bizzozero they are, when observed in the living and fresh blood, pale, circular, or slightly oval discs (Fig. 12, *b*). Their size is only $\frac{1}{3}$ to $\frac{1}{2}$ of that of the red blood corpuscles. They are called by him *blood plates*, and he supposes them to be of essential importance in the coagulation of the blood, in fact, the fibrin ferment. Hayem described them previously as being intermediate forms in the development of red blood corpuscles, and called them hæmatoplasts.

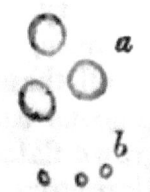

Fig. 12.—Human Blood.

a, Red blood corpuscles; *b*, blood plates of Bizzozero.

16. **Development of Blood Corpuscles.**— At an early stage of embryonal life, when blood makes its appearance it is a colourless fluid, containing only white corpuscles (each with a nucleus), which are derived from certain cells of the mesoblast. These white corpuscles change into red ones, which become flattened, and their protoplasm gets homogeneous and of a yellowish colour. All through embryonal life new white corpuscles are transformed into red ones. In the embryo of man and mammals these red corpuscles retain their nuclei for some time, but ultimately lose them. New nucleated red blood corpuscles are, however, formed by division of old red corpuscles. Such division has been observed even in adult blood of lower vertebrates (Peremeschko) as well as in mammals (marrow of bone by Bizzozero and Torre).

An important source for the new formation of red corpuscles in the embryo and adult is the red marrow of bones (Neumann, Bizzozero, Rindfleisch), in which numerous nucleated protoplasmic cells (marrow cells) are converted into nucleated red blood corpuscles. The protoplasm of the corpuscle becomes homogeneous and tinged with yellow, the nucleus being ultimately lost. The spleen is also assumed to be a place for the formation of red blood corpuscles. Again, it is assumed that ordinary white blood corpuscles are transformed into red ones, but of this there is no conclusive evidence. In all these instances the protoplasm becomes homogeneous and filled with hæmoglobin, while the cell grows flattened, discoid, and the nucleus in the end disappears.

Schäfer described intracellular (endogenous) formation of red blood corpuscles at first as small hæmoglobin particles, but soon growing into red blood corpuscles, in certain cells of the subcutaneous tissue of young animals. Malassez describes the red blood corpuscles originating by a process of continued budding from the marrow cells.

The white corpuscles appear essentially to be derived from the lymphatic organs, whence they are carried by the lymph into the circulating blood.

CHAPTER III.

EPITHELIUM.

17. **Epithelial cells** (Fig. 13A) are *nucleated protoplasmic cells, forming continuous masses* on the surface of the skin, of the lining membrane of the alimentary canal, the respiratory organs, the urinary and genital organs, the free surface of the conjunctiva, and the anterior surface of the cornea. The lining of the tubes and

alveoli of secreting and excreting glands, such as the kidney, liver, mammary gland, testis and ovary, the salivary glands, mucous, peptic, and Lieberkühn's glands, the sweat and sebaceous glands, the hair follicles, &c., consists of epithelial cells. Such is the case also with the sensory or terminal parts of the organs of the special senses. And, finally, epithelial cells occur in other organs, such as the thyroid gland, the pituitary body, &c.

The hairs and nails, the cuticle of the skin, certain parts of the rods and cones of the retina, and the rods of corti of the organ of hearing, are modified epithelial structures.

Epithelial cells are grouped together by exceedingly thin layers of an albuminous *interstitial cement substance*, which during life is of a semi-fluid nature, and belongs to the group of bodies known as globulins.

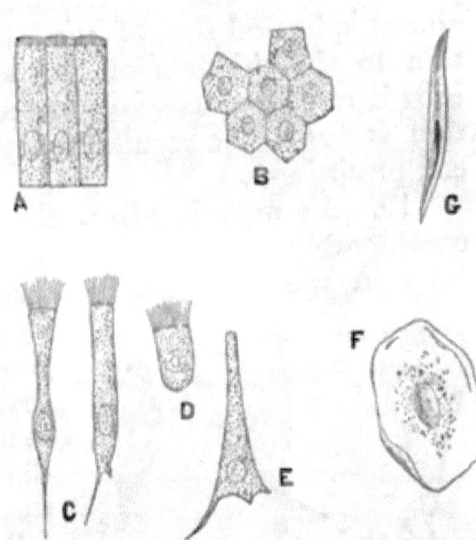

Fig. 13A.—Various kinds of Epithelial Cells.
A, Columnar cells of intestine; B, polyhedral cells of the conjunctiva; C, ciliated conical cells of the trachea; D, ciliated cell of frog's mouth; E, inverted conical cell of trachea; F, squamous cell of the cavity of mouth, seen from its broad surface; G, squamous cell, seen from its narrow side.

18. As regards **shape** we distinguish two kinds of epithelial cells—columnar and squamous. The *columnar cells* are short or long, cylindrical or prismatic, pyramidal, conical, club-shaped, pear-shaped, or spindle-shaped; their nucleus is always more or less oval, their protoplasm more or less longitudinally striated. On the free surface of the cells—*i.e.*, the

part facing a cavity, canal, or general surface—in many instances a bright thinner or thicker cuticular structure is seen, with more or less distinct vertical striation. The conical or spindle-shaped, club-shaped, and pear-shaped cells are drawn out into longer or shorter single or branched extremities.

The *squamous* or *pavement* cells are cubical, polyhedral or scaly. The nucleus of the former is almost spherical that of the latter flattened in proportion to the thinness of the scales. In polyhedral cells it can be occasionally perceived that the granulation is due to the regular honeycombed nature of the cell protoplasm.

19. As regards **size,** the epithelial cells differ considerably from one another in different parts, and even in the same part. Thus, the columnar cells, covering the surface of the villi of the small intestine, are considerably longer than those lining the mucous membrane of the uterus; the columnar cells lining the larger ducts of the kidney are considerably longer than those lining the small ducts; the polyhedral cells covering the anterior surface of the cornea are considerably smaller than those on the surface of the lining membrane of the urinary bladder; the scales lining the ultimate recesses of the bronchial tubes—the air cells—are considerably smaller than those on the surface of the membrane lining the human oral cavity and œsophagus.

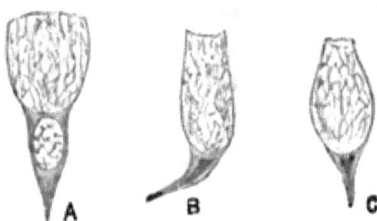

Fig. 13B.—Three Mucus-secreting Goblet Cells.

A, From the stomach of newt; B, from a mucous gland; C, from the surface of the mucous membrane of the intestine.

20. As regards **arrangement,** the epithelial cells are either arranged as a single layer or are stratified, forming several superimposed layers; in the former case

we have a simple in the latter a stratified epithelium. The simple epithelium may be composed of squamous cells—*simple squamous* or *simple pavement epithelium*—or it may be composed of columnar cells—*simple columnar epithelium*. The stratified epithelium may be *stratified pavement*, or *stratified columnar;* in the former case all or the majority of the layers consist of squamous or polyhedral-cells, in the latter all cells belong to the columnar kind. *Simple squamous epithelium* is that which lines the air cells, certain urinary tubules of the kidney (the looped tubes of Henle, the cortical parts of the collecting tubes), the acini of the milk-gland, the inner surface of the iris and choroid membrane of the eyeball. *Simple columnar epithelium* is that on the inner surface of the stomach, small and large intestine, uterus, small bronchi, ducts and acini of mucous and salivary glands, of some kidney tubules, &c. *Stratified pavement epithelium* is that on the epidermis, the epithelium lining the cavity of the mouth, pharynx, and œsophagus in man and mammals, the anterior surface of the cornea, &c.

21. The **epidermis** (Fig. 14) consists of the following layers:—(*a*) Stratum corneum: this is the superficial stratum, and it consists of several layers of horny scales, without any nucleus. The layers, which are separated from one another by narrow clefts containing air, are then in process of desquamation. This stratum is thickest on the palm of the hand and fingers, and the sole of the foot. (*b*) The stratum lucidum, composed of several dense layers of horny scales, in which traces of an exceedingly flattened nucleus may be perceived. (*c*) Then follow many layers of nucleated cells, forming the stratum or rete Malpighii or rete mucosum. The most superficial layer or layers of it are flattened scales, which are characterised by the presence around the nucleus

of globular or elliptical granules of the nature intermediate between protoplasm and keratin. Their substance is called eleidin (Ranvier); these cells form the granular layer (stratum granulosum) of Langerhans. Deeper down the cells become less flattened and more polyhedral, and the deepest form a layer of more or less columnar cells, placed vertically on the surface of the subjacent corium.

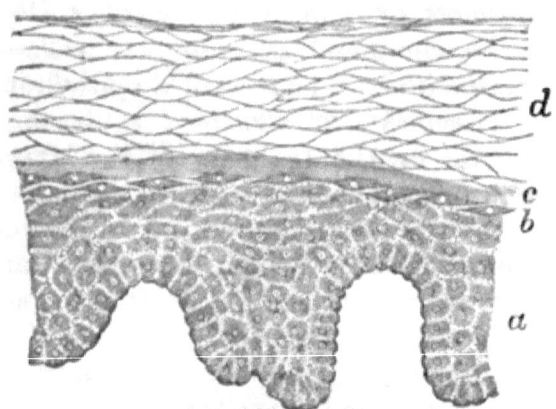

Fig. 14.—From a Vertical Section through the Epidermis.

a, The stratum Malpighii; *b*, the stratum granulosum; *c*, the stratum lucidum; *d*, the stratum corneum. (Atlas.)

The substance of the hairs, nails, claws, hoofs, consists of horny scales. (*See* chapter on Skin.)

22. The **stratified pavement epithelium** (Fig. 15) lining the cavity of the mouth, the surface of the tongue, the pharynx and œsophagus of man and mammals, and the anterior surface of the cornea, &c., is, as regards the style and arrangement of the cells, identical with the stratum Malpighii of the epider-

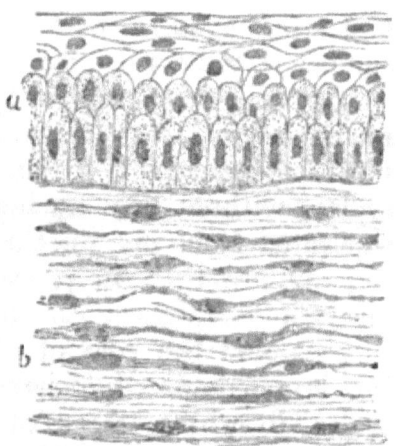

Fig. 15.—From a Vertical Section through the anterior layers of the Cornea.

a, The stratified pavement epithelium; *b*, the substantia propria, with the corneal corpuscles between its lamellæ. (Handbook.)

mis. The cell protoplasm is more transparent in the former, and the granular cells of the stratum granulosum are not always present, but they generally are in the epithelium of the tongue and of the rest of the oral cavity. The most superficial scales show more or less horny transformation.

23. **Stratified columnar epithelium** is met with on the lining membrane of the respiratory organs, as larynx, trachea, and large bronchi. It consists of several layers of columnar cells: a superficial layer of conical or prismatic cells, with a more or less pointed extremity directed towards the depth, between these are inserted spindle-shaped cells, and finally inverted conical cells.

The epithelium of the ureter and bladder is called *transitional epithelium*. It is stratified, and the most superficial layer consists of squamous cells. Underneath this is a layer of club-shaped cells, between which extend one or more layers of small spindle-shaped cells.

Amongst the columnar epithelial cells occurring in man and mammals the *ciliated cells* and the *goblet cells*, and amongst the squamous cells the *prickle cells*, deserve special attention.

24. **Ciliated cells** are characterised by possessing a bundle of very fine longer or shorter hairs or cilia on their free surface. These cilia are direct prolongations of the cell protoplasm. More correctly speaking, the cilia are continuous with the filaments or striæ of the cell protoplasm. The superficial layer of conical cells of the epithelium in the respiratory organs, the columnar cells lining the uterus and oviduct, and the columnar cells lining the tubes of the epididymis possess such cilia. In lower vertebrates the ciliated cells are much more frequently observed; in Batrachia the epithelial cells lining the mouth, pharynx, and œsophagus are ciliated.

While fresh in contact with the membrane which they line, or even after removal from it, provided the cells are still alive, the ciliated cells show a rapid synchronous whip-like movement of their cilia, the cilia of all cells moving in the same direction. The movement ceases on the death of the cell, but may become slower and may cease owing to other causes than death, such as coagulation of mucus on the surface, want of sufficient oxygen, presence of carbonic acid, low temperature, &c. Under these circumstances, removal of the impediment will generally restore the activity of the cilia. Moderate electric currents and heat stimulate the movement, strong electric currents and cold retard it. Reagents fatally affecting cell protoplasm also stop permanently the ciliary action.

25. **Goblet or chalice cells** (Figs. 13B, 16) are cells of the shape of a conical chalice. The pointed part is directed away from the free surface, and contains a compressed triangular nucleus, surrounded by a trace of protoplasm. The body of the goblet contains mucus. This latter may be in various states of formation, and may at any time be poured out of the cell. Goblet cells are most commonly met with amongst the epithelium lining the respiratory organs, the surface of the stomach and intestines, and especially in mucous glands, in whose secreting portion all cells are goblet cells.

Fig. 16.

From a Vertical Section through the Epithelium on the surface of the mucous membrane of the large intestine.

Three goblet cells are seen pouring out their mucus. The rest are ordinary columnar cells.

The protoplasm of columnar cells facing a free surface, no matter whether in simple or stratified epithelium, ciliated or non-ciliated, may undergo such alteration as will lead to the transformation of the cell into a goblet cell. This takes place during life, and, in fact, represents an important function of columnar epithelial cells—viz., the formation of mucus. In mucus-secreting glands all the epithelial cells have this function permanently, but in ordinary columnar epithelium only a comparatively small number of the cells, as a rule, undergo this change, and then only temporarily; for a cell subject to it at one time may shortly afterwards resume the original shape and aspect of an ordinary protoplasmic, cylindrical, or conical epithelial cell, and *vice versa*. If ciliated cells undergo this change, the cilia are generally first detached.

It can be shown that in this change of an ordinary columnar epithelial cell into a goblet cell, the interstitial substance of the cell reticulum increases in amount, the meshes enlarging and distending the body of the cell. The interstitial substance probably undergoes the change into mucin.

26. **Prickle Cells** (Fig. 14).—Amongst the middle and deeper layers of the stratified pavement epithelium, such as is present in the epidermis and on the surface of the oral cavity and pharynx, we meet with a close, more or less distinct and regular striation, extending from the margin of one cell to that of each of its neighbours, by means of fine transverse short fibrils which, passing from protoplasm to protoplasm, connect the surfaces of the cells.

27. Pigmented epithelial cells—*i.e.*, epithelial cells filled with black pigment particles (crystals)—are found on the internal surface of the choroid and iris of the eyeball.

In coloured skins, and in coloured patches of skin and mucous membrane, such as occur in man and animals, there is found pigment in the shape of granules lodged in the protoplasm of the deeper epithelial cells, as well as in branched cells situated between the epithelial cells of the deeper layers. Minute branched non-pigmented nucleated cells are met with in *the interstitial or cement substance* of various kinds of epithelium, simple and stratified, *e.g.*, epidermis, epithelium of oral cavity, cornea, &c.

28. Epithelial cells undergo division, **and by this means a** constant regeneration takes place. In those parts where the loss of the superficial layers of cells is conspicuous, such as the epidermis, the stratified epithelium of the tongue and oral cavity, the sebaceous follicles **of** hairs, the regeneration **goes** on more **copiously** than at places where no such conspicuous loss occurs—as, for instance, in the stomach and intestines, the secreting glands, or sense organs.

In the stratified pavement epithelium it is the cells of the deepest layer which chiefly divide; the next layer thereby becomes gradually shifted towards the surface, and more flattened, and on reaching the surface dries up owing to rapid loss of water.

29. The interstitial substance between the epithelial cells being soft and semi-fluid, and **the protoplasm of** the epithelial cells themselves being **a soft** flexible material, **it** is possible for **the cells to change their shape** and arrangement after **pressure and** tension, exerted on them by the contraction or distension of the membrane on which they are situated. Thus, for instance, the epithelium lining a middle-sized bronchus may appear at one time as composed of long, thin, columnar epithelial cells in two layers; at another, as a single layer of long columnar cells; or again as a single layer of polyhedral or short columnar cells: the first is the case **when** the bronchus is contracted, the second when it is in a medium state of distension, the third when it is much distended. Similar changes may be noticed on the epithelium lining the mucous membrane of the bladder, gland tubes, the epidermis, and various other epithelial structures.

CHAPTER IV.

ENDOTHELIUM.

30. THE free surfaces of the serous and synovial membranes, and of those of the brain and spinal cord, the posterior surface of the cornea and anterior surface of the iris, the surfaces of tendon and tendon-sheaths, the lymph sinuses or lymph sacs of amphibian animals, the cavity of the heart, of blood-vessels and of lymphatic-vessels are lined with a continuous *endothelial membrane*, composed of *a single layer of flattened transparent squamous cells*, called *endothelial cells* (Fig. 17). Each contains an *oval nucleus*, situated generally excentrically. Just as in the case of epithelium, the endothelial cell plates are joined by a fluid or semi-fluid homogeneous *interstitial* or *cement substance* of the nature of globulin.

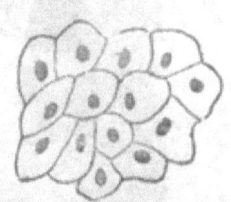

Fig. 17.—Endothelium of the Mesentery of Cat.

The outlines of the endothelial cells, and the nucleus of the latter are well shown.

When examining any of the above structures fresh the endothelial cells are not, as a rule, visible, owing to their great transparency; but by staining the structures with a dilute solution of nitrate of silver, and then exposing them to the influence of the light, the cement substance appears stained black, whereby the shape and size of the cell plates become evident. By various dyes also the nucleus of each cell plate may be brought into view.

On careful examination, and with suitable reagents, it can be shown that each endothelial cell consists of a homogeneous *ground-plate*. In it lies the nucleus, and around it is a substance which appears granular, but which is of a fibrillar nature, the fibrillæ being arranged in a network, and extend-

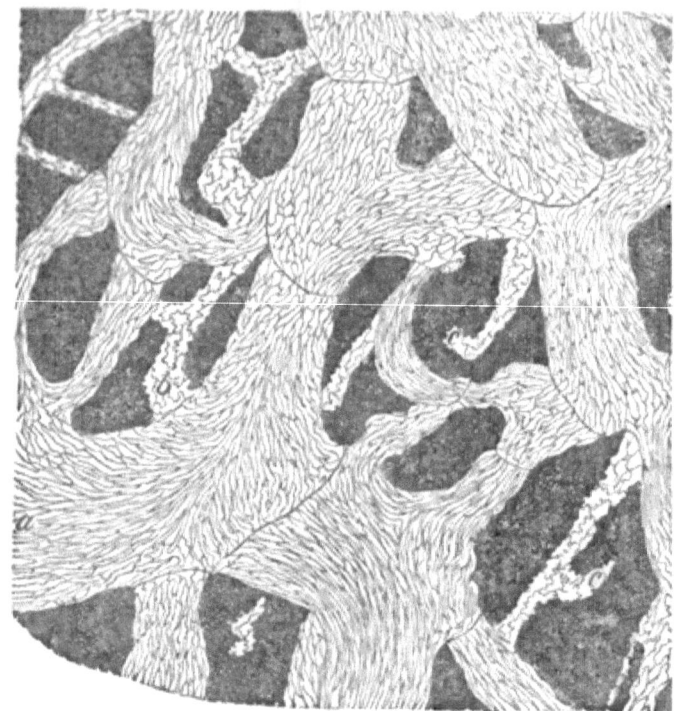

Fig. 18.—**Network of Lymphatics** in the Central Tendon of the Diaphragm of Rabbit, prepared with nitrate of silver, so as to show the outlines of the **Endothelial** Cells forming the wall of the Lymphatics.

a, Big lymphatic vessels; b, lymphatic capillaries; c, apparent ends of the capillaries. (Handbook.)

ing in many places up to the margin of the ground-plate. The nucleus is limited by a membrane, and contains a well-developed reticulum. The fibrillæ of the cell substance appear in connection with the nuclear reticulum.

31. As regards **shape**, endothelial cells differ considerably. Those of the pleura, pericardium, peritoneum, and endocardium of man and mammals

are more or less polygonal, or slightly elongated. Their outlines vary; in the lining of the lymph sacs of the frog they are much larger, and of very sinuous outline; while those of the posterior surface of the cornea

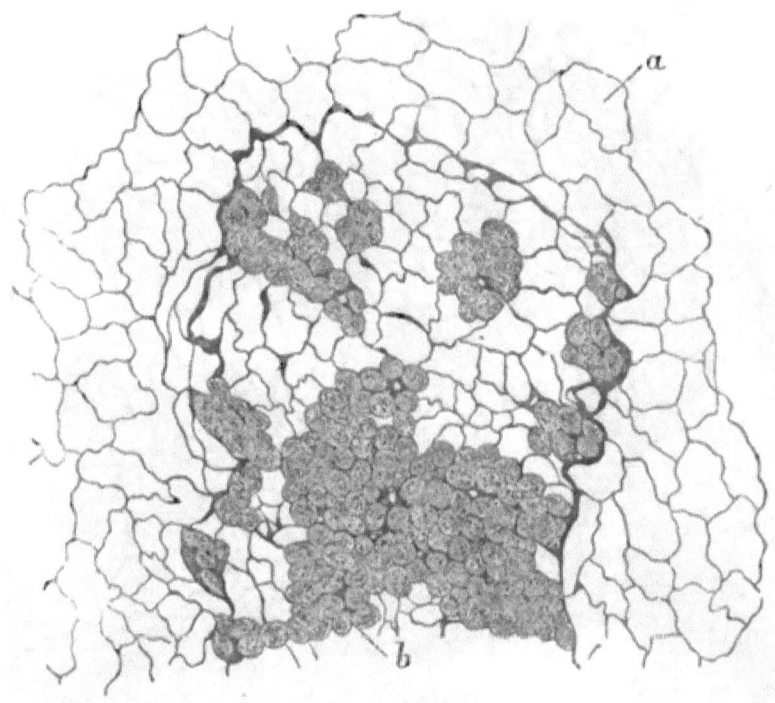

Fig. 19.—Omentum of Rabbit, stained with Nitrate of Silver. *a*, Ordinary flat endothelial cells; *b*, germinating cells. (Atlas.)

are very regular, pentagonal, or hexagonal, having straight outlines in the perfectly normal and well-preserved condition, but serrated and sinuous after they have been prepared with various reagents and in the abnormal state; the endothelial plates lining the blood-vessels and lymphatic vessels (Fig. 18) are narrow and elongated, with more or less sinuous outlines. In the lymphatic capillaries the endothelial plates are polygonal, but their outline serrated.

32. As a rule the endothelial cells are flattened,

i.e., scaly—but in some places they are polyhedral, or even short columnar. Such cells occur isolated or in small groups, or covering large and small patches, nodular, villous, or cord-like structures of the pleura

Fig. 20.—Part of Peritoneal Surface of the Central Tendon of Diaphragm of Rabbit, prepared with Nitrate of Silver.

s, Stomata; *l*, lymph-channels; *t*, tendon bundles. The surface is covered with endothelium. The stomata are surrounded by germinating endothelial cells. (Handbook.)

and omentum, on the synovial membranes, tunica vaginalis testis, &c. They are especially observable in considerable numbers in the pleura and omentum (Fig. 19) of all normal subjects (human, ape, dog, cat, and rodent animals); their number and frequency of occurrence are increased in pathological conditions (chronic inflammations, tuberculosis, cancer, &c.).

These endothelial cells are the *germinating endothelial cells*, and they can be shown to be in an active state of division. They thus produce small spherical lymphoid (amœboid) cells, which ultimately are absorbed by the lymphatics, and carried into the blood system as white blood corpuscles. On the sur-

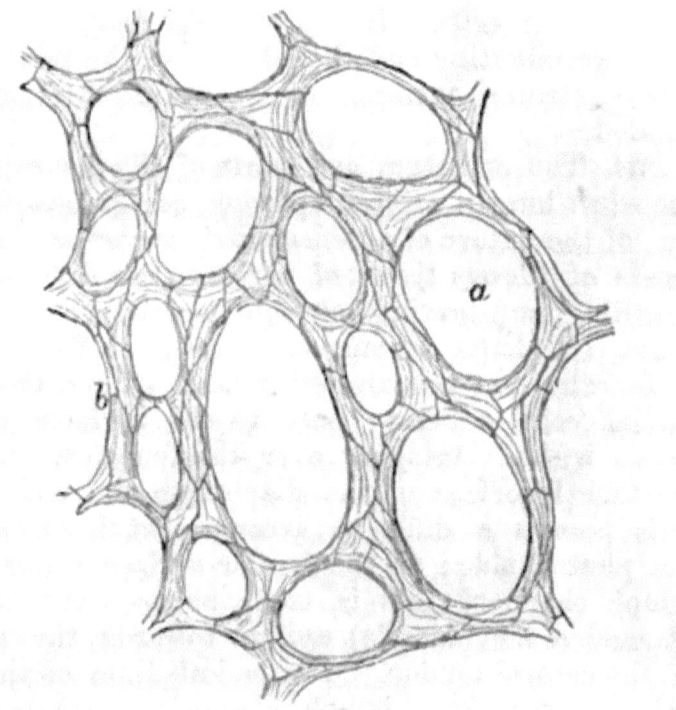

Fig. 21.—Part of Omentum of Cat, stained with Nitrate of Silver.
a, Fenestræ or holes ; *b*, trabeculae covered with endothelium. Only the outlines (silver-lines) of the endothelial cells are shown.

face of the serous membranes, especially the diaphragm (Fig. 20) and pleura, there exist minute openings, *stomata*, leading from the serous cavity into a lymphatic vessel of the serous membrane. These stomata are often lined by germinating cells.

33. In the frog, germinating cells occur in great abundance on the mesogastrium and the part of the peritoneum which separates the peritoneal cavity from the cisterna lymphatica magna. This part of

the peritoneum is called the septum cisternæ lymphaticæ magnæ, and on it occur numerous holes or stomata, by which a free communication is established between the two cavities. On the peritoneal surface of this septum the stomata are often bordered by germinating cells. In the female frog, these and other germinating endothelial cells of the peritoneum (mesogastrium, mesenterium, septum cisternæ) are ciliated.

34. The omentum and parts of the pleura are, in the adult human subject, ape, dog, cat, guinea-pig, rat, &c., of the nature of a *fenestrated membrane* (Fig 21), bands of fibrous tissue of various sizes dividing and reuniting, and leaving between them larger or smaller holes, in shape oblong or circular. These holes or fenestræ are not covered with anything, the endothelial cells adhering only to the surfaces of the bands without bridging over the fenestræ. On the peritoneal surface of the diaphragm the endothelial cells possess a different arrangement from that on the pleural side; on the former surface a number of lymph channels (that is, clefts between the bundles of tendon and muscle) radiate towards the middle of the central tendon. The endothelium of the free surface over these lymph channels is composed of much smaller cells than at the places between, so that the endothelium of the peritoneal surface of the diaphragm shows numbers of radiating streaks of small endothelial cells. Many of these small cells are not flattened, but polyhedral, and of the nature of germinating cells (Fig. 20.) The above-mentioned stomata occur amongst these small endothelial cells.

CHAPTER V.

FIBROUS CONNECTIVE TISSUES.

35. By the name of "connective tissues" we designate a variety of tissues which have these things in common—that they are developed from the same embryonal elements; that they all more or less serve as supporting tissue or framework, or connecting substance, for nervous, muscular, glandular, and vascular **tissues**; that they are capable of taking one another's place in the different classes of animals; that in the embryo and in the growing condition one may **be** changed into the other; and that in the adult **they** gradually shade off one into the other.

Connective tissues are divided into the three great groups of (1) fibrous connective tissue; (2) cartilage; (3) bone, to which may be added dentine. Each of these is subdivided into several varieties, as will appear farther on, but in all instances the *ground substance*, or *matrix*, or *intercellular substance*, is to be distinguished from *the cells*. In the fibrous connective tissue the matrix yields *glutin or gelatin*, and the cells are called *connective tissue cells*, or connective tissue corpuscles. In the cartilage the ground substance yields *chondrin*, and the cells are called *cartilage cells*. In the third group the ground substance contains inorganic lime salts, intimately connected with a fibrous matrix, and the cells are called *bone cells*.

36. The **fibrous connective tissue, or white fibrous tissue,** occurs in the skin and mucous membranes, in the serous and synovial membranes, in the membranes of the brain and spinal cord, in

tendons and tendon sheaths, in fasciæ and aponeuroses, in the intermuscular tissue, and in the tissue connecting neighbouring organs, &c. It consists of microscopic band-like or cylindrical *bundles* or fasciculi of exceedingly fine homogeneous fibrils (Fig. 22), which are known as the *elementary connective tissue fibrils*.

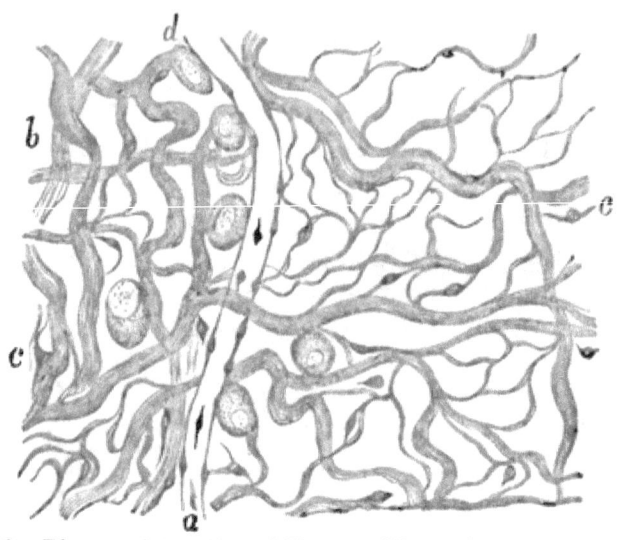

Fig. 22.—Plexus of Bundles of Fibrous Tissue from the Omentum of Rat.

a, A capillary blood vessel; *b*, bundles of fibrous tissue; *c*, the connective-tissue corpuscles; *d*, plasma cells. (Atlas.)

According to the number of these, the bundles differ in size. The bundles, and also their constituent fibrils, may be of very great length—several inches. Where the fibrous tissue forms continuous masses—as in tendon, fascia, aponeurosis, skin, and mucous membrane—the microscopic bundles are aggregated into smaller or larger groups, the *trabeculæ*, and these are again associated into groups. The fibrils are held together by an *albuminous* (globulin), semi-fluid, homogeneous *cement substance*, which is also present between the bundles forming a trabecula.

On adding an acid or an alkali to a bundle of

fibrous tissue, it is seen to swell up and to become glassy-looking, homogeneous, and gelatinous. Subjected to boiling in water, or to digestion by dilute acids, the bundles of fibrous tissue yield glutin or gelatin.

37. According to the arrangement of the bundles, the fibrous connective tissue varies in different localities. (1) In tendons and fasciæ the bundles are arranged parallel to one another. (2) In the true skin and mucous, serous, and synovial membranes, in the dura mater and tendon sheaths, the trabeculæ of bundles divide repeatedly, cross and interlace very intimately with one another, so that thereby a dense felt-work is produced. (3) In the subcutaneous, submucous, or subserous tissue, in the intermuscular tissue, in the tissue connecting with one another different organs or the parts of the same organ—*i.e.*, interstitial connective tissue—the texture of the fibrous tissue is more or less loose, the trabeculæ dividing and re-uniting and crossing one another, but leaving between them larger or smaller spaces, cellulæ or areolæ, so that the tissue assumes the character of a loose plexus, which is sometimes called "areolar" or "cellular tissue." Such tissue can be more or less easily separated into larger or smaller lamellæ, or plates of trabeculæ. (4) In the omentum and parts of the pleura of man, ape, dog, cat, and some rodents, and in the subarachnoidal tissue of the spinal cord and brain, the trabeculæ form a *fenestrated membrane*, with larger or smaller oval or circular holes or fenestræ.

38. The **connective tissue cells** or corpuscles occurring in white fibrous tissue are of several varieties. (*a*) In tendon and fasciæ the cells are called *tendon cells* or tendon corpuscles; they are flattened nucleated protoplasmic cells of a square or oblong shape (Fig. 23A), forming continuous rows (single files), situated on the surface of groups of bundles of fibrous

tissue. Between these groups are wider or narrower channels—the *interfascicular spaces*—running parallel with the long axis of the tendon (Fig. 23B). The cells in each row are separated from one another by a narrow line of albuminous cement substance, and the round nucleus of the cell is generally situated at one end, in such a way that in two adjacent cells of the growing tendon the nuclei face each other. This indicates that the individual cells undergo division. Corresponding to the margin of each row, the cells possess minute processes. The cell plate is not quite flat, but possessed of one, two, or even three membranous projections, by which it is wedged in between the individual bundles of the group to which the row of cells belongs.

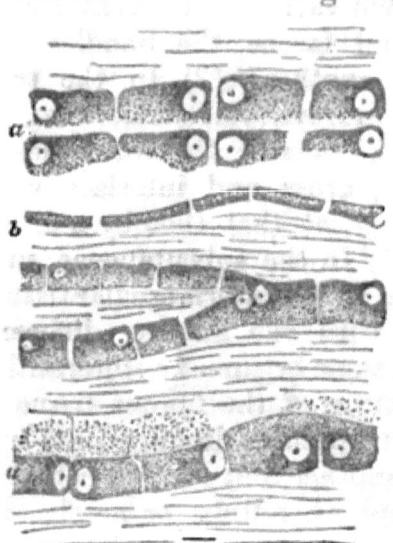

Fig. 23A.—From a Tendon of Tail of Mouse, showing the Tendon cells. (The tendon is viewed in the long axis).

a, The tendon cells seen from their broad surface; *b*, the same seen sideways. (Handbook.)

39. (*b*) In the serous membranes, cornea, subcutaneous tissue, and loose connective tissues, the cells are flattened transparent corpuscles, each with an oblong flattened nucleus, and more or less branched and connected by their processes. In the cornea they are spoken of as the *corneal corpuscles*, and are very richly branched (Fig. 25). They are situated between the lamellæ of fibrous bundles of which the ground substance of the cornea consists.

These corpuscles are also situated in the interfascicular spaces, or spaces left between the bundles of the ground substance, which are cavities in the interstitial substance cementing the bundles and trabeculæ together (von Recklinghausen). In the cornea and serous membranes these spaces possess the shape of branched lacunæ, each lacuna being the home of the body of the cell, while the branches or canaliculi contain its processes. These canaliculi form the channels by which neighbouring lacunæ anastomose with one another (Fig. 26). The cell and its processes do not fill up the lacuna and its canaliculi.

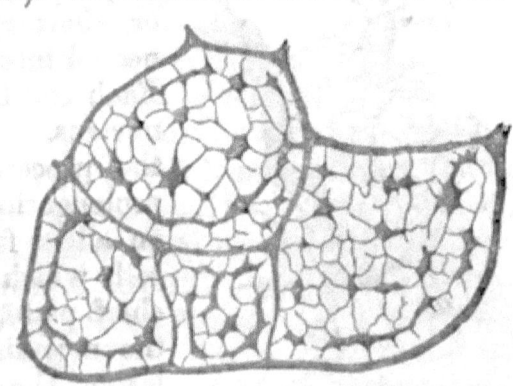

Fig. 23B.—From a Transverse Section through the Tendons of the Tail of a Mouse, stained with gold chloride.

Several fine tendons are shown here. The dark branched corpuscles correspond to albuminous cement substance stained with gold chloride; they are the channels between the bundles of fibrous tissue, constituting the tendon, and seen here as the clear spaces in cross section. In each of these channels is a row of tendon cells—not discernible here, the long axis of these rows being parallel with the long axis of the tendon. (Handbook.)

In loose connective tissue the lacuna may be of considerable size, and may contain several connective cells, which make as it were a lining for it. These in some places are very little branched, and almost form a continuous endotheloid membrane of flattened cells. Such is the *subepithelial endothelium of Debove*, occurring *underneath* the epithelium *on the surface* of the mucous membrane of the bronchi, bladder, and intestines.

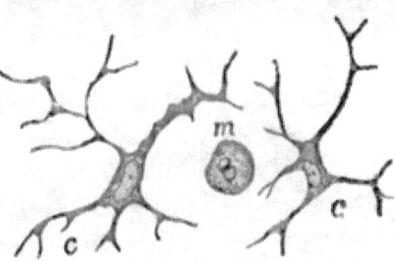

Fig. 24.—From the Tail of a Tadpole.

c, Branched connective tissue cells; m, a migratory cell. (Atlas.)

40. (c) In the true skin and mucous membranes

the connective tissue cells are also branched flattened corpuscles, and by their longer or shorter processes are connected into a network (Fig. 24). Each cell has a flattened oblong nucleus. As a rule, some of the processes are membranous prolongations coming off under an angle from the body of the cell, which is then called the chief plate, the processes being the secondary plates. By the latter the cell is wedged in between the bundles of the trabecula to which it belongs.

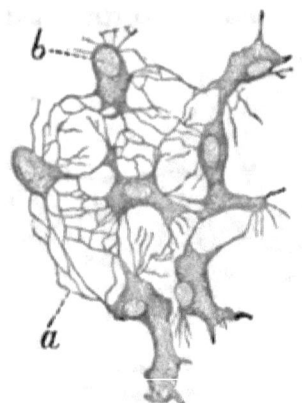

Fig. 25.—From the Cornea of Kitten, showing the Networks of the Branched Corneal Corpuscles.

a, The network of their processes; b, nucleus of the corpuscle. (Atlas.)

This character of the cells (i.e., of possessing secondary plates) is well shown by the cells of the skin and mucous membranes, but only in a very limited degree by those of the cornea and serous membranes, and somewhat better by some of those of the subcutaneous and other loose connective tissues.

In the skin and mucous membranes also the cells and their processes are situated in the interfascicular spaces.

41. The connective tissue corpuscles

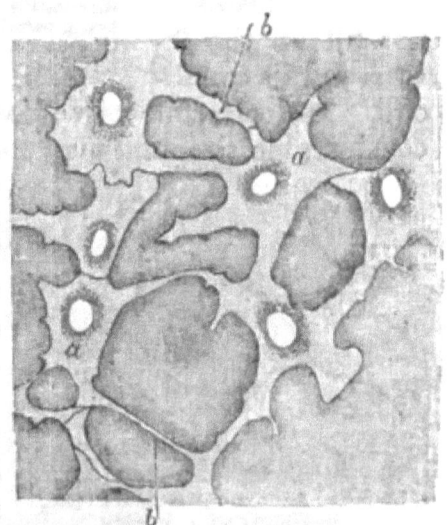

Fig. 26.—From the Cornea of Kitten, stained with Nitrate of Silver, showing the Lymph-canal System.

a, The lacunæ, each containing the nucleated cell-body, just indicated here; b, the canaliculi for the cell processes. (Atlas.)

hitherto mentioned are fixed corpuscles; they do not show movement. Kühne and Rollett ascribe to the corneal corpuscles a certain amount of contractility, inasmuch as they are said to be capable of withdrawing their processes on stimulation. When this ceases they are said again to protrude them. According to Stricker and Norris, they acquire contractility when the corneal tissue is the seat of inflammatory irritation. It can be shown that the connective tissue cells consist, like the endothelial plates, of a ground plate and a fibrillar reticulated (granular-looking) substance around the nucleus, and extending beyond the ground plate into the processes of the cell.

42. **Pigment Cells.**—In the lower vertebrates, especially fishes, reptiles, and amphibian animals, we find certain branched nucleated connective tissue corpuscles, distinguished by their size and by the protoplasm both of the cell-body and processes—but not of the nucleus—being filled with pigment granules. The pigment is either white or yellow, or more commonly dark brown to black. These cells are called pigmented connective tissue cells, or simply *pigment cells*. They are very numerous in the skin of fishes, reptiles, and amphibian animals, and also around and between the blood-vessels of the serous membranes. They are also present in man and mammals, but then they are chiefly limited to the eye-ball, where they occur in **the** proper tissue of the iris of all but albino and bright blue eyes, and in the tissue of the choroid membrane. In dark eyes of mammals a large number of these cells are found in the tissue between the sclerotic **and** choroid, as the lamina fusca, and also, but to a more limited degree, in the sclerotic. As a rule they appear to be of various kinds: such as are flattened, large plates perforated by a number of small and large holes and minute clefts; such as **possess** a more spindle-shaped body, and long, thin, **not** very richly branched

processes; and intermediate forms between the two. But on careful examination it will be seen that these

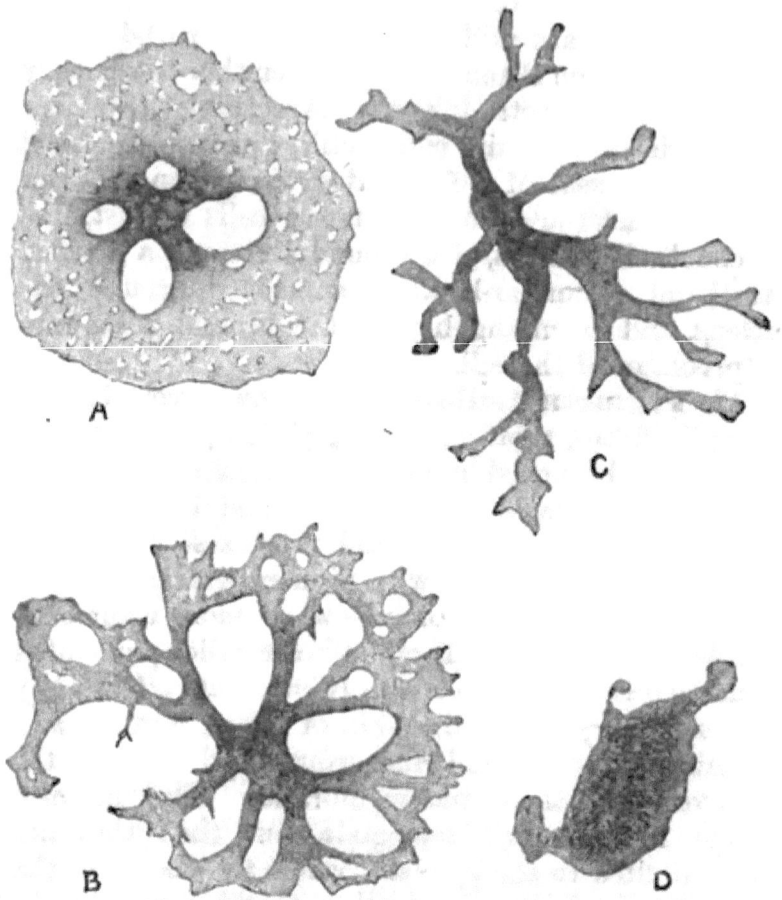

Fig. 27.—Pigment Cells of the Tail of Tadpole.
A, B, C, D represent various states; A being a cell in an uncontracted or passive state, D in a contracted or active state.

appearances are due to different states of contraction of the same kind of cells (Fig. 27).

43. In lower vertebrates the dark pigment cells show marked contractility, inasmuch as they are capable of altogether withdrawing into their body the pigmented processes. In the passive state these are

exceedingly numerous, and form a network so dense that the whole mass of cells resembles an extremely close network of pigment. In the maximum of activity the processes disappear, being withdrawn into the cell-body, which now looks like a spherical or oblong mass of black pigment. Between the states of passiveness and maximum activity there are various intermediate grades, in which the pigmented processes are of various numbers and lengths.

44. Owing to the great number of the pigment cells in the skin of fishes and amphibian animals, the state of contraction of these cells materially affects the colour of the skin. If the dark pigment cells of a particular part contract, the skin of this particular part will become lighter and brighter, the degree of lightness and brightness depending on the degree of contraction of the pigmented processes by the cells. Brücke has shown that darkness is a stimulus to the pigmented cells; they contract, and the skin becomes light. Sunlight leaves the pigmented cells in the passive state, *i.e.*, the skin becomes dark. If previously they have been contracted by darkness, on being exposed to sunlight they again return to the passive state. The contraction of the pigment cells is under the direct influence of the nervous system (Lister). Pouchet proved that the contractility of the pigment cells of the skin of certain fishes is influenced as a reflex action by the stimulation of the retina by light.

45. **Fat Cells.**—Fat cells in the ripe and fully-formed state are spherical, relatively large vesicles, each consisting (*a*) of a thin *protoplasmic membrane*, which at one point includes *an oval nucleus* flattened from side to side, and (*b*) of a substance, which is *an oil globule* filling the cavity of the vesicle (Fig. 28). These fat cells are massed together by fibrous connective tissue into smaller or larger *groups*, which in their

turn form *lobules;* these again become *lobes*, and these make continuous masses. Each group and lobule has its *afferent arteriole*, one or two *efferent veins*, and a *dense network of capillaries* between; each mesh of the capillary network holding one, two or three fat cells. (Fig. 49.) Such are the

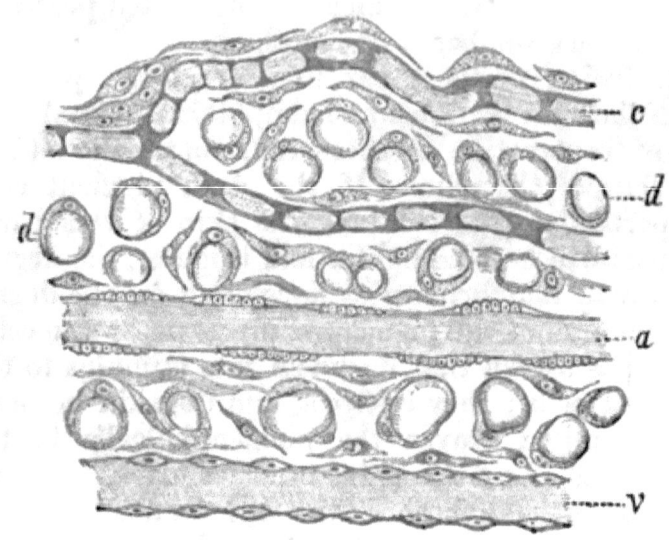

Fig. 28.—From a Preparation of the Omentum of Guinea-pig. *a,* An artery; *v,* vein; *c,* young capillary blood-vessel; *d,* fat cells. (Atlas.)

nature and arrangement of the fat tissue in the subcutaneous and submucous tissue, in the serous and synovial membranes, in the intermuscular tissue, in the loose tissue connecting organs or parts of organs. It can be shown that fat cells are derived from ordinary connective tissue cells. In some places— both in the embryo and adult—the protoplasm of the connective tissue corpuscles growing in size becomes filled with small oil globules, which, increasing in numbers, become fused with one another to larger globules; as their size thus increases the cell nucleus becomes shifted to the periphery; ultimately one large oil globule fills the cell, and what is left of the cell protoplasm surrounds this oil globule like a membrane. The cell as a whole has become in this process many times its original size.

46. It can also be shown that where at one time **only few** isolated connective tissue corpuscles are present, at another time, in the natural state of growth, and especially under very favourable conditions of nutrition, the connective tissue cells become increased by cell-multiplication so as to form groups; these continue to increase in size and to be gradually furnished with their own system of bloodvessels; the individual **cells** constituting the group become **then converted into** fat cells, and their processes are thereby **lost**.

Individual connective **tissue cells** situated in the neighbourhood of small blood-vessels are converted into fat cells under favourable conditions of nutrition.

In starvation the fat cells lose their oil globule, they become smaller and contain a serous fluid, which may ultimately also disappear. Finally, the fat cell may be reduced to a small, solid, protoplasmic, slightly branched cell.

47. In many places the fibrous connective tissue includes, besides the fixed cells, others **which** show amœboid movement. These **are of** two kinds: (1) *migratory* or *wandering cells*. These are **ident**ical with colourless blood corpuscles as regards size, shape, aspect, and general nature (Fig. 24, *m*). They wander about through the spaces of the fibrous tissue. Some of them are slightly larger, and possess one spherical relatively large nucleus. The amœboid movement of these cells is not so distinct as in the smaller variety. (2) *Plasma cells* of Waldeyer. They are larger than the former, less prone to migrating, being possessed of only slight amœboid movement, which is, however, sufficiently pronounced to be detected.

They contain always coarse granules, which are composed of a substance which is not fat, but something between protoplasm and fat. They stain deeply in dyes, and the corpuscles correspond to similar "granular" corpuscles of the blood. These "granules" may change into fat globules, and thus the plasma cell becomes transformed into a fat cell.

48. The wandering cells occur almost in all loose fibrous tissues, chiefly around or near blood-vessels; they are not numerously met with in the healthy state, but increase greatly in the state of inflammation of the part. The larger kinds are met with in certain localities only; in the sub-lingual gland of the dog and guinea-pig they occur in numbers between the gland tubes or acini. They are also found in the mucous membrane of the intestine. The plasma cells are met with chiefly in the intermuscular tissue, in the mucous and sub-mucous tissue of the intestine, in the trabeculæ of the lymphatic glands, and in the omentum.

49. **Development of fibrous tissue.**—Fibrous connective tissue is developed from embryonal connective tissue cells, *i.e.*, from spindle-shaped or branched nucleated protoplasmic cells of the mesoblast. The former are met with isolated or in bundles, as in the umbilical cord or embryonal tendon. The latter form a network, as in the fœtal skin and mucous membrane. In both instances the protoplasm of the embryonal connective tissue cells becomes gradually transformed into a bundle of elementary fibrils, with a granular-looking interstitial substance. The nucleus of the original cell finally disappears. A second mode of the formation of fibrous connective tissue is this: the embryonal connective tissue cell, while growing in substance, produces the fibrous tissue at the expense of its peripheral part. A remnant of the protoplasm persists around the nucleus.

The same modes of formation of connective tissue may be also observed in the adult under normal and pathological conditions.

50. Fibrous connective tissue is in most places associated with *elastic fibres* or yellow elastic tissue. These are of bright aspect, of variable thickness and length, branching and anastomosing so as to form networks (Fig. 29). They are straight or more or less twisted and coiled. The latter condition may be observed when the tissue is shrunk, the former when it is stretched. They do not swell up in acids or alkalies, nor yield glutin or gelatin on boiling, but contain a chemically different substance, viz., *elastin*. When broken their ends generally curl up.

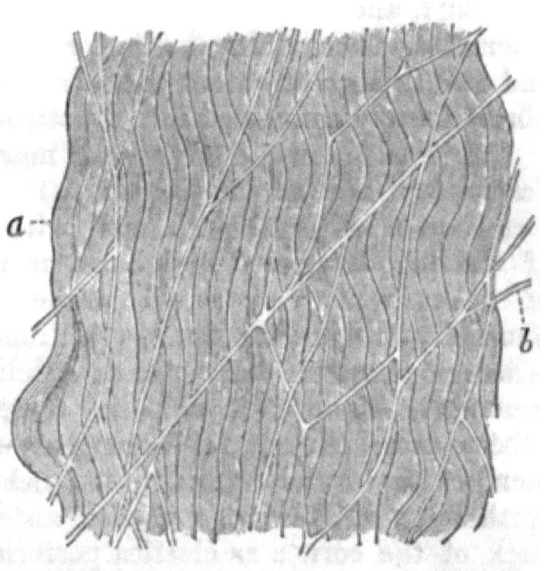

Fig. 29.—From a Preparation of the Mesentery.
a, Bundles of fibrous tissue; *b*, networks of elastic fibres. (Atlas.)

51. Elastic fibres occur in great numbers as networks extending between the bundles of fibrous tissue, in the skin and mucous membranes, in the serous and synovial membranes, and in the loose interstitial connective tissues. They are not very commonly met with in tendons and fasciæ; in the former they are seen as single fibres often twisting round the tendon bundles.

Elastic fibres forming bundles, but branched and

connected into networks within the bundle, are to be specially found in considerable numbers in the walls of the alveoli of the lung, in the ligamenta flava, in the ligamentum nuchæ of the ox—in which the fibres are exceedingly thick cylinders,—in yellow elastic cartilage (*see* below), in the endocardium and valves of the heart, and in the vascular system, particularly the arterial division. In the latter organs the intima, and also to a great extent the media, consist of elastic fibrils densely connected into a network.

52. The following are special morphological modifications of the elastic fibres: (*a*) *elastic fenestrated membranes* of Henle, as met with in the intima of the big arteries; these are in reality networks of fibres with very small meshes, and the fibres unusually broad and flat. (*b*) *Homogeneous elastic membranes*, which surround, as a delicate sheath, the connective tissue trabeculæ in some localities, *e.g.*, subcutaneous tissue. (*c*) Homogeneous-looking elastic membranes in the cornea, found behind the anterior epithelium as *Bowman's elastica anterior*, and at the back of the cornea as elastica posterior, or *Descemet's membrane;* in this latter bundles of minute fibrils have been observed. (*d*) Elastic trabeculæ forming a network, as in the ligamentum pectinatum iridis. In the embryonal state the elastic fibres are nucleated, the nuclei being the last remnants of the cells from which the fibres develop—one cell generally giving origin to one fibre. Such nucleated fibres are called Henle's nucleated fibres.

53. Special varieties of fibrous connective tissue are these :—

(1) *Adenoid reticulum.* This is a network of fine fibrils, or plates, forming the matrix of lymphatic or adenoid tissue (*see* Lymphatic Glands). The reticulum is not fibrous connective tissue nor elastic tissue; it contains nuclei in the young state, being derived from

a network of branched cells; but in the adult state the reticulum itself possesses no nuclei. Those found on it do not form an essential part of it.

(2) The *neuroglia* of Virchow is a dense network of very fine homogeneous fibrils forming the supporting tissue for the nervous elements in the central nervous system. These fibrils are supposed to be elastic fibres (Gerlach). Embedded in the network of these fibres are found branched nucleated flattened cell plates, which are the proper connective tissue cells.

(3) *Gelatinous tissue.* This occurs chiefly in the embryo, being the unripe state of fibrous connective tissue. It consists of spindle-shaped or branched connective tissue cells, separated from one another by a homogeneous transparent mucoid substance. It is met with in the umbilical cord of the embryo, and in the places where fibrous connective tissue is to be developed. After birth it is found in the tissue of the pulp of the teeth, and in the cavity of the middle ear, and in some places as precursor of fat tissue.

CHAPTER VI.

CARTILAGE.

54. Cartilage consists of a firm ground substance which yields *chondrin*, and of cells embedded in it. Most cartilages (except on the articulation surface) are covered on their free surface with a membrane of fibrous connective tissue with a few elastic fibrils. This membrane is supplied with blood-vessels, lymphatics, and nerves, and is of essential importance for the life and growth of the cartilage. This is the *perichondrium*. There are three varieties of cartilage.

55. (1) *Hyaline cartilage* (Fig. 30A). This occurs at the articular surfaces of all bones; on the borders of many short bones; at the sternal part of the ribs, as *costal cartilages;* at the margin of the sternum, scapula, and os ileum; in the rings of the trachea, the cartilages of the bronchi, the septum and lateral cartilages of the nose; and in the thyroid and cricoid cartilages of the larynx. The ground substance is hyaline, transparent, ground-glass-like, and firm. The cells are spherical or oval protoplasmic corpuscles, each with one or two nuclei. They undergo division, and although the two offspring are at first close together—half-moon-like —they gradually grow wider apart by the deposit of hyaline ground substance between them. The cells are contained in special cavities called the *cartilage lacunæ.* Each cell generally occupies one lacuna, but according to the state of division a lacuna may contain two, four, six, or eight cartilage cells; the latter are those cases in which division proceeds at a more rapid rate than the deposition or formation of hyaline ground substance between the cells.

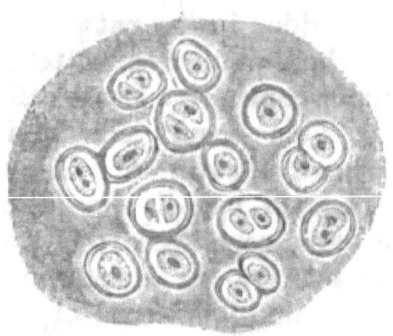

Fig. 30A.—Hyaline Cartilage of Human Trachea.

In the hyaline ground substance are seen the cartilage cells enclosed in capsules.

The part of the cartilage next to the perichondrium shows most active growth; hence the cells are here smaller, closer together, and there is less ground substance.

Each lacuna is limited by a delicate membrane, and, according to the state of the cell, is either completely or partially filled out by it. This membrane is called *the capsule* (Fig. 30A). In many cartilages, especially in growing cartilage, it is thickened by the addition of a layer or layers of hyaline ground substance; this is the most recently-formed part of the matrix, but is still distinct from the rest of the ground substance.

56. In some places, especially in articular cartilage (Tillmanns, Baber), bundles of fine connective tissue fibrils may be noticed in the hyaline ground substance.

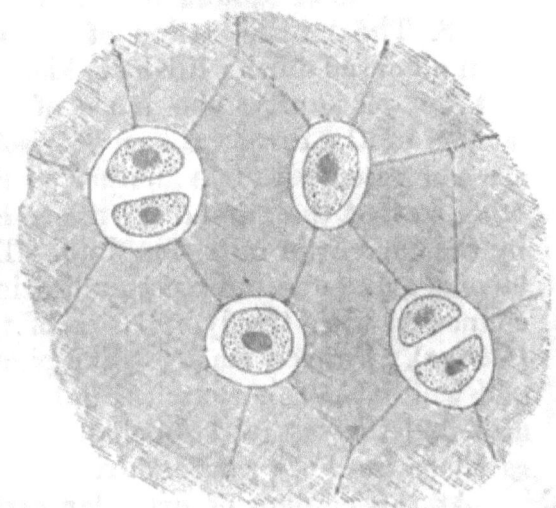

Fig. 30B.—From a Preparation of Sternal Cartilage of Newt.
The lacunæ, containing the cartilage cells, anastomose by fine channels.

57. In some cartilages, the protoplasm of the cell becomes filled with fat globules (Fig. 30c). This fact may be observed in many normal cartilages; sometimes the fat globules become confluent into one large drop, and then the cell has the appearance of a fat cell. In age, disease, and deficient nutrition, lime salts are deposited in the ground substance, beginning from the circumference of the cells. The lime matter appears in the shape of opaque granules, or irregular or angular clumps. The ground substance thereby loses its transparency, becomes opaque in transmitted, white in reflected, light, and, of course, very hard and brittle. This process is the *calcification* of cartilage. It is also met with in cartilage that is to be

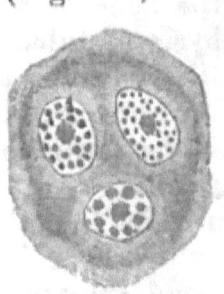

Fig. 30c.—Three Cartilage Cells filled with Fat Droplets. From the hyaline cartilage of the nasal septum of Guinea-pig.

replaced by bone, being the precursor of the formation of bone, as in the embryo (*see* below), and at the growing ends of tubular bones.

58. The multiplication of **the** cartilage cells has **been** observed during life by Schleicher and Flemming. **It takes place** after **the** mode **of** karyokinesis. The **lacunæ of** the cartilage are not isolated cavities, but are connected with one another by fine channels (Fig. 30B), so that the ground substance is easily permeable by the current of nutritive fluid. These channels and lacunæ make one intercommunicating system, and are connected with the lymphatics of the perichondrium (Budge). Formed matter—like pigment granules, red and white blood corpuscles, and pus corpuscles—may also find its way into the channels and lacunæ of the cartilage from the perichondrium.

At **the** borders **of** articular cartilage, where it is joined to the synovial membrane and the articulation-**capsule,** the cartilage cells are more or less branched, **and** pass insensibly into **the branched** connective tissue **cells** of the membrane. In fœtal hyaline cartilage **many** of the cells are spindle-shaped **or** branched.

59. In the cartilage separating the bone of the apophyses from the end of the diaphysis of tubular bones, there is a peculiar hyaline cartilage, known as the *intermediary* or *ossifying*. Its cells are arranged in characteristic vertical rows, owing to the continued division of the cells in a transverse direction.

Cartilages, or parts of cartilages, in which the cells are very closely placed, owing to the absence, **or** imperfect deposit and formation, of ground substance, are called parenchymatous.

60. (2) **Fibro-cartilage,** or connective tissue cartilage, occurs as the intervertebral discs, as the interarticular cartilages, sesamoid cartilages, and as that forming the margin of a fossa glenoidalis. *It is fibrous connective tissue arranged in bundles*, and these

again in layers. The ground substance of this cartilage is said (?) to yield chondrin and not glutin. Between the strata of the fibrous bundles are rows of more or less flattened oval protoplasmic nucleated cells, each invested in a delicate capsule(Fig. 31). They are less flattened than the cells of tendon, and the capsule distinguishes the two. Where fibro-cartilage passes into tendinous tissue, the two kinds of cells pass insensibly into one another.

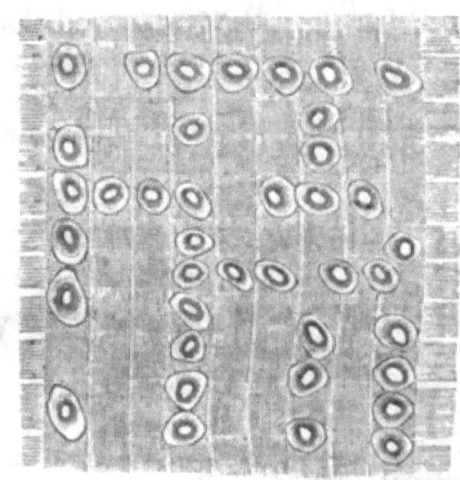

Fig. 31.—Fibro-Cartilage of an Intervertebral Ligament.
Showing the bundles of fibrous tissue and rows of cartilage cells. (Atlas.)

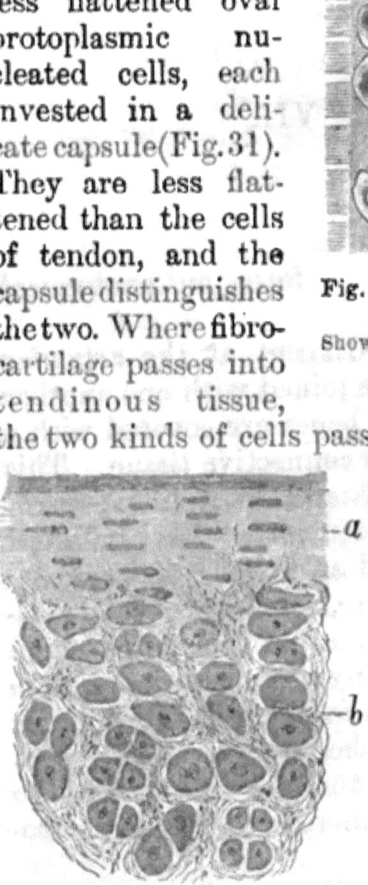

61. (3) **Yellow, or elastic cartilage.**—This variety is also called reticular; it occurs in the epiglottis, in the ear-lobe, in the Eustachian tube, in the cartilages of Wrisberg and Santorini in the larynx. In the early stage this kind is hyaline. Gradually numbers of elastic fibrils make their appearance, growing into the cartilage matrix from the perichondrium in a more or less vertical direction, and branching and anastomosing with one

Fig. 32.—From a Section through the Epiglottis.
a, Perichondrium; b, networks of elastic fibrils surrounding the cartilage cells. (Atlas.)

E

another. The final stage is reached when the ground substance is permeated by *dense networks of elastic fibrils* (Fig. 32), so arranged that spherical or oblong spaces are left, each of which contains one or two *cartilage cells*, surrounded by a smaller or larger zone of *hyaline cartilage ground substance*.

CHAPTER VII.

BONE.

62. BONE, or osseous substance, is associated with several other soft tissues to form an anatomical individual.

(a) The **periosteum.**—Except at the articular surfaces, and where bones are joined with one another by ligaments or cartilage, all bones are covered with a vascular membrane of fibrous connective tissue. This is the periosteum. It consists in most instances of an outer *fibrous layer*, composed of bundles of fibrous tissue densely aggregated, and an *inner*, or *osteogenetic layer*, which is of loose texture, consisting of a meshwork of thin bundles of *fibrous tissue*, in which numerous *blood-vessels* and many protoplasmic cells are contained. The blood-vessels form by their capillaries a network. The cells are spheroidal or oblong, each with one spherical or oval nucleus. They have to form bone-substance, and are therefore called the *osteoblasts* (Gegenbaur).

(b) The **cartilage** is hyaline cartilage, and its distribution on and connection with bone have been mentioned in §§ 55 and 59.

63. (c) The **marrow of bone** is a vascular soft tissue, filling up all spaces and cavities. It consists

of a *small amount of fibrous tissue* as a matrix, and in it are embedded *numerous blood-vessels* and *cells*. The few afferent arterioles break up into a dense network of capillaries, and these are continued as plexuses of veins, characterised by their size and exceedingly thin walls. The cells are of the same size, aspect, and shape as the osteoblasts of the osteogenetic tissue, and they are called *marrow cells*.

In origin and structure, the tissue of the *osteogenetic layer of the periosteum* and *the marrow* are identical. In the embryo, the marrow is derived from an ingrowth of the osteogenetic layer of the periosteum (*see* below), and also in the adult the two tissues remain directly continuous. As will be shown later, the marrow at the growing ends of the bones is concerned in the new formation of osseous substance in the same way as the osteogenetic layer of the periosteum is in that of the surface; and in both tissues the highly vascular condition and the cells (osteoblasts of the osteogenetic layer, and marrow cells of the marrow) are the important elements in this bone formation. Marrow is of two kinds, according to the condition of the cells. If many or most of these are transformed into fat cells, it has a yellowish aspect, and is called *yellow marrow;* if few or none of them have undergone this change, it looks red, and is called *red marrow*. In the central, or marrow, cavity of the shaft of tubular bones, and in the spaces of some spongy bones, the marrow is yellow; at the ends of the shaft, in the spongy bone substance in general, and in young growing bones, it is red.

The cells, especially those of red marrow, are the elements from which normally vast numbers of red blood-corpuscles are formed, as has been mentioned on a former page.

In marrow, particularly in red marrow, we meet with huge multinucleated cells, called *Myeloplaxes of*

Robin. They are derived by overgrowth from ordinary marrow cells, and are of importance for the absorption and formation of bone (*see* below). According to Heitzmann, Malassez, and others, they also have to do with the **formation of** blood-vessels and blood-corpuscles.

64. The **matrix of osseous substance** is dense fibrous connective tissue, *i.e.*, a tissue yielding gelatin on boiling. The cement substance between the fibrils is petrified, owing to a deposit of insoluble inorganic lime salts, chiefly carbonates and phosphates. These can be dissolved out by strong acids (hydrochloric) and are thereby converted into soluble salts. Thus the organic matrix of osseous substance—called *ossein* —may be obtained as a soft flexible material, easily cut.

The bone substance is in the adult state generally *lamellated*, the lamellæ being of microscopic thinness. Between every two lamellæ are numbers of isolated, flattened, oblong spaces—the *bone lacunæ* (Fig. 33), which anastomose by numerous fine canals with one another, and also with those of the next lamella above and below. The appearances are very similar to those presented by the lacunæ and canaliculi containing the corneal corpuscles, as described in Chapter V.

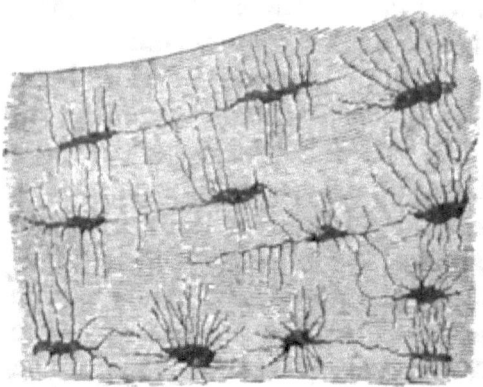

Fig. 33.—Osseous Lamellæ; oblong branched bone lacunæ and canaliculi between them. (Atlas.)

These bone lacunæ and their canaliculi are the lymph-canalicular system of osseous substance, for they are in open and free communication with lymphatic vessels of the marrow spaces and Haversian canals.

65. In the bone matrix, each lacuna contains also a nucleated protoplasmic cell, called *the bone cell*, which, however, does not fill it completely. In the young state, the cell is branched, the branches passing into the canaliculi of the lacunæ; but in the old state very few processes can be detected on a bone cell, which, with its lacuna and canaliculi, is called a *bone corpuscle*.

66. According to the arrangement of the bone substance, we distinguish *compact* from *spongy* substance. The former occurs in the shaft of tubular bones and in the outer layer of flat and short bones. Its lamellæ are arranged as: (*a*) *concentric* or *Haversian lamellæ*, directly surrounding the *Haversian canals* (Fig. 33A). These are fine canals of varying lengths pervading the compact substance in a longitudinal direction, and anastomosing with one another by transverse or oblique branches. The Haversian canals near the marrow cavity are larger than those near the periosteum. As a matter of fact, those next to the marrow cavity become gradually enlarged by absorption, until finally they are fused with the marrow cavity. Each Haversian canal contains a blood-vessel, one or two lymphatics, and a variable amount of marrow tissue. These canals open both into the marrow cavity and on the outer surface into the osteogenetic layer of the periosteum, and they form the means by which the latter remains in continuity with the marrow. They are surrounded by *numbers of concentric bone lamellæ, with the bone corpuscles between them*, and this is a system of *concentric lamellæ*. Near the external surface of the compact substance the number of lamellæ in each system is smaller than in the deeper parts. (*b*) Between these systems of concentric lamellæ are the *interstitial* or *ground lamellæ*; they run in various

directions, and in reality fill the interstices between the systems of the Haversian or concentric lamellæ. Near the external surface of long bones they have preeminently a direction parallel to the surface. These are the *circumferential lamellæ* of Tomes and de Morgan.

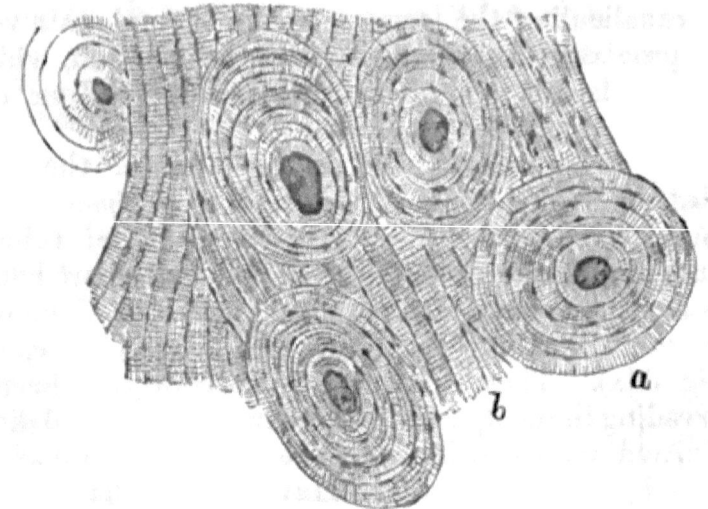

Fig. 33A.—Compact Bone Substance in Cross Section.

a, Concentric lamellæ arranged around the Haversian canals, cut across; *b*, interstitial or ground lamellæ. The bone lacunæ are seen between the bone lamellæ. (Atlas.)

The lamellæ of compact bone are perforated by perpendicular petrified fibres, the *perforating fibres* of Sharpey. They form a continuity with the fibres of the periosteum, from which they are developed.

Some of these fibres are fine, and of the nature of elastic fibres.

67. Spongy bone substance occurs in the end of the shaft, in the apophyses, in short bones, and in the diploë of flat bones. The cavities or meshes of the spongy substance are called Haversian *spaces;* they intercommunicate with one another, and are filled with marrow, which in the young and growing state is generally of the red variety. The firm parts are of the shape of spicules and septa, called *bone trabeculæ*,

of varying length and thickness, and are composed of lamellæ of bone substance.

According to the arrangement of the trabeculæ, the spongy substance is a uniform honey-combed substance, or appears longitudinally striated, as in the end of the shaft. In the latter case the marrow spaces are elongated and the trabeculæ more or less parallel, but anastomosing with one another by transverse branches.

68. **Development of bone.**—Bone is developed in the embryo, and continues to be formed also after birth as long as bone grows, either in the cartilage, or independently of this directly from the osteogenetic layer of the periosteum. The former mode is called *endochondral*, the latter *periosteal*, or *intermembranous* formation.

All bones of the limbs and of the vertebral column, the sternum, and the ribs, and the bones forming the base of the skull, are preformed in the early embryo as solid hyaline cartilage, covered with a membrane identical in structure and function with the periosteum, which at a later period it becomes. The tegmental bones of the skull, the bones of the face, the lower jaw, except the angle, are not preformed at all, only a membrane identical with the future periosteum being present, and underneath and from it the bone is gradually being deposited.

69. **Endochondral formation.**— The stage next to the one (1) in which we have solid hyaline cartilage covered with periosteum is the following (2): Starting from the "centre, or point, of ossification," and proceeding in all directions, the cartilage becomes permeated by numbers of channels (cartilage channels) containing prolongations (periosteal processes of Virchow) of the osteogenetic layer of the periosteum, *i.e.*, vessels and osteoblasts, or marrow cells. This is the stage of the *vascularisation of the cartilage*. In the next stage (3) the cartilage bordering on these

channels grows more transparent, the lacunæ becoming enlarged and the cartilage cells more transparent. The latter gradually break down, while the intercellular trabeculæ become *calcified;* the lacunæ themselves, by absorption, fusing with the cartilage channels. These latter thereby become transformed into *irregular cavities*, which are bordered by, and into which project, *trabeculæ of calcified cartilage*. The cavities are the *primary marrow cavities*, and they are filled with the *primary* or *cartilage marrow*, i.e., blood-vessels and osteoblasts, derived, as stated above, from the osteogenetic layer of the periosteum. (4) The osteoblasts arrange themselves by active multiplication as an *epitheloid layer* on the surface of the calcified cartilage trabeculæ projecting into, and bor-

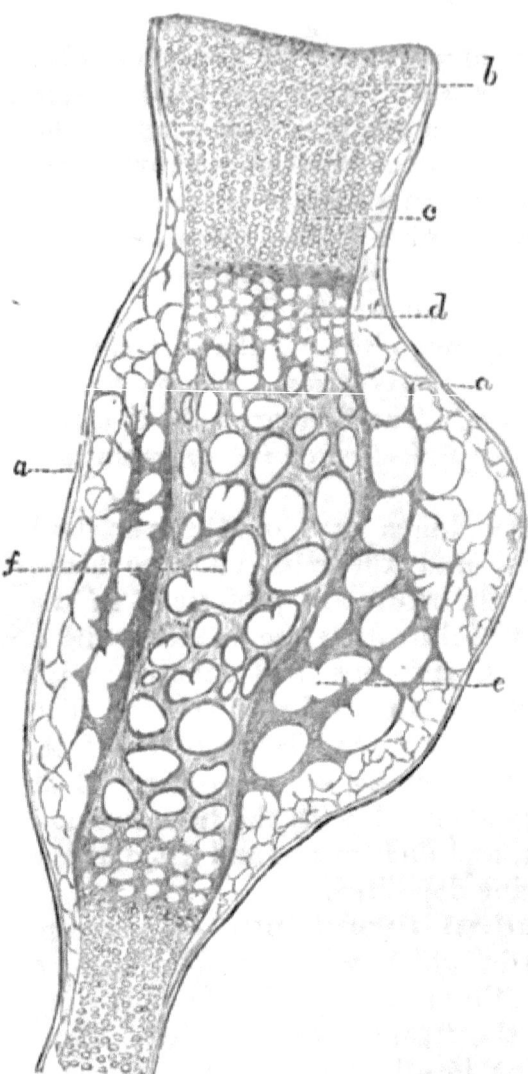

Fig. 34.—Longitudinal Section through the entire Fœtal Humerus of a Guinea-pig.

a, Periosteum ; *b*, hyaline cartilage of the epiphysis ; *c*, intermediate cartilage at the end of the shaft ; *d*, zone of calcification ; *e*, periosteal bone, spongy ; *f*, endochondral bone, spongy.

dering the primary marrow cavities. The *osteoblasts form bone substance*, and as this proceeds, *the calcified cartilage trabeculæ become gradually ensheathed and covered with a layer of osseous substance*,—the osseous matrix and branched bone corpuscles. Thus the original cartilage gradually assumes the appearance of a spongy substance, in which the cavities (primary marrow cavities) are filled with the primary marrow, and are of considerable size, while the trabeculæ bordering them are calcified cartilage covered with layers of new bone. The marrow cells, or osteoblasts, continue to deposit bone substance on the free surface of the trabeculæ, while the calcified cartilage in the centre of the trabeculæ gradually becomes absorbed.

70. The nearer the centre of ossification, the more advanced the process, *i.e.*, the more bone the less calcified cartilage is found in the trabeculæ, and the thicker the latter. At the "centre of ossification," *i.e.*, whence it started, the process is further advanced; away from it, it is in an earlier stage. At this period of embryo life, between the centre of ossification and a point nearer to the extremity of the shaft of a tubular bone, all stages described above may be met with, viz., between the solid unaltered hyaline cartilage at the end of the shaft, and the spongy bone with the unabsorbed remains of calcified cartilage in the middle of the shaft, all intermediate stages occur (Fig. 34).

71. After birth, and as long as bone grows, **we** find in the end of the shaft, and to a further degree also in the epiphysis, a continuation of the above process of endochondral formation. In fact, all bones preformed in the embryo as cartilage *grow in length before and after birth* by endochondral formation of new bone. The hyaline cartilage at their extremities (intermediate or ossifying cartilage) is the

cartilage at the expense of which the new bone is formed, by the marrow (blood-vessels and marrow-cells or osteoblasts) of the spongy substance in contact with the cartilage.

72. Following the development of a tubular bone after the above-mentioned stage 4, we find that the spongy bone once formed is not a permanent structure,

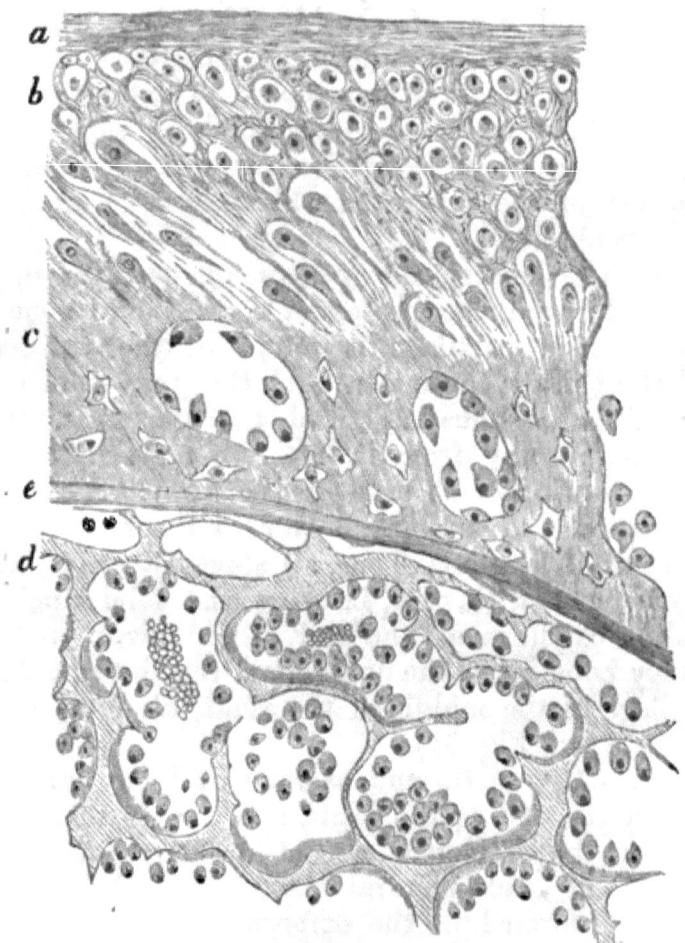

Fig. 35.—From a Transverse Section through the Tibia of Fœtal Kitten.
a, Fibrous layer of the periosteum; *b*, osteogenetic layer of the periosteum; *c*, periosteal bone; *d*, calcified cartilage not covered yet by bone; below this layer the trabeculæ of calcified cartilage covered with plates of bone—shaded darkly in the figure; *e*, boundary between periosteal and endochondral bone. (Atlas.)

but becomes gradually absorbed altogether, and this process also starts from the points of ossification. Thus a continuous cavity filled with marrow is formed, and this first appears in the region of the centre of ossification, and represents the rudiment of the future continuous central marrow cavity of the shaft. Simultaneously with or somewhat previous to this absorption of the endochondral bone, new bone—spongy bone—is deposited directly by the osteogenetic layer of the periosteum on the outer surface of the endochondral bone. This also commences at the centre of ossification and proceeds from here gradually to further points. This is the *periosteal bone* (Figs. 34, 35). It is formed without the intervention of cartilage directly by the osteoblasts of the osteogenetic layer. And as fresh layers of osteoblasts by multiplication appear on the surface of the periosteal bone, new layers of bone trabeculæ are formed, and also the old trabeculæ become increased in thickness. In the meshes or Haversian spaces of this spongy periosteal bone the same tissue is of course to be found as constitutes the osteogenetic layer of the periosteum, being derived from and continuous with it.

In these Haversian spaces concentric lamellæ of bone substance become formed by the osteoblasts, and spongy is thus transformed into compact bone, while at the same time the Haversian spaces, being narrowed in by the deposit in them of the concentric lamellæ, are transformed into the Haversian canals. When this compact bone is again absorbed—*e.g.*, that next the central marrow cavity of the shaft of a tubular bone—the concentric lamellæ are first absorbed, the Haversian canal being in this way again transformed into a Haversian space.

73. At birth all the primary endochondral bone has already disappeared by absorption from the centre of the shaft, while the bone present is all of periosteal

origin. At the extremity of the shaft, however, the spongy bone is all endochondral bone, and it continues to grow into the intermediate cartilage as stated above, as long as the bone as a whole grows (Fig. 36). Of course the parts of this spongy bone nearest to the centre of the shaft are the oldest, and ultimately disappear by absorption into the central marrow cavity. In the epiphysis the spongy bone is also endochondral bone, and its formation is connected with the deep layer of the articular cartilage.

Underneath the periosteum and on the surface of the spongy endochondral bone at the extremity of the shaft, the periosteal bone is represented only as a thin layer, extending as far as the periosteum reaches, *e.g.*, to the margin of the articular cartilage.

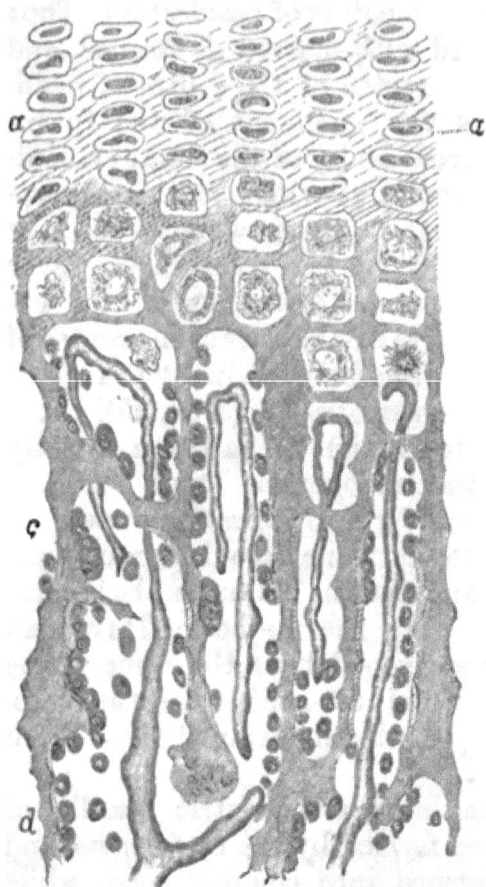

Fig. 36.—From a Longitudinal Section of Femur of Rabbit, through the part in which the intermediary cartilage joins the end of the shaft.

a, Intermediary cartilage; *b*, zone of calcified cartilage; *c*, zone, in which the calcified trabeculæ of cartilage become gradually invested in osseous substance, shaded light in the figure; the spaces between the trabeculæ contain marrow, and the capillary blood-vessels are seen here to end in loops; *d*, in this zone there is more bone formed; the greater amount the farther away from this zone. (Atlas).

74. Intermembranous formation.—All bones not preformed in the embryo as cartilage are developed directly from the periosteum in the manner of the periosteal bone just described (Fig. 37). Here also

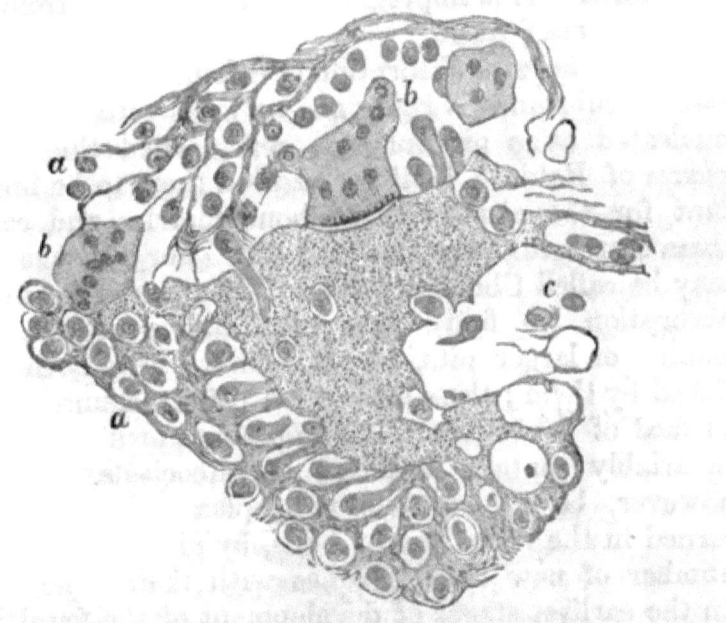

Fig. 37.—A small mass of Bone Substance in the Periosteum of the Lower Jaw of a Human Fœtus.

a, Osteogenetic layer of periosteum; *b*, multinucleated giant cells, myeloplaxes. The one in the middle of the upper margin corresponds to an osteoclast, whereas the smaller one at the left upper corner appears concerned in the formation of bone. Above *c* the osteoblast cells become surrounded by osseous substance and thus become converted into bone-cells. (Atlas.)

the new bone is at first spongy bone, which in its inner layers gradually becomes converted into compact bone.

In all instances during embryo life and after birth the growth of a bone in thickness takes place after the manner of *periosteal bone*; this is at first spongy, but is gradually converted into compact bone.

75. All osseous substance is formed in the embryo and after birth by the *osteoblast* or marrow cells (Gegenbaur, Waldeyer): each osteoblast giving origin to a zone of osseous matrix, and in the centre of this

to a nucleated protoplasmic remnant, which gradually becomes branched and then represents a bone cell. The osseous matrix is at first a soft fibrillar tissue, but is gradually and uniformly impregnated with lime salts. This impregnation always starts from the centre of ossification.

76. Wherever absorption of calcified cartilage or of osseous substance is going on, we meet with the multinucleated huge protoplasmic cells, called the *myeloplaxes* of Robin. Kölliker showed them to be important for the absorption of bone matrix, and called them therefore *Osteoclasts* (Fig. 37). For cartilage they may be called Chondroclasts. When concerned in the absorption we find these myeloplaxes situated in smaller or larger pits, which seem to have been produced by them; these absorption pits or lacunæ on the surface of bones are called *Howship's lacunæ*. They invariably contain numbers of osteoclasts. It can, however, be shown that myeloplaxes are also concerned in the formation of bone, by giving origin to a number of new osseous zones with their bone cells. In the earliest stages of development of the fœtal jaw this process is seen with great distinctness (Fig. 37).

77. **Dentine** forms the chief part of a tooth. It consists of a petrified matrix, in which are numbers of perpendicularly-arranged canals—the *dentinal tubes*—containing the *dentinal fibres*. It is in some respects similar to bone, although differing from it in certain essentials. It is similar, inasmuch as it is developed in like manner by some peculiarly transformed embryonal connective tissue—viz., by the tissue of the embryo tooth papilla—and inasmuch as cells are concerned in the production both of the petrified matrix (impregnated with lime salts), and of the processes of the cells contained in its canals—the dentine fibres. The details of structure and distribution will be described in connection with the teeth.

CHAPTER VIII.

NON-STRIPED MUSCULAR TISSUE.

78. This tissue consists of nucleated cells, which, unlike amœboid cells, are contractile in one definite direction, becoming shorter and thicker during contraction.

The *cells* are *elongated, spindle-shaped*, or *band-like* (Fig. 38A), and drawn out at each extremity into a longer or shorter, generally single but occasionally branched, tapering process. Each cell includes an *oval nucleus*,

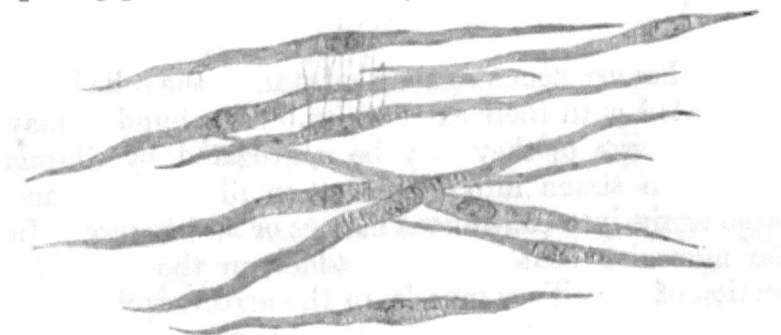

Fig. 38A.—Non-striped Muscular Fibres, isolated.
The cross-markings indicate corrugations of the elastic sheath of the individual fibres. (Atlas.)

which is flattened if the cell it belongs to is flattened. The cell-substance is a pale homogeneous-looking or finally and longitudinally striated substance.

During extreme contraction the nucleus may become more or less plicated, so that its outline becomes wavy or zig-zag.

It has been shown (Klein) in certain preparations—*e.g.*, the non-striped muscle cells of the mesentery of the newt—that each muscle cell consists of a delicate *elastic sheath,* inside of which is a *bundle* of *minute fibrils,* which cause the longitudinal striation of the cell. These fibrils are the contractile portion; and they are contractile towards the nucleus, with whose intranuclear reticulum they are intimately connected. When the cell is contracted its sheath becomes transversely corrugated (Fig. 38B).

79. The non-striped muscular cells are aggregated into smaller or larger *bundles* by an interstitial albu-

Fig. 38B.—A Non-striped Muscular Cell of Mesentery of Newt.

Showing several places where the muscular substance appears contracted, thickened. At these places the corrugations of the sheath are marked. (Atlas.)

minous homogeneous cement substance, the cells being imbricated with their extremities. The bundles may form a *plexus,* or they may be aggregated by fibrous connective tissue into larger or smaller *groups,* and these again into continuous masses or *membranes.* In the muscular coat of the bladder, in the choroidal portion of the ciliary muscle, in the arrector pili, in the muscular tissue of the scrotum, very well marked plexuses of bundles of non-striped muscular cells may be met with. In the muscularis mucosæ of the stomach and intestines, in the outer muscular coat of the same organs, in the uterus, bladder, &c., occur continuous membranes of non-striped muscular tissue.

When the muscular cells form larger bundles they are more or less pressed against one another, and, therefore, in a cross section appear of a polygonal outline.

80. Non-striped muscular tissue is found in the following places: in the muscularis mucosæ of the œsophagus, stomach, small and large intestine; in the

outer muscular coat of the lower two-thirds or half of the human œsophagus; in that of the stomach, small and large intestine; in the tissue of the pelvis and outer capsule of the kidney; in the muscular coat of the ureter, bladder, and urethra; in the tubules of the epididymis, in the vas deferens, vesiculæ seminalis and prostate; in the corpora cavernosa, and spongiosa; in the tissue of the ovary, and in the ligamentum latum; in the muscular coat of the oviduct, the uterus and vagina; in the soft or posterior part of the wall of the trachea; in the large and small bronchi, in the alveolar ducts and infundibula of the lung; in the pleura pulmonalis (guinea-pig); in the peritoneum of the frog and newt, in the upper part of the upper eye-lid, and in the fissura orbitalis; in the sphincter and dilatator pupillæ, and the ciliary muscle; in the capsule and trabeculæ of the spleen, and the trabeculæ of some of the lymphatic glands; in the arrector pili, and sweat glands of the skin, the tunica dartos of the scrotum; in the tissue of the nipple of the breast; in the large ducts of the salivary and pancreatic gland; and in the muscular coat of the gall bladder, the hepatic and cystic duct. The aorta and the arteries have a larger amount of non-striped muscular tissue, the veins and lymphatics a smaller.

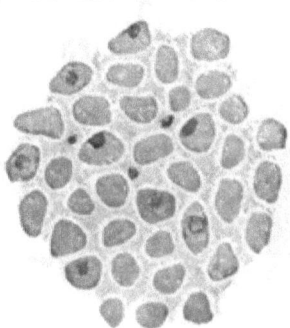

Fig. 39.—From a Transverse Section through Bundles of Non-striped Muscular Tissue of the Intestine.

The muscular cells being spindle-shaped are cut at various heights; the large corpuscles of the figure correspond to the middle, the small ones to the extremities, of the muscle-cells. (Atlas.)

81. As regards length, the muscular cells vary within considerable limits (from $\frac{1}{10}$ to $\frac{1}{500}$ of an inch), those of the intestine, stomach, respiratory, urinary, and genital organs being very long, as compared with those of the blood-vessels, which are sometimes only

twice or thrice as long as they are broad, and at the same time branched at their extremities.

Non-striped muscular tissue is richly supplied with blood-vessels, the capillaries forming oblong meshes, though their number is not so great as in striped muscle. The nerves of non-striped muscle are all derived from the sympathetic; their distribution and termination will be described in a future chapter.

CHAPTER IX.

STRIPED MUSCULAR TISSUE.

82. THIS tissue is composed of extremely long (up to $1\frac{1}{2}$-2 inches) more or less *cylindrical fibres*, of a diameter varying between $\frac{1}{200}$ to $\frac{1}{500}$ of an inch; they appear transversely striated. These are the *striped muscular fibres*. They are held together by delicate bundles of fibrous connective tissue, with the ordinary connective tissue cells—*endomysium*—so as to form larger or smaller *bundles;* these again are aggregated together by stronger bands and septa of fibrous connective tissue—*perimysium*—into *groups*, and these into the fascicles or divisions of an anatomical muscle. The fibrous connective tissue, including the perimysium tissue, is the carrier of the larger vascular and nervous branches. The endomysium contains the capillaries, which form very rich networks with elongated meshes, and are always situated between the individual muscle fibres. The capillaries and veins appear very wavy and twisted in the contracted bundles, and straighter in the uncontracted bundles. The small vessels are provided here and there with peculiar saccular dilatations, which act as a sort of safety receptacles

for the blood when, during a sudden maximal contraction, it is pressed out from some of the capillaries.

83. Each muscular fibre during contraction becomes shorter and thicker. In the living uninjured muscular fibres, spontaneously or after the application of a stimulus, a contraction starts at one point and passes over the whole muscular fibre like a wave— *contraction wave*—the progress of which is noticeable by the thickening, gradually and rapidly, shifting along the fibre, the part behind resuming its previous diameter.

84. When looked at in the fresh state, or after the action of certain re-agents, the muscular fibre shows the following parts (Fig. 40): (1) a transparent homogeneous delicate elastic sheath, the *sarcolemma*; (2) dark delicate lines stretching across the fibre at regular intervals, so as to sub-divide the space within the sarcolemma into uniform transverse compartments, the *muscular compartments* of Krause. These dark lines are the *membranes of Krause*. Under a high power they seem permeated or broken up by

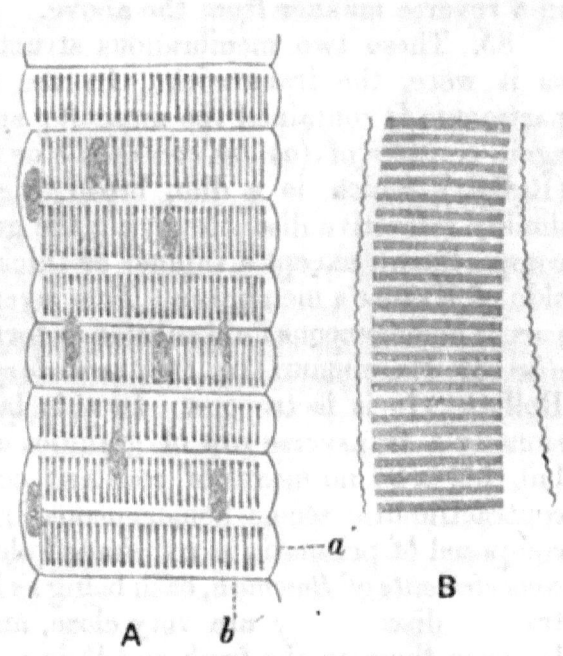

Fig. 40.—Striped Muscular Fibres of the Water-Beetle (Hydrophilus).

a, Sarcolemma; *b*, Krause's membrane. The sarcous elements are well seen. In A the oblong nuclei of the muscle-corpuscles are shown. In B the sarcolemma has become unnaturally raised from the muscular contents. The contractile discs are well shown; so also are the sarcous elements. (Atlas.)

a great number of fine, clear, longitudinal lines (*see* below), and therefore under these conditions seem to be made up of one row of granules. The membranes of Krause appear fixed to the sarcolemma, so that while a fibre contracts, or while it is contracted or shrunk, owing to the action of hardening re-agents, or merely in consequence of being detached from its fixations, its surface is not smooth, but regularly and transversely undulating, the valleys being caused by the attachment of the membranes of Krause to the sarcolemma. On stretching a fibre beyond its natural passive state, the surface becomes also uneven and undulating, but in a reverse manner from the above.

85. These two membranous structures represent, as it were, the framework. In the muscular compartments is contained the muscular substance, which again consists of (*a*) *the contractile* or *chief substance* (Rollett), which is a dim, broad, highly refractive, doubly refractive disc, occupying the greater part of a compartment, except a thinner or thicker layer at the side of Krause's membrane. This layer is (*b*) a transparent homogeneous fluid substance, forming the *lateral disc* of Engelmann, or the *secondary substance* of Rollett. It is isotropous. In this lateral disc occasionally a transverse row of granules appears present, but this is by no means of constant occurrence. The contractile disc seems homogeneous, but is in reality composed of prismatic or rod-shaped elements, the *sarcous elements of Bowman*, each being as long as the contractile disc. They are very close, and there is left between them in the fresh and living state an exceedingly minute layer of a homogeneous transparent interstitial substance, identical with that of the lateral disc. After death and shrinking of the sarcous elements, this interstitial substance is more marked, and is then easily perceived as longitudinal clear lines separating the sarcous elements in the indi-

vidual compartments. The total appearance produced is that of longitudinal striation, the sarcous elements of successive compartments forming fibrils—called the *primitive fibrils*. Sometimes in hardened muscular fibres the substance of the sarcous elements shows a middle transparent portion for the whole contractile disc; this appears to form a distinct median transparency, known as the *median disc of Hensen*.

86. Of course, each such fibril is a successive row of sarcous elements, with the corresponding portion of Krause's membranes, and the adjacent portions of the lateral discs. Generally, each fibril is thinnest at the point of Krause's membrane and lateral discs, and thicker at the part corresponding to the sarcous elements, so that in reality it is of a moniliform shape (Haycraft). This varicose condition is the more apparent the shorter and thicker the individual sarcous elements are (Fig. 43A, B and C).

These differentiations due to structure alone are sufficient to produce a transverse striation of the muscular fibres; but it must be borne in mind that a fibre when contracted or shrunk, even in the smallest degree, would show a transverse striation due to the above-mentioned undulating surface. Any other fibre with a moniliform shape would show the same transverse striation (Haycraft); and that usually observed on hardened—*i.e.*, shrunk and more or less contracted—fibres, may be accounted for in this way. Fibres stretched or prevented from shrinking generally show pronounced longitudinal striation, but also very faint cross striæ; these latter are due to the structural differences.

87. On observing a transverse section through a fresh and living muscular fibre, the muscular substance inside the sarcolemma appears as a transparent ground-glass-like substance, crossed here and there by bright

lines. These lines gradually increase in number, and to join that ultimately they form a dense network. Thus a more or less regular pattern of small polygonal fields is produced, which are styled Cohnheim's areas or fields (Fig. 41). Each corresponds to the end-view or optical section of a sarcous element prism, and is granular, as if composed of a bundle of minute fibrils. If this be the case, each sarcous element will have to be considered as a bundle of rods. The bright lines producing the Cohnheim's fields are the interstitial substance. When a muscle fibre shrinks, after death or after some hardening reagents, Cohnheim's fields' shrink into small circular areas, separated by a relatively large amount of the interstitial substance.

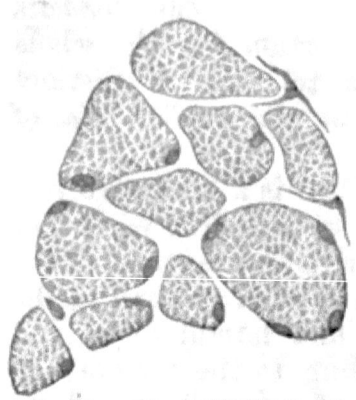

Fig. 41. — Striped Muscular Fibres in Cross Section.

Each fibre is limited by the sarcolemma; the muscular substance is differentiated into Cohnheim's areas. (Atlas.)

88. During contraction the cross striation is much narrower, the dim disc becoming shorter in the long diameter of the fibre, but broader in the transverse direction.

The broader the lateral disc in a fibre, the more apart from one another are the dim or contractile discs.

On the surface of the substance of the muscle fibres, but within the sarcolemma, are seen isolated oblong nuclei, which belong to small protoplasmic, more or less branched corpuscles—the *muscle corpuscles*. In the adult fibres these are few and far between; in the young and growing fibres they are numerous and large. Their protoplasm is the substance which, becoming converted into the muscular substance, is the material at the expense of **which new** fibres are formed, or fibres already formed become thickened, as is the case when muscle fibres are kept at constant work.

In the muscular **fibres of man and most vertebrates (except the fibres of the heart), the muscular corpuscles** are situated on the surface of the muscular substance ; but in invertebrates (especially insects and crustacea) they are often found in the central part of the fibres, and here they are occasionally seen forming almost a continuous cylindrical mass of nucleated protoplasmic cells.

89. In the **embryo** the muscular fibres are developed from spindle-shaped nucleated cells (Remak, Weissmann, Kölliker). One spindle-shaped cell with an oval nucleus grows rapidly in length and thickness, its nucleus divides repeatedly, and the offspring

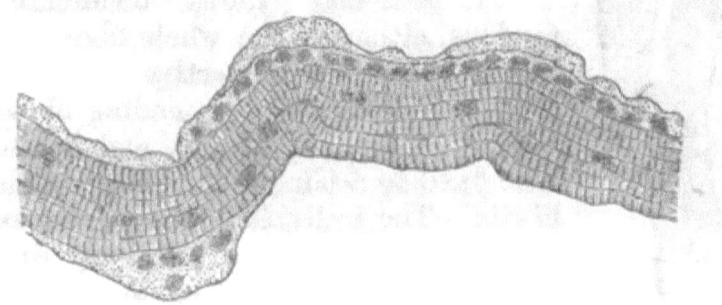

Fig. 42.—A Striped Muscular Fibre **of** the Diaphragm of a Guinea-pig.

The muscle-corpuscles are much increased in size and numbers; they are probably used here for the new formation of muscular substance. (Atlas.)

become shifted from one another as the cell continues to grow in length. The protoplasmic substance all along one side of the cell gives origin to the muscular substance—sarcous elements and lateral disc—while a small **rest of** protoplasm remains collected around the **nucleus** as the muscle corpuscle. This protoplasm continues to increase in amount, and then the increment again changes into muscular substance (Fig. 42). In this way the muscular fibre increases in thickness. Thus one spindle-shaped embryo cell gives rise to one muscular fibre, which, at first very slender, continues to grow in thickness by the active growth of the muscle corpuscles. The sarcolemma appears to be formed from cells other than muscle cells.

90. The striped muscular fibres, taken as a whole, are, as a rule, spindle-shaped, becoming gradually thinner towards their ends. They are branched in some exceptional cases—*e.g.*, in the tongue; here the extremities of the muscle fibres, passing in a transverse direction into the mucous membrane, become richly branched.

91. Muscular fibres terminate in tendons, either by the whole fibre passing into a bundle of connective tissue fibrils (Fig. 43), or by the fibre ending abruptly with a blunt, conical end, and becoming here fixed to a bundle of connective tissue fibrils. The individual fibres have only, as mentioned above, a relatively limited length, so that, following an anatomical fascicle from one point of its insertion to the other, we find some muscle fibres terminating, others originating. This takes place in the following way: the contents of a fibre suddenly terminate, while the sarcolemma, as a fine thread, becomes interwoven with the fine connective tissue between the muscular fibres.

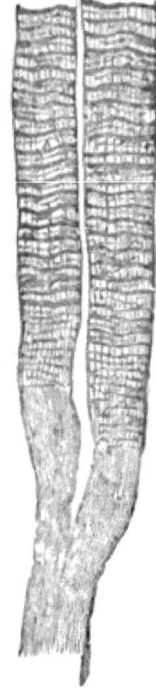

Fig. 43.—Two Striped Muscular Fibres passing into Bundles of Fibrous Tissue.
Termination in Tendon. (Handbook.)

92. The striped muscular fibres of the heart (auricles and ventricles) and of the cardiac ends of the large veins (the pulmonary veins included) differ

from other striped muscular fibres in the following respects:—(1) They possess no distinct sarcolemma. (2) Their muscle corpuscles are in the centre of the fibres, and more numerous than in ordinary fibres. (3) They are very richly branched, each fibre giving off all along its course short branches, or continually dividing into smaller fibres and forming a close network (Fig. 43A.) A transverse section through a bundle of such fibres shows, therefore, their cross sections irregular in shape and size. (4) Each nucleus of a muscle corpuscle occupies the centre of one prismatic portion; each fibre and its branches thus appear composed of a single row of such prismatic portions, and they seem separated from one another —at any rate in an early stage—by a septum of a transparent substance.

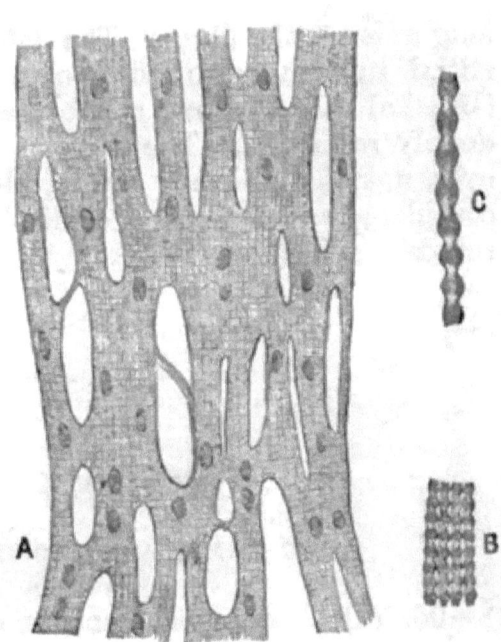

Fig. 43A.—Striped Muscular Fibres of the Heart of Mouse.

A, Showing the branching of the fibres and their anastomosis in networks; B, part of a thin fibre, highly magnified, showing the moniliform primitive fibrillæ; C, one primitive fibrilla more highly magnified.

93. Muscular fibres seem either markedly pale or markedly red (Ranvier); in the former (*e.g.*, quadratus lumborum, or adductor magnus femoris of rabbit) the transverse striation is more distinct and the muscular corpuscles less numerous, than in the latter (*e.g.*, semi-tendinosus of rabbit,

diaphragm). Here the longitudinal striation appears very distinct, but these differences are not constant in other muscular fibres of other animals (E. Meyer).

94. Brücke has shown that striped muscular fibres are doubly refractive, like uniaxial positive crystals (rock crystal), the **optical axis** coinciding with the long axis of the fibres. The lateral disc **and interstitial sub**stance are isotropous, the sarcous elements (Brücke) and Krause's membrane (Engelmann) being doubly refractive. The sarcous elements are, however, not the ultimate optical elements, but must be considered as composed of disdiaclasts, the real doubly refractive elements (Brücke).

CHAPTER X.

THE HEART AND BLOOD-VESSELS.

95. (A) THE **heart** consists of an outer serous covering (*the visceral pericardium*), an inner lining (*the endocardium*), and between the two the *muscular wall* (Fig. 44). Underneath the pericardium and endocardium is a loose connective tissue, called the subpericardial and subendocardial tissue respectively.

The free surface of both the pericardium and endocardium has an endothelial covering, like other serous membranes—*i.e.*, a single layer of transparent nucleated cell plates of a more or less polygonal or irregular shape. The ground-work of these two membranes is fibrous connective tissue, forming a dense texture, and in addition there are many elastic fibres composing networks. Capillary blood-vessels, lymphatic vessels, and small branches of nerve-fibres are met with everywhere. The subpericardial and subendocardial tissues consist of loosely connected trabeculæ of fibrous connective tissue, forming a continuity with the intermuscular connective tissue of the muscular part of the heart. The former contains in many places groups of fat cells.

96. On the free surface of the papillary muscles, in some parts of the surface of the trabeculæ carneæ, and at the insertion of the valves, the endocardium is thickened by tendinous connective tissue. The valves themselves are folds of the endocardium, and contain in their essential parts fibrous connective tissue, to which, especially in the semilunar valves, numerous elastic fibres are added. The muscular tissue of the wall of the auricle penetrates a short way into the auriculo-ventricular valves.

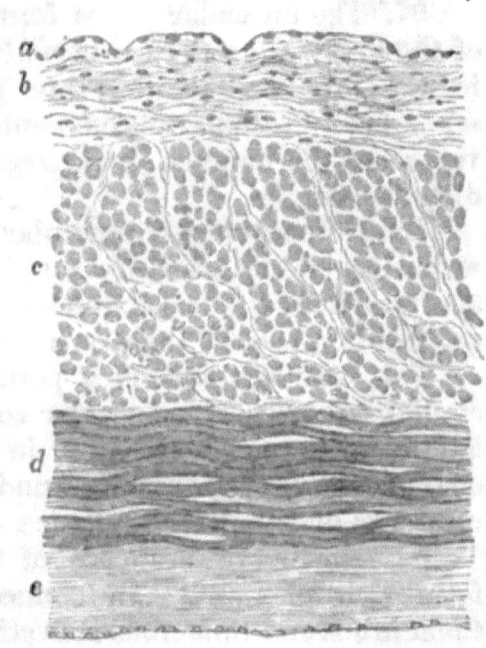

Fig. 44.—Transverse Section through the Auricle of the Heart of a Child.

a, Endothelium lining the endocardium; *b*, endocardium; *c*, muscular bundles cut transversely; *d*, muscular bundles cut longitudinally; *e*, pericardial covering.

All the cordæ tendineæ and the valves are of course covered on their free surfaces with endothelium.

Special tracts of muscle fibres occur in the subendocardial tissue.

The *fibres of Purkinje* are peculiar fibres occurring in the subendocardial tissue in some mammals and birds (not in man). They are thin, transversely striped, muscular fibres, the central part of which is a continuous mass of protoplasm, with nuclei at regular intervals, the same as is the case with some skeletal muscular fibres of insects.

97. The muscular fibres forming the proper wall of the heart, the structure of which has been described in the previous chapter, are grouped in bundles separated by vascular fibrous connective tissue. In the ventricles the bundles are aggregated into more or less distinct lamellæ.

Like other striped muscular fibres, those of the wall of the heart are richly supplied with blood-vessels and lymphatics. The endocardium and valves and the pericardium possess their own systems of capillaries.

The lymphatics form a pericardial and an endocardial network, the muscular substance of the heart having numerous lymphatics in the shape of lymph clefts between the muscular bundles, and also typical networks of tubular lymphatics.

98. The nerve branches of the **plexus cardiacus** form rich plexuses. In connection with some of them are found numerous collections of ganglion cells or ganglia. These are very numerous in the nerve plexus of the auricular septum of the frog's heart (Ludwig, Bidder), and in the auriculo-ventricular septum of the frog (Dogiel). In man and mammals numerous ganglia are found on the subpericardial nerve branches, chiefly at the point of junction of the large veins with the heart, and at the boundary between the auricles and the ventricles.

99. (B) The **arteries** (Fig. 45) consist of: (*a*) an *endothelial layer* lining the lumen of the vessel; (*b*) an *intima* consisting of elastic tissue; (*c*) a *media*, containing a large proportion of non-striped muscular cells arranged chiefly in a transverse, *i.e.*, circular, manner; and (*d*) an *adventitia* composed chiefly of fibrous

connective tissue, with an admixture of networks of elastic fibres.

(a) The endothelium is a continuous single layer of flattened elongated cell plates.

(b) The intima in the aorta and large arteries is a very complex structure, consisting of an innermost layer of fibrous connective tissue, which is the "*inner longitudinal fibrous layer*" of Remak, outside of which is a more or less longitudinally-arranged elastic membrane. This is laminated, and composed of *fenestrated elastic membranes of Henle* (see a former chapter). The greater the artery the thicker the intima. In microscopic arteries the intima is a thin fenestrated membrane, the fibres having distinctly a longitudinal arrangement.

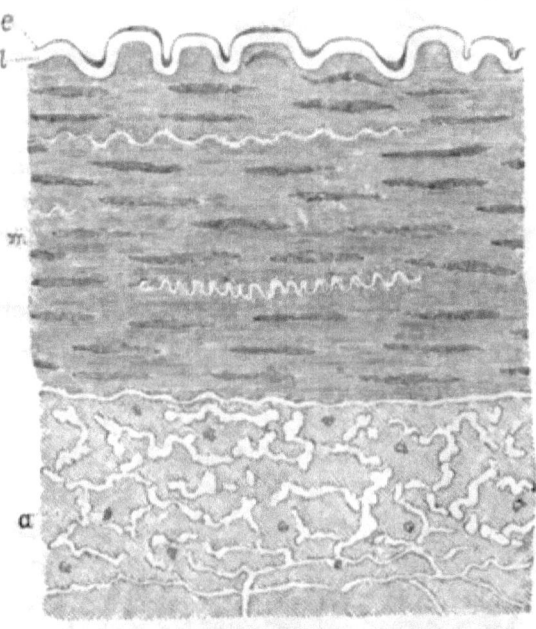

Fig. 45.—From a Transverse Section through the Inferior Mesenteric Artery of the Pig.

e, Endothelial lining; *i*, elastic intima; *m*, muscular media; *a*, adventitia with numerous elastic fibrils, cut in transverse section. (Atlas.)

(c) The media is the chief layer of the wall of the arteries (Fig. 46). It consists of transversely arranged elastic lamellæ (fenestrated membranes and networks of elastic fibres), and between them smaller or larger bundles of circularly arranged muscular cells. The larger the artery the more is the relation of elastic and muscular tissue of the media in favour of the former, in the smaller arteries the reverse is the case. In microscopic branches of arteries the media consists almost entirely of circular non-striped muscle cells with only few elastic fibres.

100. In the last branches of the microscopic arteries, the muscular media becomes discontinuous, inasmuch as the (circular) muscular cells are arranged not as a continuous membrane, but as groups of small cells (in a single layer) in a more or less alternate fashion.

When the media contracts, the intima is placed in longitudinal folds.

The aorta has, in the innermost and in the outermost parts of the media, numbers of longitudinal and oblique muscle cells. According to Bardeleben, all large and middle-sized arteries have an inner longitudinal muscular coat.

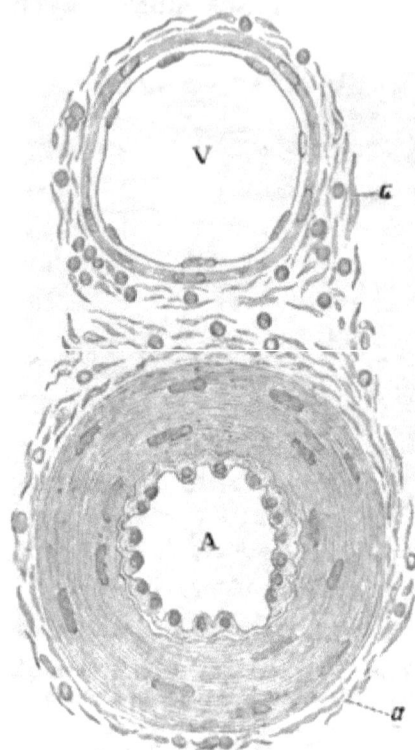

Fig. 46.—Transverse Section through a Microscopic Artery and Vein in the Epiglottis of a Child.

A, The artery, showing the nucleated endothelium, the circular muscular media, and at a the fibrous-tissue adventitia; v, the vein, showing the same layers; the media is very much thinner than in the artery. (Atlas.)

101. Between the media and the next outer layer there is, in larger and middle-sized arteries, a special elastic membrane, the *elastica externa* of Henle. (*d*) The adventitia is a relatively thin fibrous connective tissue membrane. In large and middle-sized arteries there are numbers of elastic fibres present,

especially in the part next to the media; they form networks, and have pre-eminently a longitudinal direction.

The larger the artery the more insignificant is the adventitia as compared with the thickness of media.

In microscopic arteries (Fig. 47), the adventitia is represented by thin bundles of fibrous connective tissue and branched connective tissue cells.

Large and middle-sized arteries possess their own system of blood-vessels (vasa vasorum), situated chiefly in the adventitia and media; lymphatic vessels and lymphatic clefts are also present in these coats.

102. (c) The **veins** differ from the arteries in the greater thinness of their wall. The intima and media are similar to those of arteries, only thinner, both absolutely and relatively. The media contains in most veins circularly arranged muscular fibres; they form a continuous layer, as in the arteries, and there is between them generally more fibrous connective tissue than elastic. The adventitia is usually the thickest coat, and it consists chiefly of fibrous connective tissue (Fig. 46). The smallest veins—*i.e.*, before passing into the capillaries—are composed of a lining endothelium, and outside this are delicate bundles of connective tissue forming an adventitia. The valves of the veins are folds, consisting of the endothelium lining the surface, of the whole intima, and of part of the muscular media.

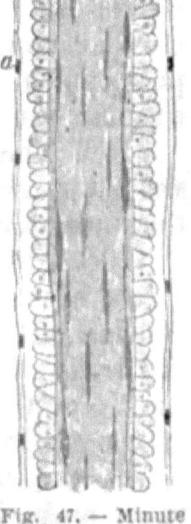

Fig. 47.— Minute Microscopic Artery.
e, Endothelium; *i*, intima; *m*, muscular media, composed of a single layer of circularly-arranged non-striped muscular cells; *a*, adventitia. (Atlas.)

103. There are many veins that have no muscular fibres at all, *e.g.*, vena jugularis—interna and externa —the vena subclavia, the veins of the bones and

retina, and of the membranes of the brain and cord. Those of the gravid uterus have only longitudinal muscular fibres. The vena cava, azygos, hepatica, spermatica interna, renalis and axillaris, possess an inner circular and an outer longitudinal coat. The vena iliaca, cruralis, poplitea, mesenterica, and umbilicalis possess an inner and outer longitudinal and a middle circular muscular coat. The intima of the venæ pulmonales in man is connective tissue containing circular bundles of non-striped muscular cells (Stieda).

104. The trunk of the venæ pulmonales possesses striped muscular fibres, these being continuations of the muscular tissue of the left auricle.

105. Hoyer showed that a direct communication exists between arteries and veins without the intervention of capillaries—as in the matrix of the nail, in the tip of the nose and tail of some mammals, in the tip of the fingers and toes of man, in the margin of the ear lobe of dog and cat and rabbit.

In the cavernous tissue of the genital organs veins make large irregular sinuses, the wall of which is formed by fibrous and non-striped muscular tissue.

106. (D) The **capillary blood-vessels** are minute tubes of about $\frac{1}{2000}$ to $\frac{1}{3000}$ of an inch in diameter. Their wall is *a single layer* of transparent *elongated endothelial plates*, separated by thin lines of *cement substance* (Fig. 48); each cell has an oval nucleus. In fact, the wall of the capillaries is merely a continuation of the endothelial membrane lining the arteries and veins.

In some places the capillaries possess a special *adventitia* made up of branched nucleated connective tissue cells (hyaloidea of frog, choroidea of mammals), or of an endothelial membrane (*pia mater* of brain and cord, retina and serous membranes), or of adenoid reticulum (lymphatic glands, His).

The smallest capillaries are found in the central

nervous system, the largest in the marrow of bone. The capillaries form always networks, the richness and

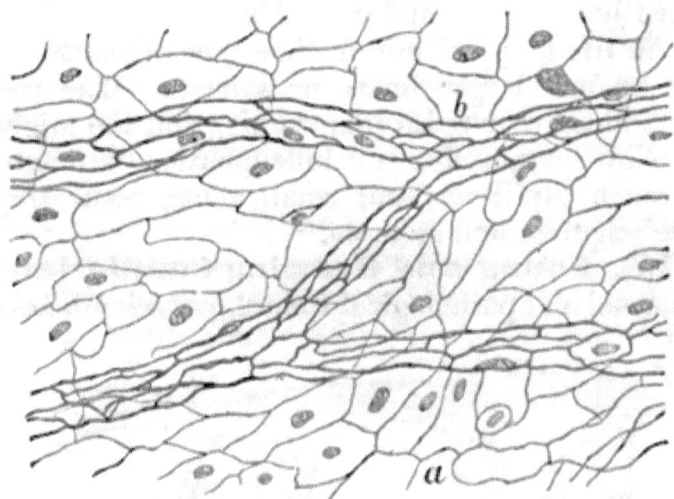

Fig. 48.—From a Preparation of the Peritoneum, stained with Nitrate of Silver.

a, The endothelium on the free surface of the membrane; b, the capillary blood-vessels in the membrane; their wall is a layer of endothelium. (Handbook.)

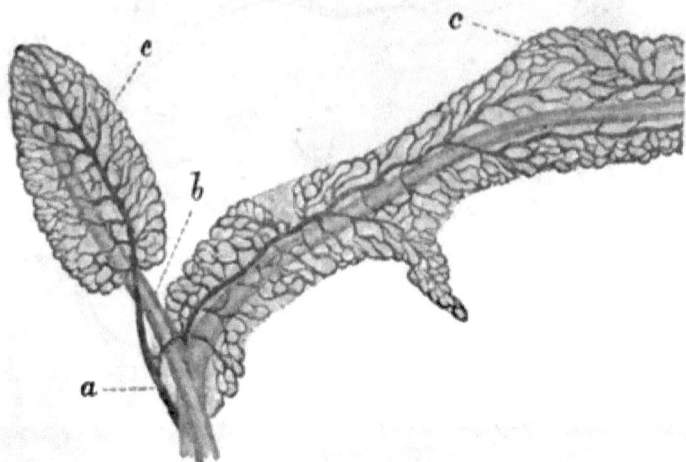

Fig. 49.—Young Fat Tissue of the Omentum, its Blood-vessels injected.
a, Artery; b, vein; c, network of capillaries. (Handbook.)

arrangement of which vary in the different organs, according to the nature and arrangement of the elements of the tissue (Fig. 49).

G

107. If capillaries are abnormally distended, as in inflammation, or otherwise injured, the cement substance between the endothelial plates is liable to give way in the shape of minute holes, or *stigmata*, which may become larger holes, or *stomata*. The passage of red blood corpuscles (diapedesis), and the migration of white corpuscles in inflammation through the unbroken capillaries and small veins, occur through these stigmata and stomata.

108. **Young and Growing Capillaries,** both of normal and pathological tissues, possess *solid thread-*

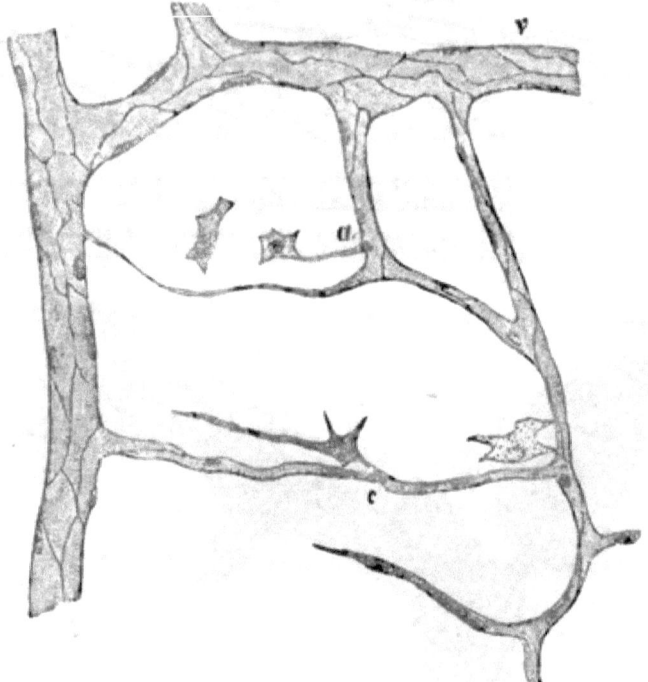

Fig. 50.—From a Preparation of Omentum of Rabbit, after staining with Nitrate of Silver.

v, A minute vein; *a*, solid protoplasmic prolongations of the wall of a capillary, connected with connective tissue corpuscles; *c*, a solid young sprout. (Atlas.)

like shorter or longer *nucleated protoplasmic processes* (Fig. 50), into which the canal of the capillary is gradually prolonged, so that the thread becomes converted into a new capillary branch. Such growing capillaries are capable of contraction (Stricker).

All blood-vessels, arteries, veins, and capillaries, in their early stages, both in embryonal and adult life, are of the nature of minute tubes, the wall of which consists of a simple endothelial membrane. In the case of the vessel becoming an artery or vein, cells are added to the outside of the endothelium, thus forming the elastic, muscular, and fibrous connective tissue elements of the wall.

109. In the first stage, both in the embryo and in the adult, the vessel is represented by a solid nucleated protoplasmic cell, elongated or spindle-shaped or branched. Such a cell may be an isolated cell of the connective tissue independent of any pre-existing vessel, or it may be a solid protoplasmic outgrowth of the endothelial wall of an existing capillary vessel (Fig. 51). In both cases it becomes hollowed out by a process of vacuolation; isolated vacuoles appear at first, but they gradually become confluent, and thus a young vessel is formed, at first very irregular in outline, but gradually

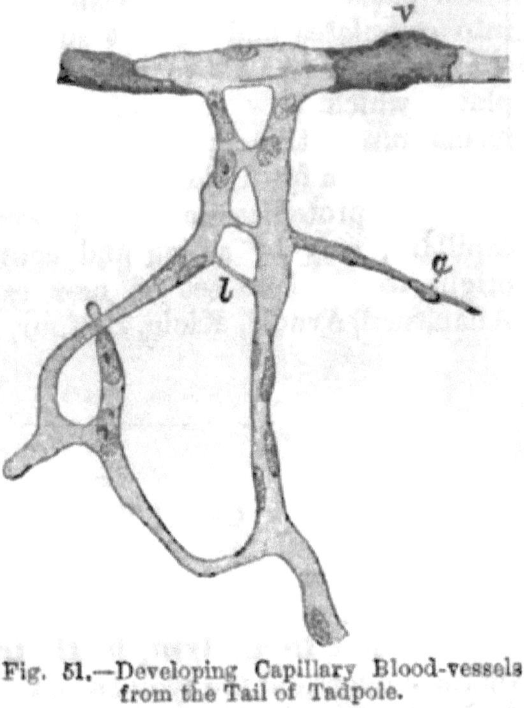

Fig. 51.—Developing Capillary Blood-vessels from the Tail of Tadpole.

v, Capillary vein with clumps of pigment in the wall; *a*, nucleated protoplasmic sprout; *l*, solid anastomosis between two neighbouring capillaries. (Atlas.)

acquiring more and more of a tubular form. In the case of an isolated cell, its protoplasmic processes grow by degrees to the nearest capillary, to the wall of which they become fixed, and the cavity of the cell finally opens through such processes into that of the capillary vessel.

The wall of young capillaries is granular-looking protoplasm (the original cell substance), and in it are disposed, in more or less regular fashion, oblong nuclei, derived by multiplication from the nucleus of the original cell. In a later stage, a differentiation takes place in the protoplasmic wall of the capillary into cell-plates and cement substance, in such a way that each of the above nuclei appertains to one cell-plate, which now represents the final stage in the formation of the capillary. Both in the embryo and in the adult a few isolated nucleated protoplasmic cells, or a few protoplasmic solid processes of an existing capillary, may by active and continued growth give origin to a whole set of new capillaries (Stricker, Affanasieff, Arnold, Klein, Balfour, Ranvier, Leboucq).

CHAPTER XI.

THE LYMPHATIC VESSELS.

110. The **large lymphatic trunks,** such as the thoracic duct, and the lymphatic vessels passing to and from the lymphatic glands, are thin-walled vessels, similar in structure to arteries. Their lining endothelium is of the same character as in an artery, and so are the elastic intima and the media with its circular muscular tissue, only they are very much thinner than in an artery of the same calibre. The adventitia is an

exceedingly thin connective tissue membrane with a few elastic fibres. The valves are semi-lunar folds of the endothelium and intima.

111. The **lymphatics** in the tissues and organs form rich plexuses. They are tubular vessels, the wall of which is, like that of a capillary blood-vessel, a single layer of endothelial plates (Fig. 52). The lymphatic may be, and often is, many times wider than a blood capillary. The endothelial plates are elongated, but

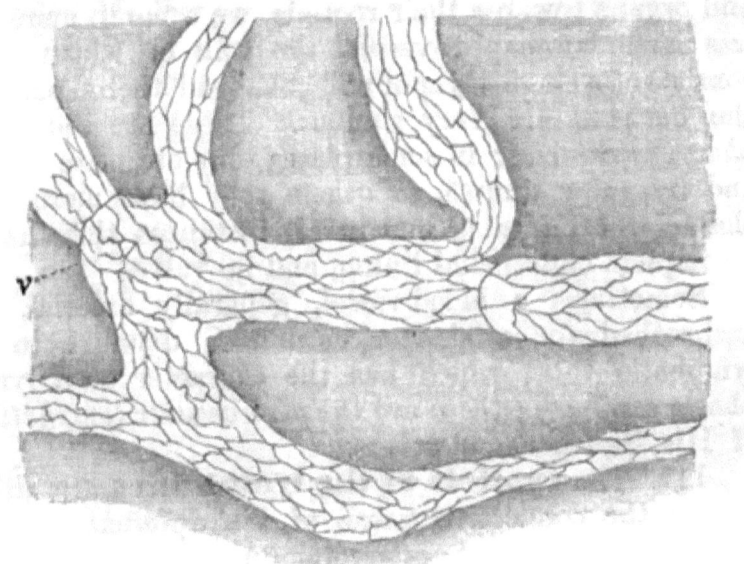

Fig. 52.—Lymphatic Vessels of the Diaphragm of the Dog, stained with Nitrate of Silver.
The endothelium forming the wall of the lymphatics is well shown; v, valves.
(Atlas.)

not so long as in a blood capillary, with more or less sinuous outlines, but this depends on the amount of shrinking of the tissue in which the vessel is embedded; when there is no shrinking in the tissue or in the vessel, the outlines of the cells are more or less straight.

The lymphatics are supported by the fibrous connective tissue of the surrounding tissue, which does not, however, form part of their wall.

112. The outline of the vessel is not straight, but more or less moniliform, owing to the slight dilatations present below and at the *semi-lunar valves;* these are folds of the endothelial wall, and they are met with in great numbers. The vessel appears slightly dilated immediately below the valve, that is, on the side farthest from the periphery, or rootlet, whence the current of lymph starts.

113. Tracing the lymphatic vessels in the tissues and organs towards their rootlets, we come to more or less irregular-shaped vessels, the wall of which also consists of a single layer of polygonal endothelial plates; the outlines are very sinuous. These are the *lymphatic capillaries;* in some places they are mere clefts and irregular sinuses, in others they have more the character of a tube, but in all instances they have a complete endothelial lining, and no valves.

Sometimes a blood-vessel, generally arterial, is ensheathed for a shorter or longer distance in a lymphatic tube, which has the character of a lymphatic capillary; these are the *perivascular lymphatics* of His, Stricker, and others.

114. The **rootlets of the lymphatics** are situated in the connective tissue of the different organs in the shape of an intercommunicating system of crevices, clefts, spaces, or canals, existing between the bundles, or groups of bundles, of the connective tissue. These rootlets are generally without a complete endothelial lining, but are identical with the spaces in which the connective tissue corpuscles are situated; where these are branched cells anastomosing by their processes into a network—such as the cornea, or serous membranes—we find that the rootlets of the lymphatics are the lacunæ and canaliculi of these cells—the typical lymph-canalicular system of von Recklinghausen. (Fig. 53). The endothelial cells forming the wall of the lymphatic capillaries are directly continuous with the

connective cells situated in the rootlets. In tendons and fasciæ the minute lymphatics lie between the bundles, and have the shape of continuous long clefts

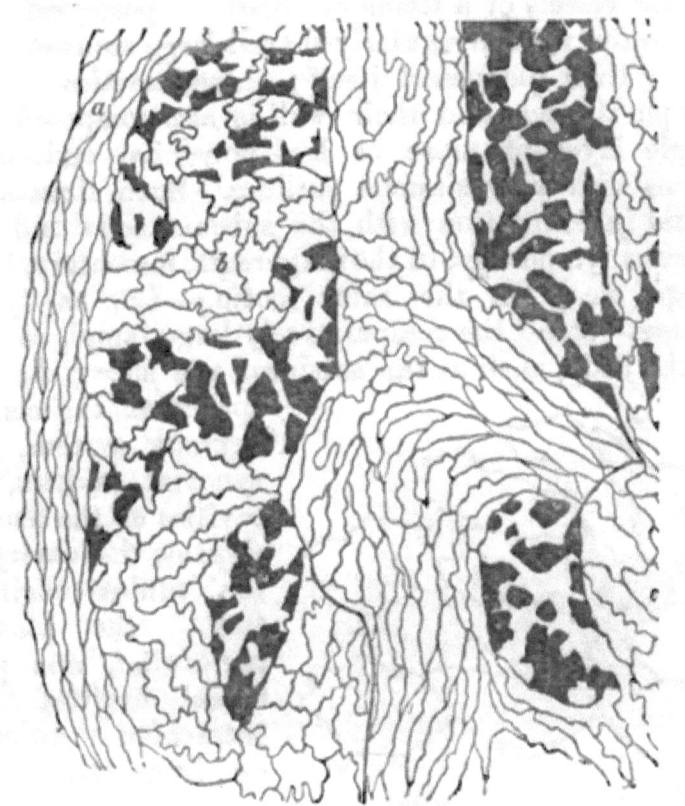

Fig. 53.—From a Silver-stained Preparation of the Central Tendon of the Rabbit's Diaphragm. Showing the direct connection of the Lymph-canalicular System of the Tissue with the Lymphatic Capillaries.

a, Lymphatic vessel; b, lymphatic capillary lined with 'sinuous' endothelium.
(Handbook.)

or channels; in striped muscular tissue they have the same character, being situated between the muscular fibres.

The passage of plasma from the minute arteries and capillary blood-vessels into the lymph-rootlets situated in the tissues, and thence into the lymphatic capillaries and lymphatic vessels, represents the natural current of lymph irrigating the tissues.

115. **Lymph cavities.**—In some places the lymphatic vessels of a tissue or organ are possessed of, or connected with, irregularly-shaped large sinuses, much wider than the vessel itself; these cavities are the lymph sinuses, and their wall is also composed of a single layer of more or less polygonal endothelial plates with very sinuous outlines. Such sinuses are found in connection with the subcutaneous and submucous lymphatics, in the diaphragm, mesentery, liver, lungs, &c. On the same footing—*i.e.*, as lymph sinuses—stand the comparatively large lymph cavities in the body, such as the subdural and subarachnoidal spaces of the central nervous system, the synovial cavities, the cavities of the tendon-sheaths, the cavity of the tunica vaginalis testis, the pleural, pericardial, and peritoneal cavities. In batrachian animals, *e.g.*, frogs, the skin all over the trunk and extremities is separated from the subjacent fasciæ and muscles by large bags or sinuses—the *subcutaneous lymph sacs*. These sinuses are shut off from one another by septa. Between the trunk and the extremities, and on the latter, the septa generally occur in the region of the joints. In female frogs in the mesogastrium smaller or larger

Fig. 54.—Stomata, lined with Germinating Endothelial Cells, as seen from the Cisternal Surface of the Septum Cisternæ Lymphaticæ Magnæ of the Frog. (Handbook.)

cysts lined with ciliated endothelium are sometimes found. Behind the peritoneal cavity of the frog, along and on each side of the vertebral column, exists a similar large lymph sinus, called the cisterna lymphatica magna.

116. **In all** instances **these cavities are** in direct communication with the lymphatics of the surrounding parts by holes or open mouths (*stomata*), often lined by a special layer of polyhedral endothelial cells—germinating cells (Figs. 54, 55). Such stomata are numerous on the peritoneal surface of the central tendon of the diaphragm, in which are found straight lymph channels between the tendon bundles, and these channels communicate here and there with the free surface by stomata. A similar arrangement exists on the costal pleura, the omentum, and the cisterna lymphatica magna of the frog. (See Chapter IV.)

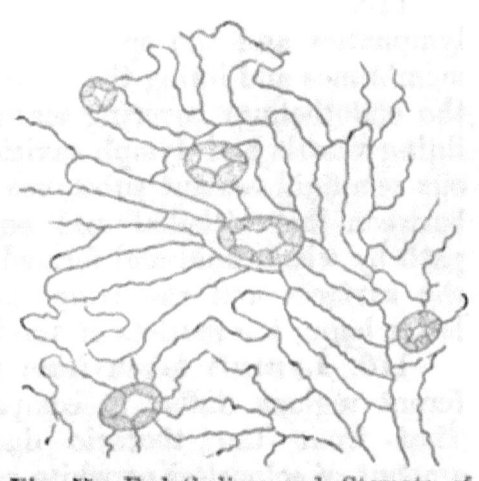

Fig. 55.—Endothelium and Stomata of the Peritoneal Surface of the Septum Cistern Lymphatic Magnæ of the Frog. (Handbook.)

117. The **serous membranes** consist of a matrix of fibrous connective tissue with networks of fine elastic fibres; they contain networks of blood capillaries and numerous lymphatic vessels arranged in (superficial and deep) plexuses. Those of the pleura costalis—or rather, intercostalis—and of the diaphragm and pleura pulmonalis, are most numerous. They are important in the process of absorption from the pleural and peritoneal cavity respectively. Lymph and lymph

corpuscles, and other formed matter, are readily taken up by the stomata (see Fig. 20) and brought into the lymphatics, and in this the respiratory movement of the intercostal muscles, of the diaphragm, and of the lungs respectively, produces the result of the action of a pump.

118. There is a definite relation between the lymphatics and the epithelium covering the mucous membranes and lining the various glands and between the endothelium covering serous membranes and that lining vessels and lymph cavities—viz., the albuminous semifluid cement substance (*see* former chapters) between the epithelial and endothelial cells is the path by which fluid and formed matter pass between the surfaces and the lymph-canalicular system, the latter being the rootlets of the lymphatics.

119. **Lymph** taken from the lymphatics of different regions differs in composition and structure. That from the thoracic duct contains a large amount of colourless or white corpuscles—lymph corpuscles—each of which is a protoplasmic nucleated cell similar in aspect and nature to a white blood corpuscle. They are of various sizes, according to the stage of ripeness. The smaller contain one, some of the larger contain two and three, nuclei. The latter show more pronounced amœboid movement than the small ones. A few red corpuscles are also met with. Granular and fatty matter is present in large quantities during and after digestion.

In the frog (and also in other lower vertebrates, *e.g.*, reptiles) there exist certain small vesicular lymph cavities, about an eighth of an inch in diameter, which show rhythmic pulsation; they are called *lymph hearts*. On each side of the os coccygis and underneath the skin is a pulsating *posterior* lymph heart. The *anterior* lymph heart is oval, and situated on each side between the processus transversus of the third and fourth vertebra; it is slightly smaller than the posterior one. The lymph hearts have an efferent vessel, which is a vein, and from them the venous system

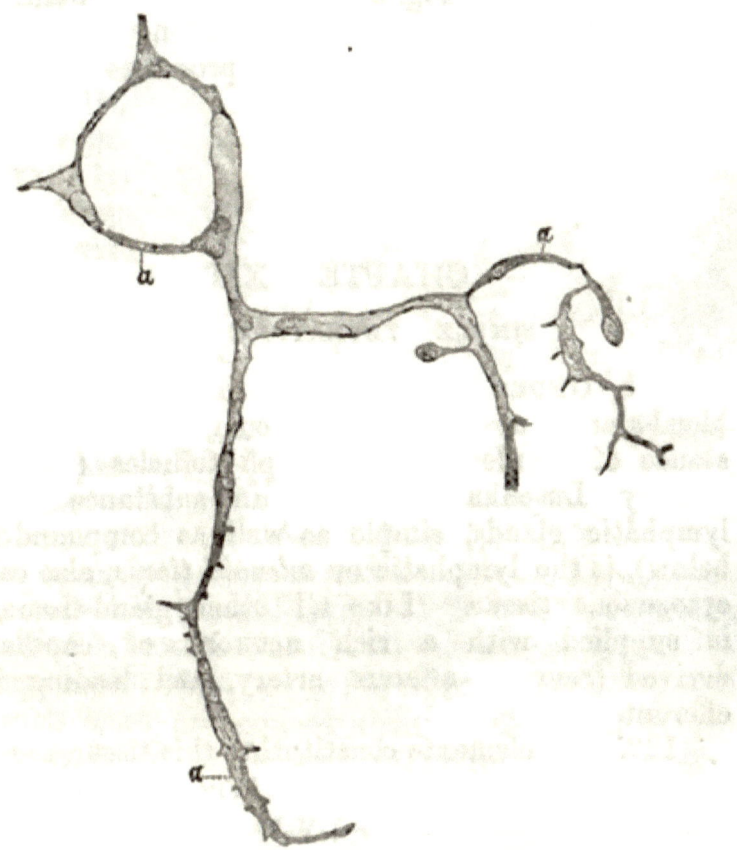

Fig. 56.—Developing Lymph-capillaries in the Tail of Tadpole.
a, Solid nucleated protoplasmic branches not yet hollowed out. (Atlas.)

of the neighbourhood can be easily injected, whereas the reverse is not possible. They are lined with an endothelium like the lymph sacs, and in their wall they possess plexuses of striped, branched, muscular fibres. The nerve fibres terminate in these striped muscular fibres in the same manner as in those of other localities. (Ranvier.)

120. Lymphatic vessels are developed and newly formed under normal and pathological conditions in precisely the same way as blood-vessels. The accompanying woodcut (Fig. 56) shows this very well. We have also here to do with the hollowing out of (connective tissue) cells and their processes previously solid and protoplasmic.

CHAPTER XII.

SIMPLE LYMPHATIC GLANDS.

121. UNDER this name are to be considered the blood-glands of His, or the conglobate gland substance of Henle, or the lymph follicles (Kölliker, Huxley, Luschka). The ground-substance of all lymphatic glands, simple as well as compound (*see* below), is the lymphatic or *adenoid tissue*, also called cytogenous tissue. Like all other gland-tissue, it is supplied with a rich network of capillaries derived from an afferent artery, and leading into efferent veins.

122. The elements constituting this tissue are:—

(*a*) *The adenoid reticulum* (Fig. 57), a network of fine homogeneous fibrils, with numerous plate-like enlargements.

(*b*) *Small, transparent, flat, endotheloid cell-plates*, each with an oval nucleus. These cell-plates are fixed on the reticulum, of which at first sight they seem to form part. Their oval nucleus especially appears to belong to a nodal point—*i.e.*, to one of the enlargements of the reticulum; but by continued shaking of a section of any lymphatic tissue, the oval nuclei and their cell-plates can be got rid of, so that only the reticulum is left, without any trace of a nucleus.

(c) *Lymph-corpuscles* completely fill the meshes of the adenoid reticulum. These can be easily shaken out of the reticulum. They are of different sizes; some — the young ones — are small cells, with a comparatively large nucleus; others — the ripe ones — are larger, have a distinct protoplasmic cell body, with one or two nuclei. They all show on a warm stage amœboid movement, but in the large ones it is much more pronounced than in the small ones.

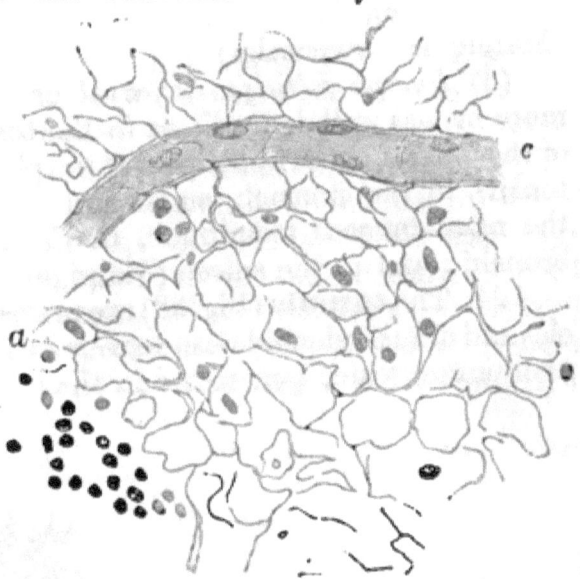

Fig. 57.—Adenoid Reticulum shaken out; most of the Lymph-corpuscles are removed. From a Lymphatic Gland.

a, The reticulum; *c*, a capillary blood-vessel. (Atlas.)

The capillary blood-vessels supplying the adenoid tissue receive a more or less distinct special investment from the adenoid reticulum; this is the capillary adventitia.

123. The adenoid tissue occurs as:

(1) *Diffuse adenoid tissue*, without any definite grouping or arrangement. This is the case in the sub-epithelial layer of the mucous membrane of the trachea, in the mucous membrane of the false vocal cords and the ventricle of the larynx, in the posterior part of the epiglottis, in the soft palate and tonsils, at the root of the tongue, in the pharynx, in the mucosa of the small and large intestine, including the villi

of the former; and in the mucous membrane of the nasal cavity and vagina.

(2) *Cords, cylinders*, or *patches* of adenoid tissue; as in the omentum and pleura, and in the spleen (Malpighian corpuscles).

(3) *Lymph follicles*, *i.e.*, oval or spherical masses more or less well defined; as in the tonsils, at the root of the tongue, in the upper part of the pharynx (pharynx-tonsil), in the stomach, small and large intestine; in the nasal mucous membrane, in the large and small bronchi; and in the spleen (Malpighian corpuscles).

124. The **tonsils** (Fig. 58) are masses of lymph follicles and diffuse adenoid tissue covered with a thin mucous membrane, which penetrates in the shape of longer or

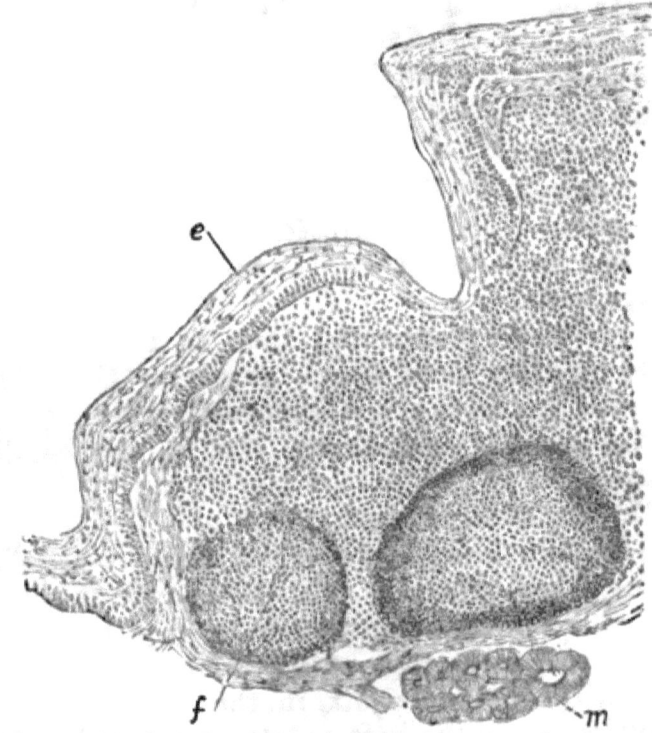

Fig. 58.—Vertical Section through part of the Tonsil of Dog.

e, Stratified pavement epithelium covering the free surface of the mucous membrane. The tissue of the mucous membrane is infiltrated with adenoid tissue. *f*, lymph follicles; *m*, mucous gland of the submucous tissue. (Atlas.)

shorter folds into the substance within. Numbers of mucus-secreting glands situated outside the layer of lymph follicles discharge their secretion into the pits— the *crypts*—between the folds. The free surface of the tonsils and the crypts is covered or lined respectively with the same stratified epithelium that lines the oral cavity. Numbers of lymph corpuscles constantly, in the perfectly normal condition, migrate through the epithelium on to the free surface, and are mixed with the secretions (mucus and saliva) of the oral cavity. The so-called mucous or salivary corpuscles of the saliva, taken from the oral cavity, are such discharged lymph corpuscles. They become swollen up by the water of the saliva, and assume a spherical shape. They finally disintegrate.

Similar relations, only on a smaller scale, obtain at the root of the tongue.

The *pharynx tonsil* of Luschka, occurring in the upper part of the pharynx, is in all essential respects the same as the palatine tonsil. Owing to large parts of the mucous membrane of the upper portion of the pharynx being covered with ciliated columnar epithelium, some of the crypts in the pharynx tonsil are also lined with it.

125. The **lenticular glands** of the stomach are single lymph follicles.

The **solitary glands** of the small and especially the large intestine are single lymph follicles.

The **agminated glands** of the ileum are groups of lymph follicles. The mucous membrane containing them is much thickened by their presence, and represents a *Peyer's patch* or a *Peyer's gland*.

126. In most instances the capillary blood-vessels form in the lymph follicles meshes, arranged in a more or less radiating manner from the periphery towards the centre; around the periphery there is a network of small veins. A larger or smaller portion of the

circumference of the follicles of the tonsils, pharynx, intestine, bronchi, &c., is surrounded by a lymph sinus leading into a lymphatic vessel. The lymphatic vessels and lymph sinuses in the neighbourhood of lymphatic follicles or of diffuse adenoid tissue are almost always found to contain numerous lymph corpuscles, thus indicating that these are produced by the adenoid tissue and absorbed by the lymphatics.

127. The **Thymus gland** consists of a framework and the gland substance. The former is fibrous connective tissue arranged as an outer capsule, and in connection with it are septa and trabeculæ passing into the gland and subdividing it into lobes and lobules, which latter are again subdivided into the *follicles* (Fig. 59A). The follicles are very irregular in shape, most of them being oblong or cylindrical streaks of adenoid tissue. Near the capsule they are well defined from one another, and present a polygonal outline; farther inwards they are more or less fused. Each shows a central transparent *medulla* and a peripheral less transparent *cortex* (Watney). At the places where two follicles are fused with one another the medulla of both is continuous. The matrix is adenoid reticulum, the fibres of the medullary part being coarser and shorter, those of the cortical portion of the follicle finer and longer. The meshes of the of the reticulum in the cortical part of the follicles are filled with the same lymph-corpuscles as occur in the adenoid tissue of other organs, but in the medullary part they are fewer, and the meshes are more or less completely occupied by the enlarged but transparent endotheloid plates. These conditions cause the greater transparency of the medulla. In some places the endotheloid cells are granular, and include more than one nucleus; some are even multinucleated giant cells.

128. There occur in the medulla of the follicles, larger or smaller, more or less concentrically-arranged

nucleated protoplasmic cells, which are the *concentric bodies of Hassall* (Fig. 59B). They are met with al-

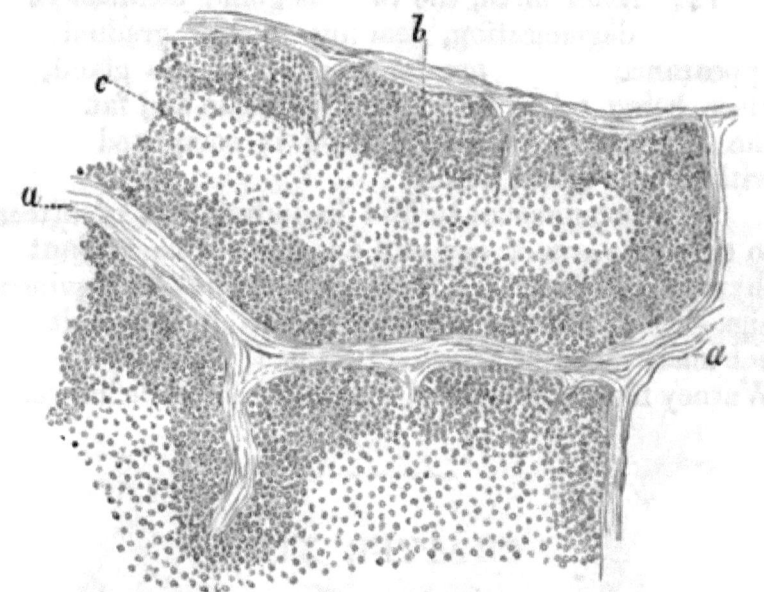

Fig. 59A.—Section through the Thymus Gland of a Fœtus.
a, Fibrous tissue between the follicles; *b*, the cortical portion of the follicles; *c*, the medullary portion.

ready in the early stages of the life of the thymus, and cannot therefore be connected with the involution of the gland, as maintained by Afanassief, according to whom the **concentric corpuscles** are formed in blood-vessels which thereby become obliterated. According to Watney they are concerned in the formation of blood-vessels and connective tissue.

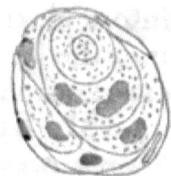

Fig. 59B.—Two Concentric or Hassall's Corpuscles of the Thymus. Fœtal Gland.

The lymphatics of the interfollicular septa and trabeculæ always contain numbers of lymph corpuscles. The blood capillaries of the follicles are more richly distributed in the

H

cortex than in the medulla, and they radiate from the periphery towards the central parts.

129. After birth, the thymus gland commences to undergo degeneration, leading to the gradual disappearance of the greater portion of the gland, its place being taken by connective tissue and fat. But the time when the involution is completed varies within very broad limits.

It is not unusual to find in individuals of fifteen to twenty years of age still an appreciable amount of thymus gland tissue. In some animals—*e.g.*, guinea-pigs—the involution of the gland even in the adult has not made much progress. In the thymus of the dog Watney found cysts lined with ciliated epithelial cells.

CHAPTER XIII.

COMPOUND LYMPHATIC GLANDS.

130. THE compound or true lymphatic glands are directly interpolated in the course of lymphatic vessels. Such are the mesenteric, portal, bronchial, splenic, sternal, cervical, cubital, popliteal, inguinal, lumbar, &c., glands. Afferent lymphatic vessels anastomosing into a plexus open at one side (in the outer capsule) into the lymphatic gland, and at the other (the hilum) emerge from it as a plexus of efferent lymphatic tubes.

131. Each true lymphatic gland is enveloped in a *fibrous capsule* which is connected with the interior and the hilum by connective tissue *trabeculæ* and *septa*. The trabeculæ having advanced a certain distance, about one-third or one-fourth, in a manner more or less radiating towards the centre, branch into minor trabeculæ, which in the middle part of the gland anastomose with one another so as to form a plexus with small

meshes. Thus the peripheral third or fourth of the gland is subdivided by the septa and trabeculæ, into relatively large spherical or oblong compartments, while the middle portion is made up of relatively small cylindrical or irregularly-shaped compartments (Fig. 60). The former region is the *cortex*, the latter the *medulla* of the gland. The compartments of the cortex anastomose with one another and with those of the medulla, and these latter also form one intercommunicating system.

The fibrous capsule, the septa and trabeculæ are the carriers of the vascular trunks; the trabeculæ consist of fibrous connective tissue and of a certain

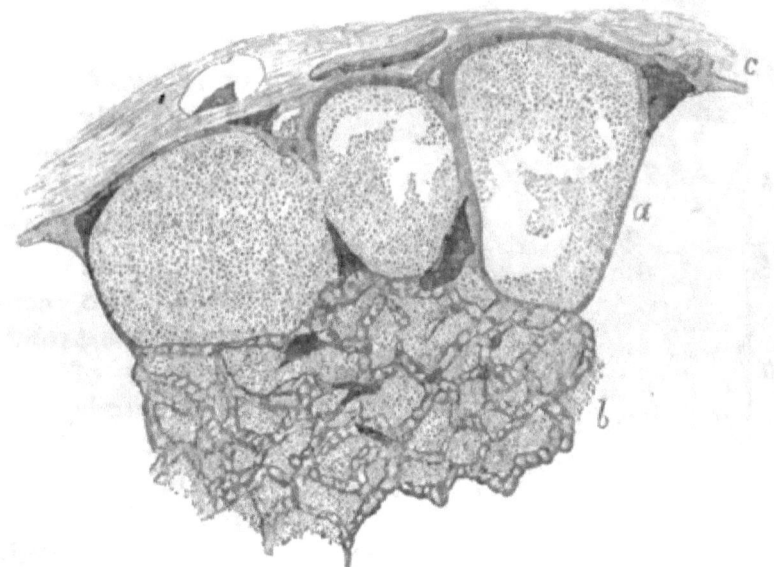

Fig. 60.—From a Vertical Section through a Lymphatic Gland, the Lymphatics of which had been injected.

c, The outer capsule, with lymphatic vessels in section; *a*, the cortical lymph follicles; around them are the cortical lymph sinuses; *b*, the medullary; injected lymph sinuses between the masses of adenoid tissue. (Atlas.)

amount of non-striped muscular tissue, which is conspicuous in some animals—*e.g.*, pig, calf, rabbit, guinea-pig—but is scarce in man.

Sometimes coarsely granular connective tissue cells

(plasma cells) are present in considerable numbers in the trabeculæ.

132. The compartments contain masses of adenoid tissue, without being completely filled with it; those of the cortex contain oval or spherical masses—*the lymph follicles of the cortex;* those of the medulla cylindrical or irregularly-shaped masses—*the medullary cylinders.* The former anastomose with one another and with the latter, and the latter amongst themselves, a condition easily understood from what has been said above of the nature of the compartments containing these lymphatic structures. The follicles and medullary cylinders consist of *adenoid tissue* of exactly the same character as that described in the previous chapter. And this tissue also contains the last ramifications of the blood-vessels, *i.e.*, the last branches of the arteries, a rich network of capillary blood-vessels, and the first or smaller branches of the veins. The capillaries and other vessels receive also here an adventitious envelope from the adenoid reticulum.

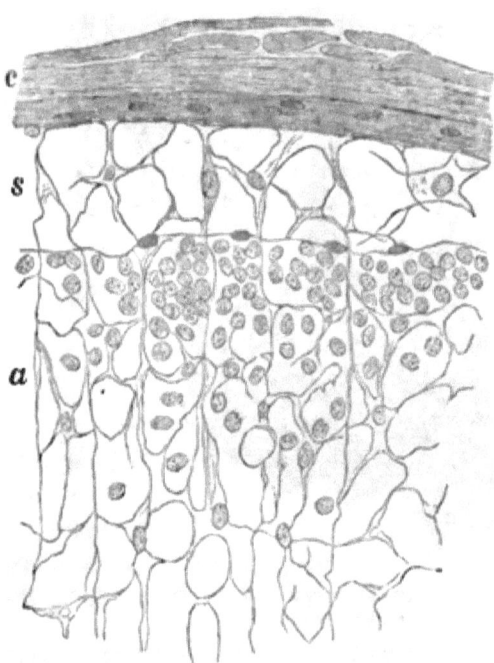

Fig. 61.—From a Section through a Lymphatic Gland.

c, The outer capsule; *s*, cortical lymph sinus; *a*, adenoid tissue of cortical follicle. Numerous nuclei, indicating lymph corpuscles. (Atlas.)

133. The cortical follicles and the medullary

cylinders do not completely fill out the compartments made for them by the capsule and trabeculæ respectively, but a small peripheral zone of each compartment is left free; this is a *lymphatic sinus*. In the cortex it is spoken of as a *cortical* (Fig. 61), in the medulla as a *medullary, lymph sinus* (Fig. 62). The

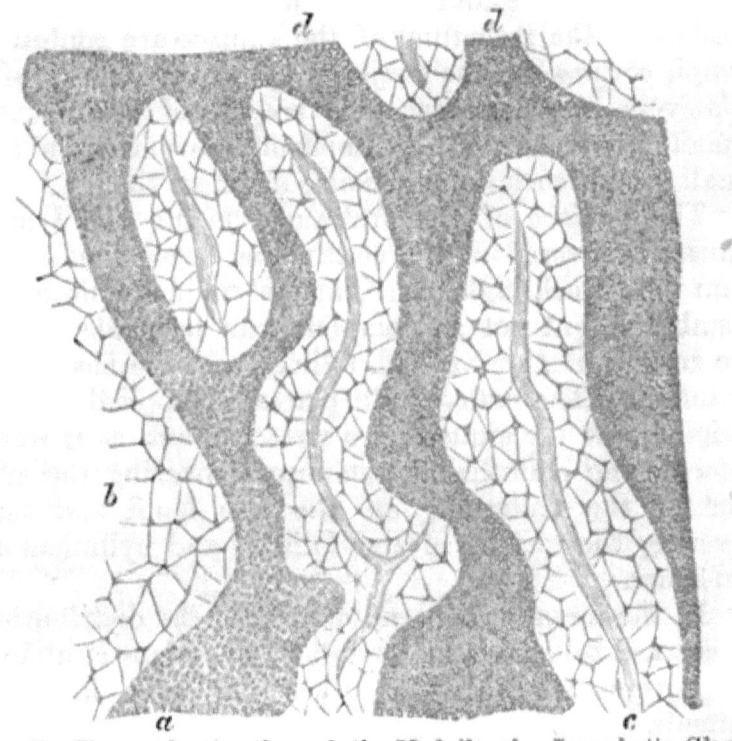

Fig. 62.—From a Section through the Medulla of a Lymphatic Gland.
a, Transition of the medullary cylinders of adenoid tissue into the cortical follicles: *b*, the lymph sinuses occupied by a reticulum; *c*, the fibrous tissue trabeculæ; *d*, the medullary cylinders. (Atlas.)

former is a space between the outer surface of the cortical lymph follicle and the corresponding part of the capsule or cortical septum, the latter between the surface of a medullary cylinder and the trabeculæ. From what has been said of the relation of the compartments, it follows that the cortical and medullary lymph sinuses form one intercommunicating system.

These are not empty free spaces, but are filled with a coarse reticulum of fibres, much coarser than the adenoid reticulum; to it are attached large transparent cell plates—endotheloid plates. In some instances (as in the calf) these cell-plates of the medullary sinuses contain brownish pigment granules, which give to the medulla of the gland a dark brown aspect. In the meshes of the reticulum of the sinuses are contained lymph corpuscles, the majority of which consist of a relatively large protoplasmic body, and one or two nuclei; they show lively amœboid movement; a few small lymph corpuscles are also amongst them.

The surface of the trabeculæ facing the lymph sinuses is covered with a continuous layer of endothelium (von Recklinghausen), and a similar endothelial membrane, but not so complete, can be made out on the surface of the cortical follicles and the medullary cylinders. The endotheloid plates, as applied to the reticulum of the sinuses, are stretched out, as it were, between the endothelial membrane covering the surface of the trabeculæ on the one hand and that covering the surface of the follicles and cylinders on the other.

In the mesenteric gland of the pig the distribution of cortical follicles and medullary cylinders is almost the reverse from that of other glands or in other animals.

134. The *afferent lymphatic vessels* having entered the outer capsule of the gland, and having formed there a dense plexus, open directly into the cortical lymph sinuses. The medullary lymph sinuses lead into lymphatic vessels, which leave the gland per hilum as the efferent vessels.

Both afferent and efferent vessels are supplied with valves.

135. The course of the lymph through a lymphatic gland is then simply this—from the afferent vessels,

situated in the capsule, into the cortical lymph sinuses, from these into the medullary sinuses, and from these into the efferent lymphatics. Owing to the presence of the reticulum in the sinuses the current of the lymph will proceed only very slowly and with difficulty, as if it were passed through a spongy filter. Hence a large number of formed corpuscles, pigment, inflammatory or other elements, passing into the gland by the afferent vessels are easily arrested and deposited in the sinuses, and there readily swallowed by the amœboid corpuscles lying in the meshes.

Passing a stream of water through the gland, the contents of the meshes of the reticulum of the sinuses— *i.e.*, the lymph corpuscles—are of course the first things washed out (von Recklinghausen), and on continuing the stream some of the lymph corpuscles of the follicles and cylinders are also washed out. Hence it is probable also that by the normal stream of lymph passing through the gland, lymph corpuscles are drained, as it were, from the follicles and cylinders into the sinuses. The amœboid movement of the lymph corpuscles, especially of the large and ripe ones, will greatly facilitate their passage from the follicles and cylinders into the lymph sinuses.

CHAPTER XIV.

NERVE-FIBRES.

136. THE nerve-fibres conduct impulses to or from the tissues and organs on the one hand, and the nerve-centres on the other, and accordingly we have to consider in each nerve-fibre the peripheral and central termination and the conducting part. The latter, *i.e.*,

the nerve-fibres proper, in the cerebro-spinal nerves are grouped into bundles, and these again into anatomical nerve-branches and nerve-trunks. Each anatomical cerebro-spinal nerve consists, therefore, of *bundles* of nerve-fibres (Fig. 63). The general matrix by which these bundles **are held** together is fibrous connective tissue called the *Epineurium* (Key and Retzius); this epineurium is the carrier of the larger and smaller blood-vessels with which **the** nerve-trunk is supplied, of a plexus of lymphatics, of groups of fat-cells, and sometimes of numerous plasma **cells.**

137. The *nerve-bundles* (Fig. 64) are of various sizes, according to the number and size of the nerve-fibres they contain. They are well-defined by a sheath of their

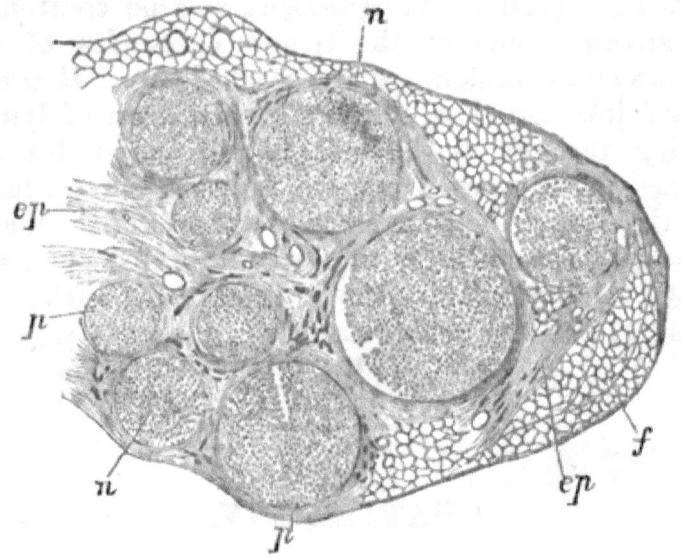

Fig. 63.—From a Transverse Section through the Sciatic Nerve of the Dog.
ep, Epineurium; *p*, perineurium; *n*, nerve fibres constituting **a nerve-bundle in cross** section; *f*, fat tissue surrounding the nerve. (**Atlas.**)

own, called *perineurium* (Key and Retzius). This perineurium consists of bundles of fibrous connective tissue arranged in lamellæ, every two lamellæ being separated from one another by smaller or larger lymph spaces, which form an intercommunicating system, and anastomose with the lymphatics of the epineurium whence they can be injected. Between the lamellæ, and in the spaces, are situated flattened endotheloid connective tissue corpuscles.

The nerve-bundles are either *single* or *compound*. In the former the nerve-fibres contained in a bundle are not sub-divided into groups, in the latter the bundles are sub-divided by thicker and thinner septa of fibrous connective tissue connected with the perineurium. When a nerve-bundle divides—as when a trunk repeatedly branches, or when it enters on its peripheral distribution—each branch of the bundle receives a continuation of the lamellar perineurium. The more branches the perineurium has to supply, the more reduced it becomes in thickness. In some of these minute branches the perineurium is reduced to a single layer of endothelial cells. When one of these small bundles breaks up into the single nerve-fibres, or into small groups of them, each of these has also a continuation of the fibrous tissue of the perineurium. In some places this perineural continuation is only a very delicate endothelial membrane, at others it is of considerable thickness, and still shows the laminated nature. Such thick sheaths of single nerve-fibres, or of small groups of them, represent what is called *Henle's sheath*.

138. The nerve-fibres are held together within the bundle by connective tissue, called the *Endoneurium* (Fig. 64). This is a homogeneous ground substance in which are embedded fine bundles of fibrous connective tissue, and connective tissue corpuscles, and capillary blood-vessels arranged so as to form a network with elongated meshes. Between the perineurium and the nerve-fibres are found here and there lymph spaces; similar spaces separate the individual nerve-fibres, and have been injected by Key and Retzius.

When nerve-trunks anastomose so as to form a plexus—*e.g.*, in the brachial, or sacral plexus—there occurs an exchange and re-arrangement of nerve-bundles in the branches. A similar condition obtains in the ganglia of the cerebro-spinal nerves. Nerve-

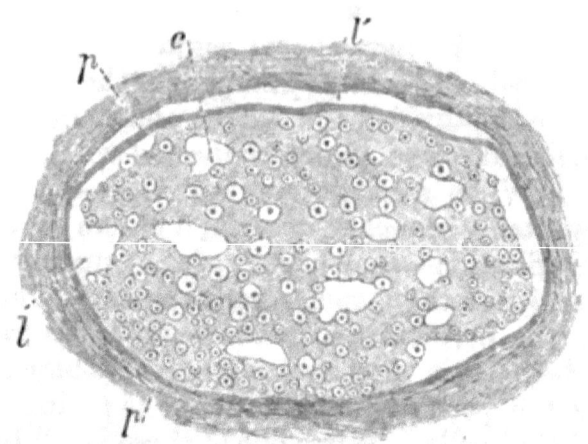

Fig. 64.—Transverse Section through a Nerve-bundle in the Tail of Mouse.

p, The perineurium ; *e*, the endoneurium separating the medullated nerve-fibres seen in cross section ; *l*, lymph spaces in the perineurium ; *l'*, lymph spaces in the endoneurium. (Atlas.)

trunks and nerve-branches passing through a lymph cavity, such as the subdural spaces, or the subcutaneous lymph sacs, or the cisterna lymphatica magna in the frog, receive from the serous membrane an outer endothelial covering.

139. The nerve-fibres in the nerve-bundles of the cerebro-spinal nerves, with the exception of the olfactory nerve, are *medullated nerve-fibres*. These are doubly or darkly contoured smooth cylindrical fibres, varying in diameter between $\frac{1}{2000}$ and $\frac{1}{12000}$ of an inch. Within the same bundle of a nerve—*e.g.*, of the brachial or sacral plexus—there occur fibres which are several times thicker than others, and it is probable that they are derived from different sources. Schwalbe has shown that the thickness of the nerve-

fibre stands in a certain relation to the distance of its periphery from the nerve-centre and to functional activity.

A medullated nerve-fibre in the fresh condition is a bright glistening cylinder, showing a dark double contour. Either spontaneously after death, or after re-agents—as water, salt solution, dilute acids—or after pressure and mechanical injury, the outline of the nerve-fibre becomes irregular; smaller, or larger, glistening dark-bordered droplets and masses appear and gradually become detached. These droplets and masses are derived from the fatty substance constituting the medullary sheath or white substance of Schwann (see below). When a nerve-fibre within the bundle undergoes degeneration during life, either after section of the nerve or after other pathological changes, or in the natural course of its existence (S. Mayer), the medullary sheath breaks up into similar smaller or larger globules or particles, which gradually become absorbed.

140. Each medullated nerve-fibre (Figs. 64A, 66) consists of the following parts: (*a*) the central *axis cylinder*. This is the essential part of the fibre, and is a cylindrical or bandlike, pale, transparent structure, which in certain localities (near the terminal distribution, in the olfactory nerves, in the central nervous system), and especially after certain re-agents, shows itself composed of very fine homogeneous or more or less beaded fibrillæ—the *elementary or primitive fibrillæ* (Max Schultze)—held together by a small amount of a faintly granular interstitial substance. The longitudinal striation of the axis cylinder is due to its being composed of primitive fibrillæ. The thickness of the axis cylinder is in direct proportion to the thickness of the whole nerve-fibre. The axis cylinder is said to be enveloped in its own hyaline more or less *elastic sheath*, composed of neurokeratin.

141. (*b*) The *medullary sheath* or white substance of Schwann, also called the medulla of the nerve-fibre. This is a glistening bright fatty substance surrounding the axis cylinder, as an insulating hollow cylinder surrounds an electric wire. The medullary sheath gives to the nerve-fibre its double or dark contour. Between the axis cylinder and the medullary sheath there is a small amount of albuminous fluid which appears greatly increased when the former, owing to shrinking, stands farther apart from the latter.

142. (*c*) The *sheath of Schwann*, or the *neurilemma*, surrounds closely the medullary sheath, and forms the outer boundary of the nerve-fibre. It is a hyaline delicate membrane. From place to place there is present between the neurilemma and the medullary sheath, and situated in a depression of the latter, an oblong nucleus, surrounded by a thin zone of protoplasm. These nucleated corpuscles are the *nerve corpuscles* (Fig. 64A), and are analogous to the muscle corpuscles, situated between the sarcolemma and the striated muscular substance. They are not nearly so numerous as the muscle corpuscles.

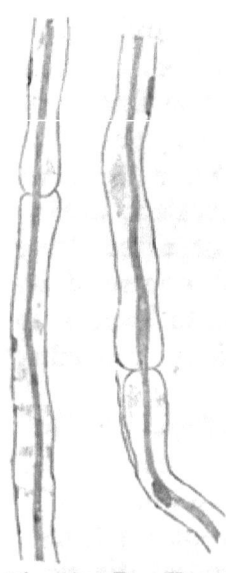

Fig. 64A.—Two Nerve Fibres, showing the nodes or constrictions of Ranvier and the axis cylinder. The medullary sheath has been dissolved away. The deeply-stained oblong nuclei indicate the nerve corpuscles within the neurilemma. (Atlas.)

143. The neurilemma produces at certain definite intervals annular constrictions, the *nodes* or *constrictions of Ranvier* (Figs. 64A, 65, 66), and at these nodes of Ranvier the medullary sheath, but not the axis cylinder and its special sheath, suffers a discontinuity and sharply terminates. The portion of the nerve-fibre situated between two nodes is

the *internodal segment.* Each internodal segment has generally one, occasionally more than one, nerve corpuscle. The medullary cylinder of each internodal segment is made up of a number of conical sections (Fig. 66A) imbricated at their ends (Schmidt, Lantermann) (Fig. 66), and each such

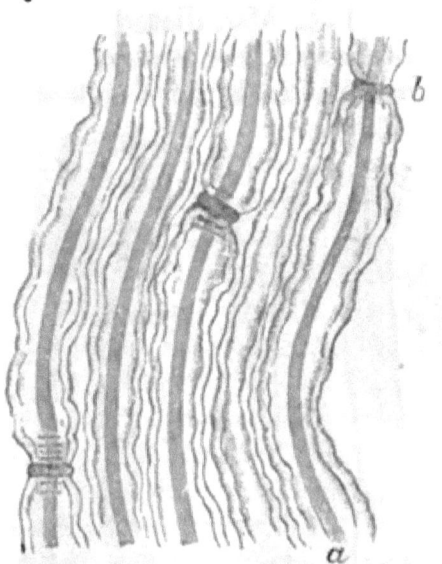

Fig. 65.—Medullated Nerve-fibres, after staining with Nitrate of Silver.

a, The axis cylinder; *b*, Ranvier's constriction. (Key and Retzius.)

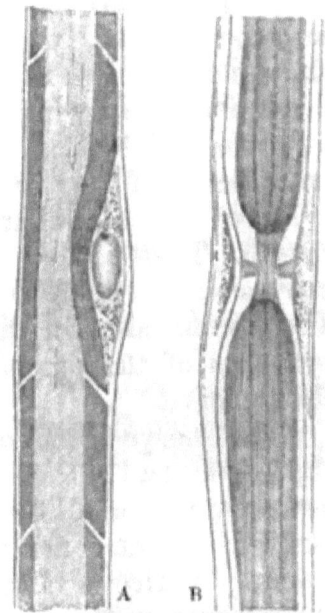

Fig. 66.—Medullated Nervefibres.

A, A medullated nerve-fibre, showing the subdivision of the medullary sheath into cylindrical sections imbricated with their ends; a nerve corpuscle with an oval nucleus is seen between the neurilemma and the medullary sheath; B, a medullated nerve fibre at a node or constriction of Ranvier; the axis cylinder passes uninterruptedly from one segment into the other, but the medullary sheath is interrupted. (Key and Retzius.)

section is again made up of a large number of rod-like structures (Fig. 67) placed vertically on the axis cylinder.

These rods are, however, connected into a network. The network itself is very likely the neurokeratin of Ewald and Kühne, whereas the interstitial substance of the network is probably the fatty substance leaving the nervefibre in the shape of droplets when pressure or reagents are applied to the fresh nerve-fibre.

144. Medullated nerve-fibres *without any neurilemma*, and consequently without any nodes of Ranvier, with a thick more or less distinctly laminated medullary sheath, form the white substance of the brain and spinal cord. In these organs, in the hardened and fresh state, numerous nerve-fibres may be noticed, which show more or less regular varicosities, owing to local accumulations of fluid between the axis cylinder and medullary sheath. These are called *varicose nerve-fibres*. They occur also in the branches of the sympathetic nerve.

The nerve-fibres of the *optic* and *acoustic* nerve are medullated, but without any neurilemma; varicose fibres are common in them.

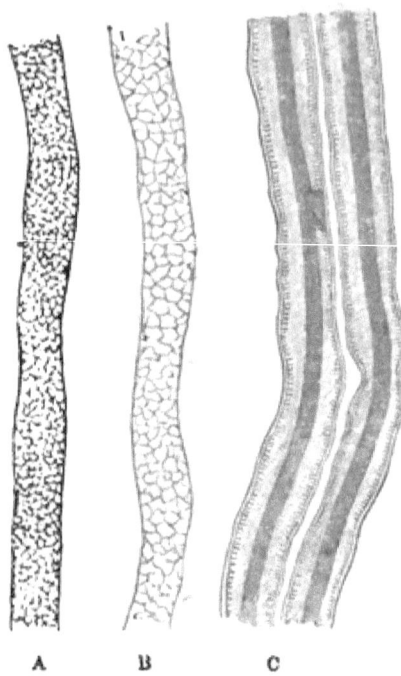

Fig. 67.—Medullated Nerve-fibres.
A, B, showing on a surface view the reticulated nature of the medullary sheath; C, two nerve-fibres showing the axis cylinder, the medullary sheath with their vertically-arranged minute rods, and the delicate neurilemma or outer hyaline sheath. (Atlas.)

145. Medullated nerve-fibres occasionally in their course divide into two medullated fibres. Such division is very common in medullated nerve-fibres supplying striped muscular fibres, especially at or near the point of entrance into the muscular fibres (see below). But also in other localities division of nerve-fibres may be met with. The electric nerve of the electric fishes (malapterurus, gymnotus, silurus, etc., elec-

tricus) shows such divisions to an extraordinary degree, one huge nerve-fibre dividing at once into a bundle of minor fibres. Division of a medullated fibre takes place generally at a node of Ranvier. The branches taken together are generally thicker than the undivided part of the fibre, but in structure they are identical with the latter.

146. When medullated nerve-fibres approach their peripheral termination, they change sooner or later, in so far as their medullary sheath suddenly ceases; and now we have a *non-medullated*, or Remak's nerve-fibre. Each of these consists of an axis cylinder, a neurilemma, and between the two a nucleated nerve corpuscle from place to place. Non-medullated nerve-fibres always show the fibrillar nature of their axis cylinder. The olfactory nerve-branches are entirely made up of non-medullated nerve-fibres. In the branches of the sympathetic most fibres are non-medullated.

The non-medullated fibres undergo always repeated divisions. They form *plexuses*, large fibres branching into smaller ones, and these again joining (Fig. 68). Generally at the nodal points of these plexuses there are triangular nuclei, indicating the corpuscles of the neurilemma.

147. Finally the non-medullated nerve-fibres lose their neurilemma, and then we have *simple axis cylinders*. These branch and ultimately break up into the constituent *primitive nerve-fibrillæ*, which occasionally show regular varicosities (Fig. 69). Of course, of a neurilemma or the nuclei of the nerve corpuscles there is nothing left. These primitive fibrillæ branch and anastomose with one another, and thus form a *network*. The density of this network depends on the number of primitive fibrils and the richness of their branching. These primitive fibrils and their networks represent the

peripheral termination, and this mode of termination occurs in the nerve-fibres of common sensation, as in many of the nerve-fibres of the skin, cornea, and

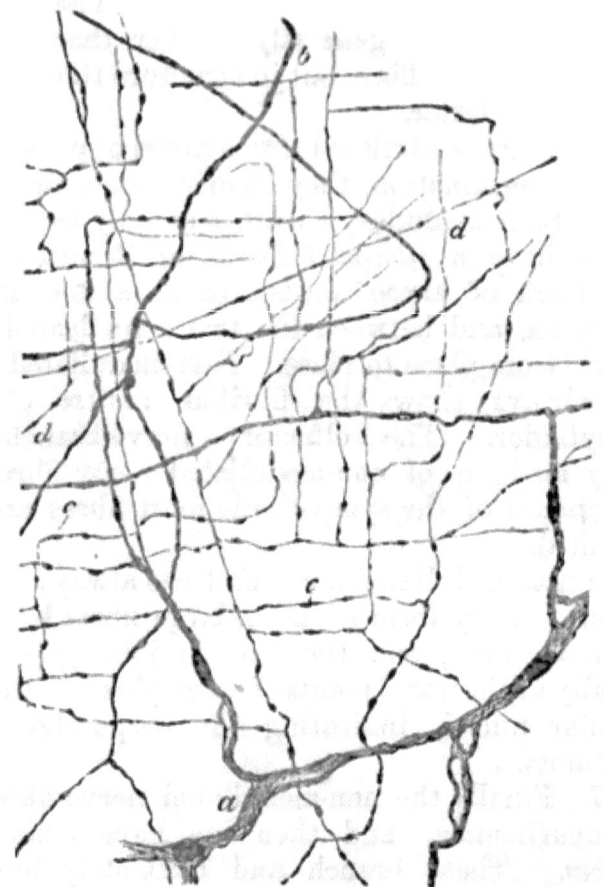

Fig. 68.—Plexus of Fine Non-medullated Nerve-fibres of the Cornea.
a, A thick non-medullated nerve-fibre; *b*, a fine one; *c*, *d*, elementary fibrils, anastomosing into a network.

mucous membranes. In all these cases the peripheral termination, *i.e.*, the primitive fibrils and their networks are found intra-epithelial (Fig. 70), *i.e.* situated in the stratum Malphigii of the epidermis, in the epithelial parts of the hair follicle, in the anterior epithelium of the cornea, and in the epithelium of the mucous membranes. The primitive nerve-fibrils lie in the interstitial substance *between* the epithelial cells, as

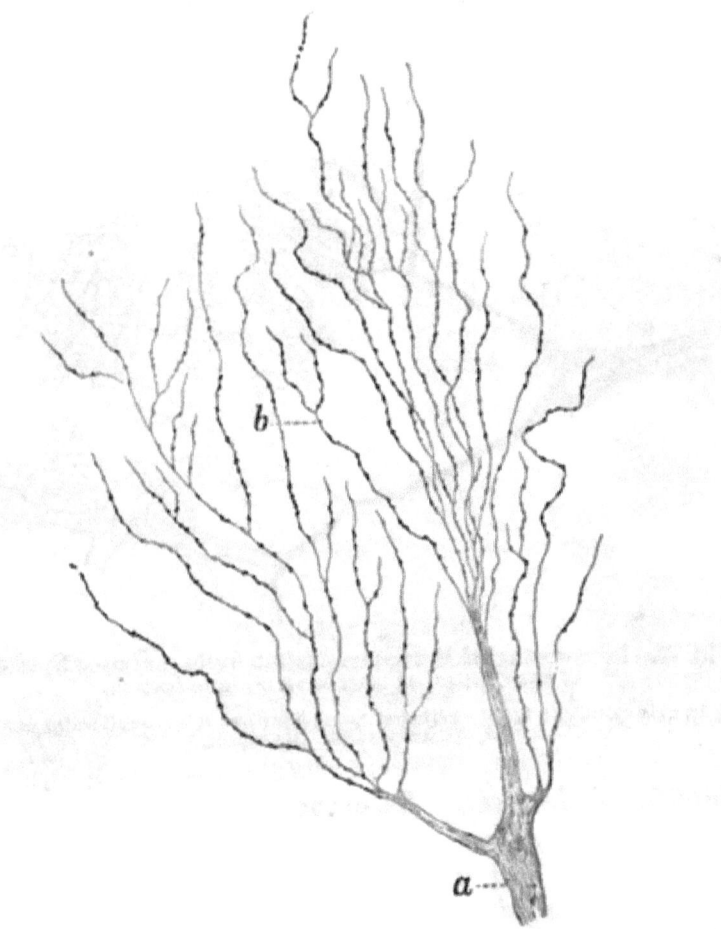

Fig. 69.—Nerve-fibres of the Cornea.

a, An axis cylinder splitting up into its constituent primitive fibrillæ near the anterior epithelium of the cornea; *b*, primitive fibrillæ.

intra-epithelial networks and as primitive fibrils which appear to terminate with free ends.

148. Tracing then a nerve-fibre, say one of common sensation, from the periphery towards the centre, we have isolated *primitive fibrils* or networks of them;

they form by association *simple axis cylinders*, which vary in thickness according to the number of their constituent primitive fibrils. These simple axis cylin-

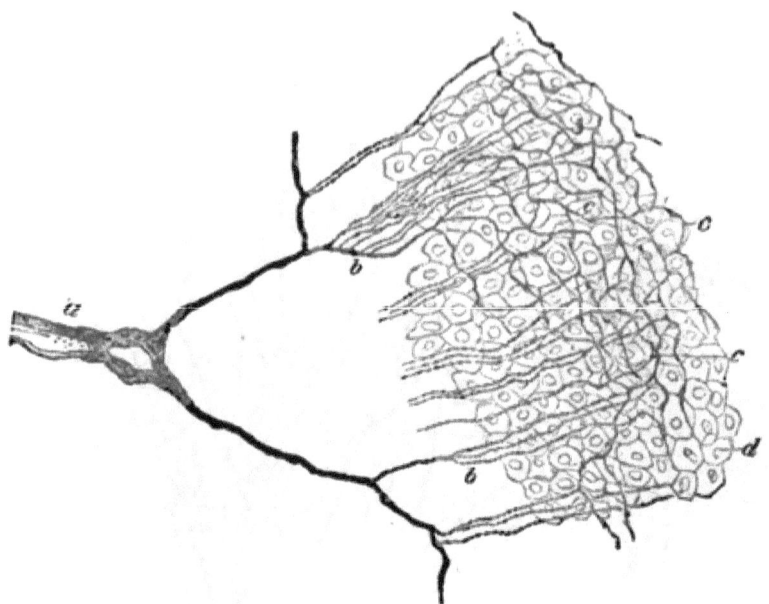

Fig. 70.— Intra-epithelial Nerve-termination in the Anterior Epithelium of the Cornea, as seen in an oblique section.

a, An axis cylinder; *b*, sub-epithelial nerve-fibrillæ; *c*, intra-epithelial network; *d*, epithelial cells. (Handbook.)

ders form plexuses. By association they make larger axis cylinders, and these becoming invested with a neurilemma, and with the nuclei of nerve corpuscles, form *non-medullated nerve-fibres*. These also form plexuses. A medullary sheath makes its appearance between the neurilemma and the axis cylinder, and thus forms a *medullated nerve-fibre*.

CHAPTER XV.

PERIPHERAL NERVE-ENDINGS.

149. In the preceding chapter the termination of the nerves of common sensation, as isolated primitive fibrillæ, and as *networks* of these, has been described in the epithelium of the skin and mucous membranes, and in the anterior epithelium of the cornea. Besides these there are other special terminal organs of sensory nerves, probably concerned in the perception of some special quality or quantity of sensory impulses. They are all connected with a medullated nerve-fibre, and are situated, not in the epithelium of the surface, but in the tissue, at greater or lesser depth. Such are the **Pacinian** corpuscles, the Herbst corpuscles, the end-bulbs of Krause in the tongue and conjunctiva, the genital end-corpuscles or end-bulbs in the external genital organs, the corpuscles of Meissner, or tactile corpuscles, in the papillæ of the skin of the volar side of the fingers, the touch-cells of Merkel in the beak and tongue of duck, &c.

150. The **Pacinian corpuscles.** — These are also called Vater's corpuscles. They occur in large numbers on the subcutaneous nerve-fibres of the palm of the hand and foot of man, in the mesentery of the cat, along the tibia of the rabbit, on the genital organs of man (corpora cavernosa, prostate). Each corpuscle is oval, more or less pointed, and in some places easily perceptible to the unaided eye (palm of the human hand, mesentery of the cat), the largest being about $\frac{1}{20}$th of an inch long and $\frac{1}{30}$th of an inch broad; in other places they are of microscopic size only. Each possesses a *stalk*, to which it is attached,

and which is a single *efferent medullated nerve-fibre* (Fig. 71), differing from an ordinary medullated nerve-fibre merely in the fact that outside the neurilemma of the nerve-fibre there is present a thick laminated connective tissue sheath, which is the sheath of Henle—continuous with the perineural sheath of the nerve branch with which the nerve-fibre is in connection. This medullated nerve-fibre within its sheath possesses generally a very wavy outline. The corpuscle itself is composed of a large number of lamellæ, or capsules, more or less concentrically arranged around a *central elongated or cylindrical clear space*. This space contains in its axis from the proximal end, *i.e.*, the one nearest to the stalk, to near the opposite or distal end, a continuation of the nerve-fibre in the shape of a *simple axis cylinder*. But this axis cylinder does not fill out the central space, since there is, all round the faintly and longitudinally striated axis cylinder, a good deal of space left filled with a transparent substance, in which, in some instances, rows of spherical nuclei may be perceived along the margin of the axis cylinder. At or near the distal end of the central space *the axis cylinder divides in two or more branches*, and these *terminate* in pear-shaped, oblong, spherical, or irregularly-shaped granular-looking *enlargements*.

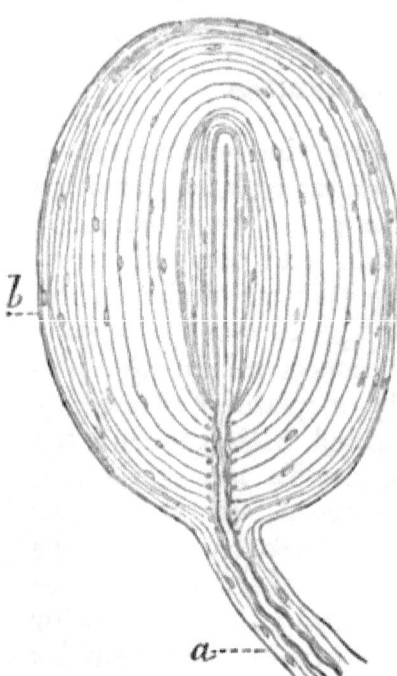

Fig. 71.—A Pacinian Corpuscle, from the Mesentery of the Cat.
a, The medullated nerve-fibre; *b*, the concentric capsules.

151. The concentric *capsules* forming the corpuscle itself are disposed in a different manner at the periphery and near the central space from that in which they are disposed in the middle parts, viz., in the former localities they are much closer together, being thinner than in the latter. On looking, therefore, at a Pacinian corpuscle in its longitudinal axis, or in cross section, we always notice the striation (indicating the capsules) to be closer in the former than in the latter places. Each capsule consists of—(a) a hyaline, probably elastic, *ground substance*, in which are embedded here and there (b) *fine bundles of* connective *tissue fibres;* (c) on the inner surface of each capsule, *i.e.*, the one directed to the central axis of the Pacinian corpuscle, is a single layer of *nucleated endothelial plates*. The oblong nuclei visible on the capsules at ordinary inspection are the nuclei of these endothelial plates. There is no fluid between the capsules, but these are in contact with one another (Huxley). Neighbouring capsules are occasionally connected with one another by thin fibres.

152. In order to reach the central space of the corpuscle, the medullated nerve-fibre has to perforate the capsule at one pole; thus a canal is formed in which is situated the medullated nerve-fibre, and as such, and in a very wavy condition, it reaches the proximal end of the central space. This part of the nerve-fibre may be called the intermediary part. The lamellæ of the sheath of Henle pass directly into the peripheral capsules of the corpuscle.

Immediately before entering the central space, the nerve-fibre divests itself of all parts except the axis cylinder, which, as stated above, passes into the central space of the Pacinian corpuscle. In some cases a minute artery enters the corpuscle at the pole, opposite to the one for the nerve-fibre; it penetrates the peripheral capsules, and supplies them with a few capillary vessels.

153. The **corpuscles of Herbst** are similar to the Pacinian corpuscles, with this difference, that they are smaller and more elongated, that the axis cylinder of the central space is bordered by a continuous row of nuclei, and that the capsules are thinner and more closely placed (Fig. 72). This applies especially to those near the central space, and here between these central capsules we miss the nuclei indicating the endothelial plates. Such is the nature of Herbst's corpuscles in the mucous membrane of the tongue of the duck, and to a certain degree also in those of the rabbit, and in tendons.

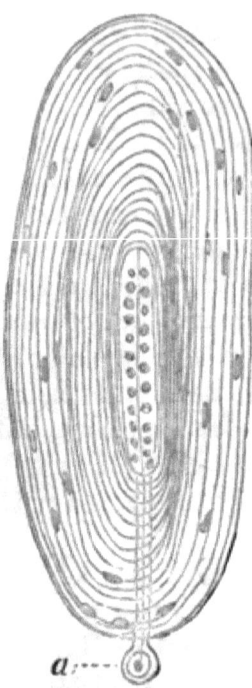

Fig. 72.—A Herbst's Corpuscle, from the Tongue of Duck.

a, The medullated nerve-fibre cut away.

154. The **tactile corpuscles, or corpuscles of Meissner**, occur in the papillæ of the corium of the volar side of the fingers and toes in man and ape; they are oblong, straight, or slightly folded. In man they are about $\frac{1}{250}$ to $\frac{1}{300}$ of an inch long, and $\frac{1}{500}$ to $\frac{1}{200}$ of an inch broad. They are connected with a medullated nerve fibre—generally one, occasionally, but rarely, two—with a sheath of Henle. The nerve-fibre enters the corpuscle, but usually before doing so it winds round the corpuscle as a medullated fibre once or twice or oftener, and its Henle's sheath becomes fused with the fibrous capsule or sheath of the tactile corpuscle. The nerve-fibre ultimately loses its medullary sheath and penetrates

into the interior of the corpuscle, where the axis cylinder branches; its branches retain a coiled course all along the tactile corpuscle (Fig. 73), anastomose with one another, and terminate in slight enlargements, pear-shaped or cylindrical. These enlargements, according to Merkel, are touch cells. The matrix, or main part of the tactile corpuscle consists, besides the fibrous sheath with nuclei and numerous elastic fibres, of fine bundles of connective tissue, and of a number of nucleated small cells.

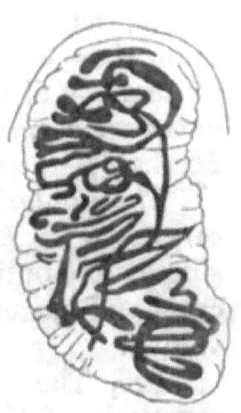

Fig. 73.—A Tactile Corpuscle of Meissner from the Skin of the Human Hand. Showing the convolutions of the nerve-fibre. (E. Fischer and W. Flemming.)

155. The **end-bulbs of Krause**.—These occur in the conjunctiva of calf and man, and are oblong or cylindrical minute corpuscles situated in the deeper layers of the conjunctiva, near the corneal margin. A medullated nerve-fibre, with Henle's sheath, enters the corpuscle (Fig. 74). This possesses a nucleated capsule, and is a more or less laminated (in man more granular-looking) structure, numerous nuclei being scattered between the laminæ. Of the nerve-fibre, as a rule, only the

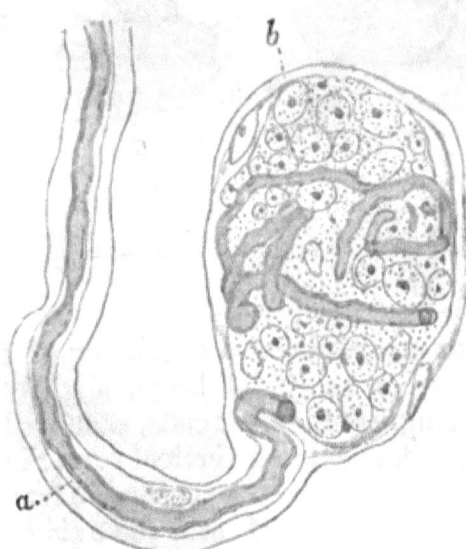

Fig. 74.—An End-bulb of Krause.
a, Medullated nerve-fibre; b, the capsule of the corpuscle.

axis cylinder is prolonged into the interior of the corpuscle. Occasionally the medullated nerve-fibre passes, as such, into the corpuscle, being at the same time more or less convoluted. Having passed to near the distal extremity, it branches, and terminates with small enlargements (Krause, Longworth, Merkel, Key and Retzius).

The *end-bulbs in the genital organs*, or the *genital corpuscles of Krause*, are similar in structure to the simple end-bulbs. They occur in the tissue of the cutis and mucous membrane of the penis, clitoris, and vagina.

156. The **corpuscles of Grandry** or *touch-corpuscles of Merkel*, in the tissue of the papillæ in the beak and tongue of birds, are oval or spherical corpuscles of minute size, possessed of a very delicate

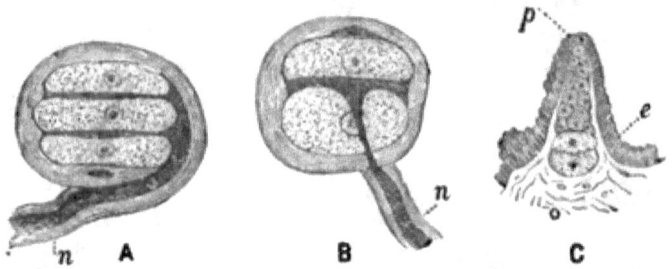

Fig. 75.—Corpuscles of Grandry in the Tongue of Duck.

A, Composed of three cells; B, composed of two cells; C, showing the development of a Grandry's corpuscle from the epithelium covering the papilla; e, epithelium; n, nerve-fibre. (Izquierdo.)

nucleated membrane as a capsule, and consisting of a series (two, three, four, or more) of large, slightly-flattened, granular-looking, transparent cells, each with a spherical nucleus, and arranged in a vertical row (Fig. 75). A medullated nerve-fibre enters the corpuscle from one side, and losing its medullary sheath, the axis cylinder branches, and its branchlets terminate, according to some (Merkel, Henle), in the cells of the corpuscle (touch-cells of Merkel); according to others (Key and Retzius, Ranvier, Hesse, Izquierdo), in the transparent substance between the touch-cells, thus forming the 'disc tactil' of Ranvier or the Tastplatte of Hesse. Neither theory seems to me to answer to the facts of the case, since I find that the branchlets of the axis cylinder terminate, not in the touch-cells, nor as the disc tactil, but with minute swellings in the interstitial substance between the touch-cells, in a manner very similar to what is the case in the conjunctival end bulbs. According to Merkel, single or small groups of touch-cells occur in the tissue of the papillæ, and also in the epithelium, in the skin of man and mammals.

157. In **articulations**—*e.g.*, the knee-joint of the rabbit—Nicoladoni described numerous nerve branches, from which fine nerve-fibres are given off. Some of these terminate in a network, others on blood-vessels, and a third group enter Pacinian corpuscles. Krause described in the synovial membranes of the joints of the human fingers medullated nerve-fibres which end in peculiar tactile corpuscles, called by him "articulation-nerve corpuscles."

158. The **nerve branches supplying non-striped muscular tissue** are derived from the sympathetic system. They are composed of non-medullated fibres, and the branches are invested in an endothelial sheath,—perineurium. The branches divide into single or small groups of axis cylinders, which reunite into a plexus—the *ground plexus of Arnold*. Small fibres coming off from the plexus supply the individual bundles of non-striped muscle cells, and they form a plexus called the *intermediary plexus* (Fig. 76). The fibres joining this plexus are smaller or larger bundles of primitive fibrillæ; in the nodes—*i.e.*, the points, of meeting of these fibres are found angular nuclei. From the intermediary

plexus pass off isolated or small groups of primitive fibrillæ, which pursue their course in the interstitial

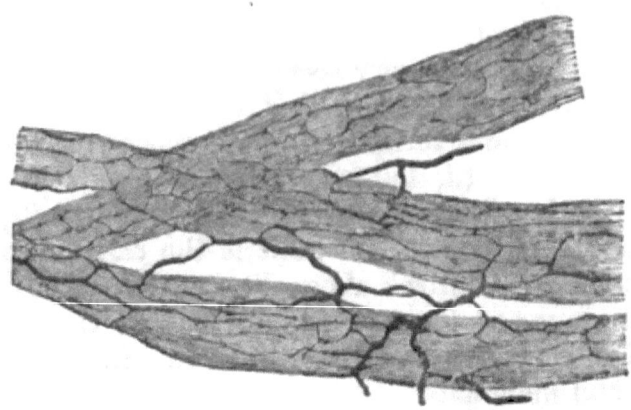

Fig. 76.—Bundles of Non-striped Muscular Tissue surrounded by Plexuses of Fine Nerve-fibres. (Handbook.)

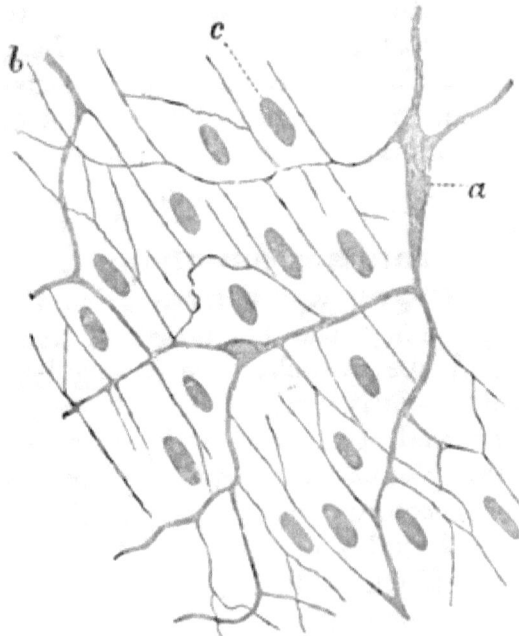

Fig. 76A.—Termination of Nerves in Non-striped Muscular Tissue.

a, Non-medullated fibre of the intermediary plexus; *b*, fine intermuscular fibrils; *c*, nuclei of muscular cells. (Atlas.)

substance between the muscle cells; these are the *intermuscular fibrils* (Fig. 76A). According to Frankenhaüser and Arnold, they give off finer fibrils, ending in the nucleus (or nucleolus). According to Elischer, the primitive fibrils terminate on the surface of the nucleus with a minute swelling.

In many localities there are isolated ganglion cells in connection with the intermuscular fibres.

159. The nerves of **blood-vessels** are derived from the sympathetic, and they terminate in arteries and veins in essentially the same way as in non-striped muscular tissue, being chiefly present in those

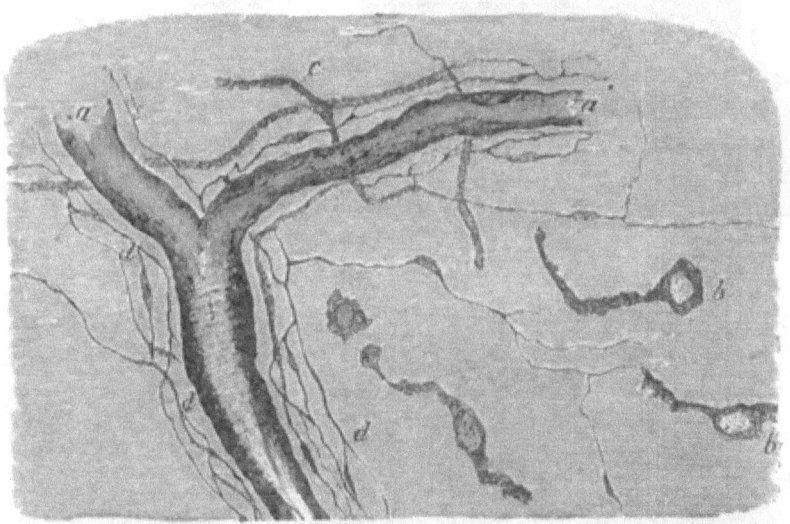

Fig. 77.—Plexus of Fine Non-medullated Nerve-fibres surrounding Capillary Arteries in the Tongue of Frog, after staining with Chloride of Gold.
a, Blood-vessel; b, connective tissue corpuscles; c, thick non-medullated fibres; d, plexus of fine nerve-fibres. (Handbook.)

parts (media) which contain the non-striped muscular tissue. But there are also fine non-medullated nerve fibres, which accompany capillary vessels—capillary arteries and capillary veins—and in some places they give off elementary fibrils, which form a network around the vessel (Fig. 77). In some localities the vascular nerve branches are provided with small groups of ganglion cells.

160. In **striped muscle** of man and mammals, reptiles, and insects, the termination of nerve fibres takes place, according to the commonly accepted view of Kühne, in the following manner :—A medullated

nerve-fibre, generally derived from one that has divided, enters at almost a right angle a striped muscular fibre, the neurilemma becoming fused with

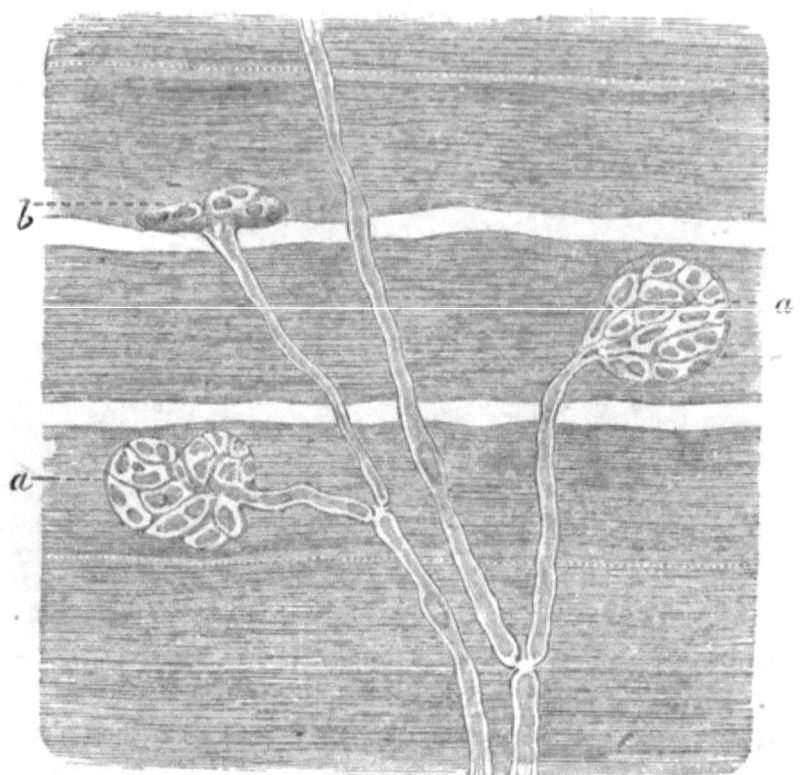

Fig. 78.—From a Preparation of Striped Muscular Fibres of the Snake, showing the termination of the Medullated Nerve-fibres. (After a preparation of Mr. A. Lingard.)

a, The nerve endplate seen from the broad side; b, the same seen from the narrow side. Each endplate is a network connected with the axis cylinder of a medullated nerve-fibre, and contains numerous nuclei of various sizes and shapes.

the sarcolemma, and the nerve-fibre, either at the point of entrance or soon after loses its medullary sheath, so that only the axis cylinder passes on. This latter divides simultaneously into a number of smaller fibres, which soon break up into a network of fine fibrils, this ultimate network being embedded in a more or less granular-looking plate, provided with a number of oblong nuclei (Fig. 78). The whole structure represents the *nerve endplate*. When the muscular fibre contracts, this endplate naturally assumes the shape of a prominence—Doyère's nerve-mount. Each muscular fibre has at least one nerve endplate, but occasionally has several in near proximity. An endplate is generally supplied by one, sometimes, however, by two, nerve-fibres. The contraction wave generally starts from the endplate. In Batrachia the nerve-fibre does not, as a rule, terminate in the shape of a granular endplate, but having penetrated the sarcolemma ramifies into several axis cylinders, each of which again branches; all branches have a more or less longitudinal direction, and are provided, either terminally or in

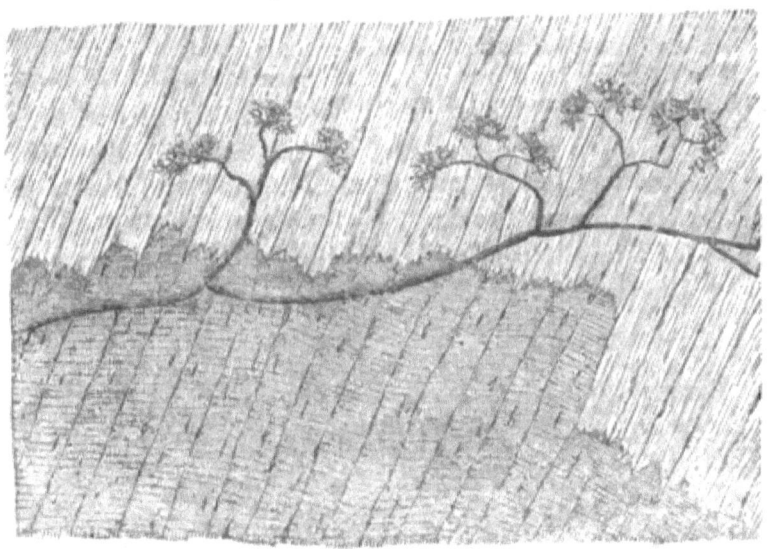

Fig. 79.—Termination of Medullated Nerve-fibres in Tendon, near the Insertion of the Striped Muscular Fibres.

The nerve-fibres terminate in peculiar reticulated endplates of primitive fibrillæ. (Golgi.)

their course, with oblong nuclei. Arndt has shown that both kinds of terminations occur in Batrachia. These two sorts of nerve endings lie underneath the sarcolemma and on the surface of the muscular substance proper. But besides this intramuscular termination, there is a plexus of nerve-fibres, which is situated outside the sarcolemma—*i.e.*, intermuscular; this has been seen by Beale, Kölliker, Krause, and others. Arndt considers these intermuscular fibres as sensory nerves.

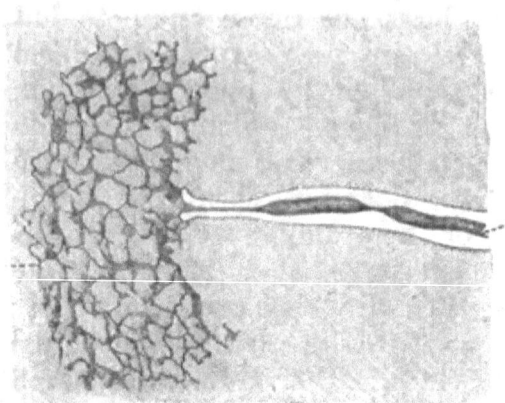

Fig. 80.—One of the Reticulated Endplates of the previous figure, more highly magnified.

a, The medullated nerve fibre ; *b*, the reticulated endplate. (Golgi.)

161. **Tendons** are supplied with special nerve endings, studied by Sachs, Rollett, Gempt, Rauber, and particularly Golgi, whose work on this subject is very minute. These terminations are especially numerous near the muscular insertion. They

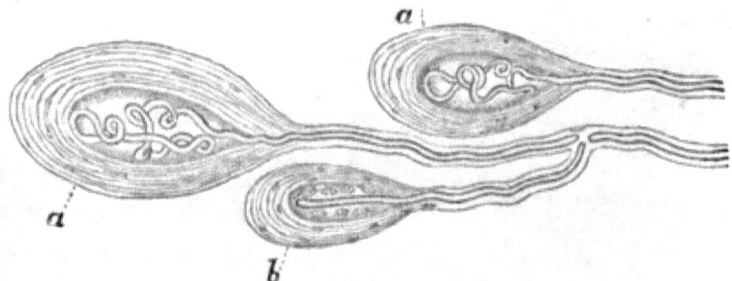

Fig. 81.—Termination of Medullated Nerve-fibres in Tendon.

a, End-bulbs with convoluted medullated nerve-fibre; *b*, end-bulb similar to a Herbst's corpuscle. (Golgi.)

are of the following kinds :—(*a*) A medullated nerve fibre branches repeatedly, and the axis cylinder breaks up into a small plate composed of a network of fine primitive nerve-fibrils (Fig. 79). (*b*) This network is occasionally embedded in a granular-looking material, and thereby a similar organ as the nerve endplate of muscular fibres is produced (Fig. 80). (*c*) A medullated nerve-fibre terminates in an end-bulb (Fig. 81), similar to those of the conjunctiva, or of a Herbst's corpuscle.

CHAPTER XVI.

THE SPINAL CORD.

162. **The spinal cord is** enveloped in three distinct membranes. The outermost one is the dura mater. This is composed of more or less distinct lamellæ of fibrous connective tissue with the flattened connective tissue cells and networks of elastic fibres. The outer and inner surface of the dura mater is covered with a layer of endothelial plates.

163. Next to the dura mater is the arachnoid membrane. This also consists of bundles of fibrous connective tissue. The outer surface is smooth and covered with an endothelial membrane facing the space existing between it and the inner surface of the dura mater; this space is the sub-dural lymph space. The inner surface of the arachnoidea is a fenestrated membrane of trabeculæ of fibrous connective tissue, covered on its free surface—*i.e.*, the one facing the sub-arachnoidal lymph space—with an endothelium.

164. The innermost membrane is the pia mater. Its matrix is fibrous connective tissue, and it is lined on both surfaces with an endothelial membrane. Between the arachnoid and pia mater extends, from the fenestrated portion of the former, a spongy plexus of trabeculæ of fibrous tissue, the surfaces of the trabeculæ being covered with endothelium. By this

spongy tissue—the *sub-arachnoidal tissue* (Key and Retzius)—the sub-arachnoidal space is subdivided into a labyrinth of lacunæ. On each side of the cord, between the anterior and posterior nerve roots, extends a spongy fibrous tissue, called *ligamentum denticulatum*, between the arachnoidea and pia. By it the sub-arachnoidal space is subdivided into an anterior and posterior division.

165. The sub-dural and sub-arachnoidal spaces do not communicate with one another. (Luschka, Key and Retzius.)

The dura mater, as well as the arachnoidea, sends prolongations on to the nerve roots; and the sub-dural and sub-arachnoidal spaces are continued into lymphatics of the peripheral nerves.

All three membranes contain their own system of blood-vessels and nerve-fibres.

166. The cord itself (Fig. 82) consists of an outer or cortical part composed of medullated nerve-fibres; this is the *white matter*, and an inner core of *grey matter*. On a transverse section through the cord the contrast of colour between the white mantle and the grey core is very conspicuous. The relation between the white and grey matter differs in different parts; it gradually increases in favour of the former as we ascend from the lumbar to the upper cervical portion. The grey matter presents in every transverse section through the cord more or less the shape of a capital H; the projections being the *anterior* and *posterior horns* or cornua of grey matter, and the transverse stroke being the *grey commissure*. In the centre of this grey commissure is a cylindrical canal lined with a layer of columnar epithelial cells; this is the *central canal;* the part of the grey commissure in front of this canal is the *anterior*, the rest the *posterior, grey commissure*. The shape of the whole figure of the grey matter

differs in the different regions, and this difference is chiefly brought about by the length and thickness of the grey commissure. In a section through the cervical region the grey commissure is thick and short; in the dorsal region it becomes thinner and

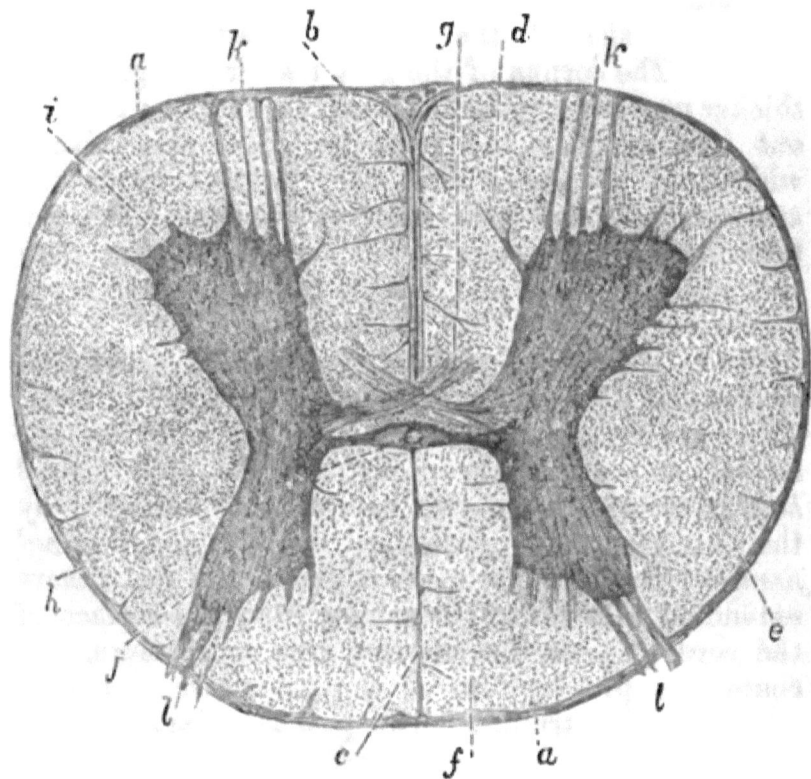

Fig. 82.—Transverse Section through the Spinal Cord of Calf.

a, Pia mater; *b*, prolongation of pia mater into the anterior longitudinal fissure; *c*, posterior longitudinal fissure; *d*, anterior column of white matter; *e*, lateral column of same; *f*, posterior column of same; *g*, anterior white commissure; *h*, central canal; *i*, anterior horn of grey matter; *j*, posterior horn of grey matter; *k*, anterior nerve roots; *l*, posterior nerve roots.

longer; and in the lumbar region it is comparatively very thin and long. Besides this, of course the relative proportions of grey and white matter, as mentioned before, indicate the region from which the particular part of the cord has been obtained. In the lower cervical and lumbar region where the nerves of the brachial and sacral plexus respectively join the cord, this latter possesses a swelling, and the grey matter is there increased in amount, the swelling being in fact due to an accumulation of grey matter, in which an additional number of nerve-fibres originates; but the general shape of the grey matter is there retained.

167. The cornua of the grey matter are generally thicker near the grey commissure; they become thinned out into anterior and posterior edges respectively, which are so placed that they point towards the antero-lateral and postero-lateral fissures. The anterior horns are generally thicker and shorter than the posterior ones, and, therefore, the latter reach nearer to the surface than the former.

168. The white matter is composed chiefly of medullated nerve-fibres running a longitudinal course. They are arranged into columns, *one anterior, one lateral,* and *one posterior column for each lateral half of the cord;* the two halves being indicated by the *anterior* and *posterior median longitudinal fissure.* The anterior median fissure is a real fissure extending in a vertical direction from the surface of the cord to *near* the anterior grey commissure. It contains a prolongation of the pia mater and in it large vascular trunks. The posterior fissure is not in reality a space, but is filled up by neuroglia. It extends as a continuous mass of neuroglia in a vertical direction from the posterior surface of the cord to the posterior grey commissure. The exit of the anterior or motor nerve roots, and the entrance of the posterior or sensitive nerve roots are indicated by the *anterior lateral* and *posterior lateral* fissures respectively. These are not real fissures in the same sense as the anterior median fissure, but correspond more to the posterior median fissure, being in reality filled up with neuroglia tissue, into which extends a continuation

from the pia mater with large vascular trunks. The white matter between the anterior median and anterior lateral fissure is the anterior column, that between the anterior lateral and posterior lateral fissure is the lateral column, and that between the posterior lateral and posterior median fissure is the posterior column.

169. Besides the septa situated in the two lateral

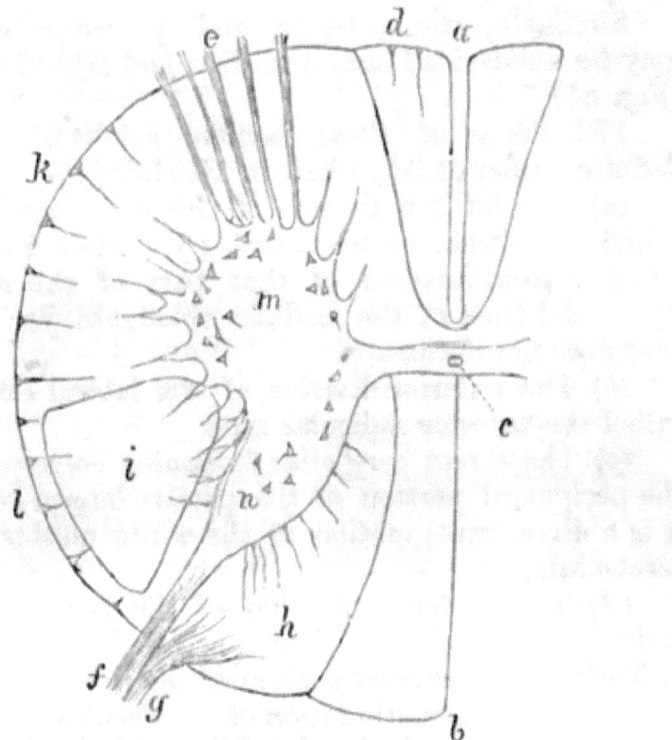

Fig. 83.—Diagram of a Transverse Section through the Cord in the region of the Cervical Swelling.

a, Anterior longitudinal fissure; *b*, posterior longitudinal fissure: the part of the white matter next to it is the fasciculus of Goll; *c*, central canal; *d*, direct pyramidal fasciculus; *e*, roots of anterior or motor nerves; *f*, lateral bundle of posterior nerve roots; *g*, median bundle of same; *h*, the cuneate fasciculus; *i*, the crossed pyramidal fasciculus or the fasciculus of Türk; *k*, the anterior radicular zone of the lateral column of white matter; *l*, the direct cerebellar fasciculus; *m*, the anterior horn of grey matter; *n*, the posterior horn of same.

fissures respectively, there are other smaller septa, neuroglia tissue and prolongations of the pia mater, which pass in a vertical and radiating direction into the white matter of the columns, and they are thus subdivided into a number of smaller portions; one such big septum is sometimes found corresponding to the middle of the circumference of one half of the cord. This is the median lateral fissure, and the lateral column is subdivided by it into an anterior and posterior division.

Similarly, the anterior and posterior columns may be subdivided into a median and lateral division (Fig. 83).

170. Some of these various subdivisions bear definite names (Türk, Charcot, Flechsig) :—

(*a*) The median division of the anterior column is called the direct or uncrossed pyramidal fasciculus, being a continuation of that part of the anterior pyramidal tract of the medulla oblongata (see below) that does not decussate.

(*b*) The anterior division of the lateral column is called the anterior radicular zone.

(*c*) The direct cerebellar fasciculus corresponds to the peripheral portion of the postero-lateral column; it is a direct continuation of the white matter of the cerebellum.

(*d*) The posterior division of the lateral column inside the cerebellar fasciculus is called the fasciculus of Türk, or the crossed portion of the pyramidal fasciculus, it being a continuation of the decussated part of the anterior pyramidal tract of the medulla oblongata.

(*e*) The lateral division of the posterior column, with the exception of a small peripheral zone, is the cuneiform or cuneate fasciculus, or posterior radicular zone.

(*f*) The median division of the posterior column is called the fasciculus of Goll.

This part is connected directly with the median bundle of the posterior nerve roots. (*See* below.)

These various divisions can be traced from the medullata oblongata into the cervical, and a greater or smaller portion of the dorsal part of the cord, but farther down most of them are lost as separate tracts, except the fasciculus of Türk.

171. The **ground substance** (Fig. 84) of both the white and grey matter—*i.e.*, the substance in which nerve-fibres, nerve cells, and blood-vessels are embedded—is a peculiar kind of connective tissue, which is called by Virchow *neuroglia* and by Kölliker supporting tissue. It consists of three different kinds of elements: (*a*) a homogeneous transparent semi-fluid *matrix*, which in hardened sections appears more or less granular; (*b*) a network of very delicate fibrils—*neuroglia fibrils*—which are similar in some respects, but not quite identical with elastic fibres.

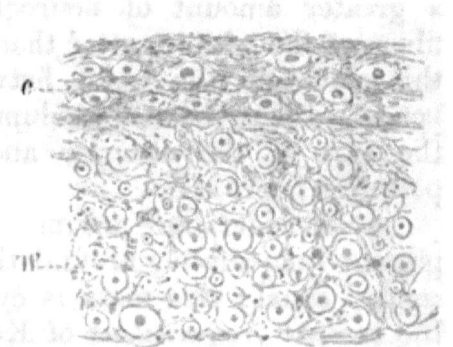

Fig. 84.—From a Transverse Section through a most Peripheral Part of the White Matter of the Cord.

c A special peripheral condensation of neuroglia; *w*, white matter with the medullated nerve-fibres shown in cross section, and neuroglia between them. (Atlas.)

In the columns of the white matter the fibrils extend pre-eminently in a longitudinal direction, in the grey matter they extend uniformly in all directions, and in the septa between the columns they extend chiefly in the direction vertical to the long axis of the cord.

(*c*) Small branched nucleated cells intimately woven into the network of neuroglia fibrils. These cells are the *neuroglia cells*. The greater the amount of neuroglia in a particular part of the white or grey matter the more numerous are these three elements.

172. In both the white and grey matter the

neuroglia has a very unequal distribution; but there are certain definite places in which there is always a considerable amount—a condensation, as it were, of neuroglia tissue. These places are: underneath the pia mater—*i.e.*, on the outer surface of the white matter—there most of the neuroglia fibrils have a horizontal direction; near the grey matter there is a greater amount of neuroglia between the nerve-fibres of the white matter than in the middle parts of this latter; in the septa between the columns and between the divisions of columns of white matter; at the exit of the anterior and the entrance of the posterior nerve roots.

A considerable accumulation of neuroglia is present immediately around the epithelium lining the central canal; this mass is cylindrical, and is called the *central grey nucleus* of Kölliker. The epithelial cells lining the central canal are conical, their bases facing the canal, their pointed extremity being drawn out into a fine filament intimately interwoven with the network of neuroglia fibrils. In the embryo and young state, the free base of the epithelial cells has a bundle of cilia, but in the adult they are lost.

Another considerable accumulation of neuroglia exists in the posterior portion of the posterior grey horns, as the *substantia gelatinosa* of Rolando.

173. **The white matter** (Fig. 85) is composed, besides neuroglia, of medullated nerve-fibres varying very much in diameter, and forming the essential and chief part of it. They possess an axis cylinder and a thick medullary sheath more or less laminated, but are devoid of a neurilemma and its nerve corpuscles. Of course, no nodes of Ranvier are observable. In specimens of white matter of the posterior columns, where the nerve-fibres have been isolated by teasing or otherwise, many fine medullated fibres are met with which show the varicose appearance mentioned

in a former chapter. The medullated nerve-fibres, or rather the matrix of their medullary sheath, contains neurokeratin. The nerve-fibres of the white matter run chiefly in a longitudinal direction, and they are separated from one another by the neuroglia. Here and there in the columns of white matter are seen connective tissue septa with vessels, by which the nerve-fibres are grouped more or less distinctly in divisions.

Fig. 85.—From a Transverse Section through the White Matter of the Cord. Showing the transversely-cut medullated nerve-fibres, the neuroglia between them with two branched neuroglia cells. (Atlas.)

174. Although most of the nerve-fibres constituting the columns of white matter are of a longitudinal direction—*i.e.*, passing upwards or downwards between the grey matter of the cord on the one hand, and the brain and medulla oblongata on the other—there are nevertheless a good many nerve-fibres and groups of nerve-fibres which have an oblique or even horizontal course.

(1) The anterior median fissure does not reach the anterior grey commissure, for between its bottom and the latter there is *the white commissure*. This consists of bundles of medullated nerve-fibres passing in a horizontal or slightly oblique manner between the grey matter of the anterior horn of one side, and the anterior white column of the opposite side (Fig. 82,*g*).

(2) Numerous medullated fibres are derived from the grey matter, and they pass in a horizontal or oblique direction into the white matter, especially in considerable numbers into that of the lateral columns.

Having entered the white matter, they take a longitudinal direction. Most of these fibres enter the white matter in the septa and septula, by which the nerve-fibres of the white matter of the columns are subdivided, and having passed in a horizontal direction in the septa and septula, some for a shorter, others for a longer distance, they enter the columns and pursue a longitudinal course.

175. (3) The medullated nerve-fibres which leave the cord by the *anterior nerve roots* are comparatively thick fibres, which pass out of the anterior portion of the grey matter of the anterior horns in bundles; they pass through the white matter in an oblique direction by septa, and emerge in the anterior lateral fissure above mentioned.

(4) The medullated nerve-fibres entering the cord by the *posterior nerve roots* are thinner than those of the anterior nerve roots; they pass into the cord by the posterior lateral fissure. Having entered, they divide into two bundles, one *median* and another *lateral*. The fibres of the former pass in an oblique direction into the white matter of the posterior columns—the cuneiform fasciculus (*see* above);—and, having run in these in a longitudinal direction, again leave them, sooner or later, and enter, in a horizontal or slightly oblique direction, the grey matter of the posterior horns. The fibres of the lateral bundle, on the other hand, pass directly from the posterior nerve root into the hindmost portion of the grey matter of the posterior horn. The nerve-fibres of the posterior roots entering the grey matter divide repeatedly, and show very markedly the varicose appearance.

176. The **grey matter** consists, besides the uniform network of neuroglia fibres and neuroglia cells, of *nerve-fibres* and of nerve-cells, or *ganglion cells*.

The nerve-fibres are of three kinds—medullated

fibres, simple axis cylinders of various sizes, and primitive nerve-fibrillæ.

The *medullated nerve-fibres* run a more or less horizontal course, and they belong to different sources:—

(1) Medullated nerve-fibres *connected directly*—*i.e.*, by the axis cylinder process (*see* below)—*with ganglion cells of the anterior horns;* they leave the anterior horns by the septa and septula, and they form the anterior nerve roots.

(2) Medullated nerve-fibres which form *the anterior white commissure;* they are traceable from the anterior column of one side into the grey matter of the anterior horns of the opposite side, as has been mentioned above; some of these, at any rate, are distinctly and directly traceable to ganglion cells.

(3) Medullated nerve-fibres **derived** *indirectly from the median bundle of the posterior* **nerve root**—*i.e.*, coming out of the cuneiform fasciculus of the posterior column and medullated nerve-fibres *derived directly from the lateral bundle of the posterior nerve-root.* Both these kinds of nerve-fibres can be traced for longer or shorter distances in the grey matter of the posterior horns; on their way they undergo numerous divisions into very fine medullated fibres.

(4) Medullated nerve-fibres *passing from the grey matter into the white matter of the lateral column.* Some of these are nerve-fibres that pass simply through the grey matter of the anterior horns from an anterior nerve root; others are derived directly from ganglion cells forming the columns of Clarke in the dorsal region (*see* below). But the majority are derived from that part of the grey matter intermediate between the anterior and posterior horn.

177. The *simple axis cylinders* are found very numerously in the grey matter of all parts; they are of many various sizes, and run in all directions, many

of them, especially the larger ones, are only the first part of the medullated nerve-fibres, being the axis cylinder process of a ganglion cell, which process, after a shorter or longer course in the grey matter, becomes ensheathed in a medullary sheath, and represents one of the above medullated fibres. But there are also numerous fine axis cylinders, which are the last outrunners of the nerve-fibres entering the grey matter by the posterior roots. They are seen everywhere, *isolated* and *running in smaller or larger bundles*.

178. The *primitive nerve fibrillæ* form the greater part of the grey matter, in fact the matrix of the grey matter of all parts being composed, besides the network of neuroglia fibrils, of an exceedingly fine and *dense network of primitive fibrillæ* (Gerlach). These are the nervous groundwork into which pass, and from which originate, nerve-fibres. The nerve-fibres which are derived from the posterior roots having entered the grey matter of the posterior horn undergo repeated divisions, and ultimately become connected with this network of primitive fibrillæ. Numerous nerve-fibres take their origin in the network of primitive fibrillæ, and leave the grey matter as medullated nerve-fibres, which pursue a longitudinal course in the anterior and especially in the lateral column of the white matter.

179. The **ganglion cells** (Fig. 86) of the grey matter are of various sizes and shapes, the branched, or stellate, or multipolar shape being predominant; some have a more or less spindle-shaped or bipolar body, but each extremity may be richly branched. Each has a relatively large nucleus bordered by a membrane, and in it is a reticulum with one or two nucleoli. The largest ganglion cells occur in the anterior horns, likewise in the Clarke's column of the dorsal region; the smallest in the posterior horns. The ganglion cells

are much more numerous in the anterior horn than in the posterior, where they are relatively scarce.

In the former they are all stellate or multipolar, and form definite groups: (*a*) an *anterior group*,

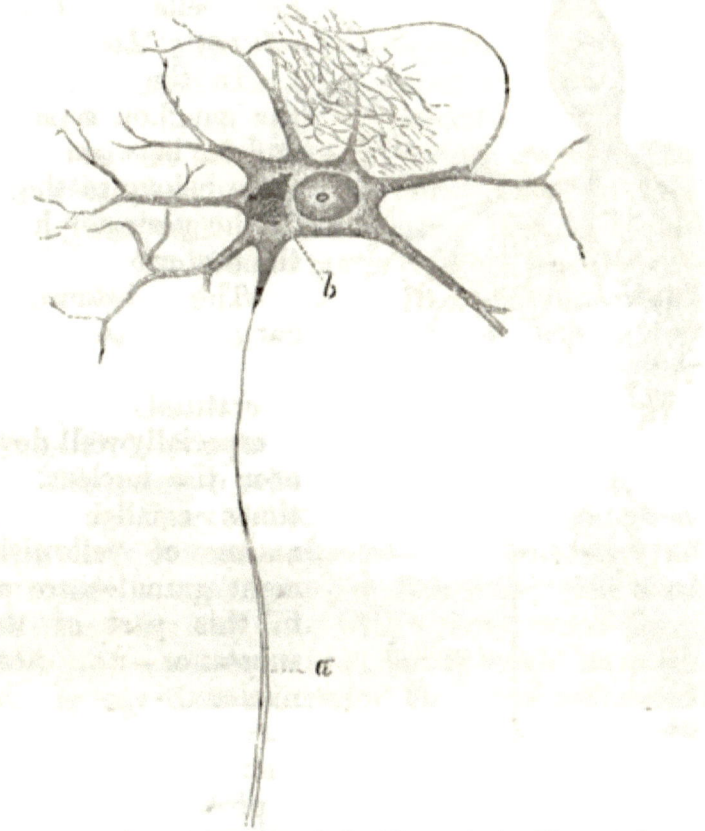

Fig. 86.—An Isolated Ganglion Cell of the Anterior Horn of the Human Cord.

a, Axis cylinder process; *b*, pigment. The branched processes of the ganglion cell break up into the fine nerve network shown in the upper part of the figure. (Gerlach, in Stricker's "Manual of Histology.")

(*b*) a *median* or *inner group*, and (*c*) a *lateral group*. The cells of the lateral group are the largest, those of the inner or median group are the smallest of the three. The lateral group of ganglion cells extends in the cervical region for a longer or shorter distance into the white matter of the lateral column.

180. In the dorsal region of the cord there exists near the grey commissure a special cylindrical group of large multipolar ganglion cells, which form the *column of Lockhart Clarke*.

In the posterior horns the ganglion cells are few and far between. Most of them belong to the portion of the posterior horn near the posterior commissure.

The substance of the ganglion cells is fibrillated, but there exists a granular interstitial material, which is especially well developed near the nucleus. Sometimes smaller or larger masses of yellowish pigment granules are present in this part of the cell substance—*i.e.*, near the nucleus.

181. The fibrillated substance of the ganglion cells is prolonged on to the processes. There are always one or two that are thicker than the others. At a longer or shorter distance from the cell the *processes branch dendritically* into a large number of fibres, which eventually break up into the fine network of primitive fibrillæ, forming the nervous groundwork of

Fig. 87.—An Isolated Multipolar Ganglion Cell of the Grey Matter of the Cord.

The dendritically-branched processes break up into the fine nerve network into which is seen to pass a fine nerve-fibre derived from a posterior nerve root. (Gerlach, in Stricker's Manual.)

the grey matter (Fig. 87). The ganglion cells of the anterior horn and the cells of Clarke's column have, in addition to these branched processes, generally one unbranched pale process (occasionally, but rarely, this is double), which takes its origin in the cell substance with a thin neck. This is the *axis cylinder process of Deiters*; it becomes invested sooner or later in a medullary sheath, and then represents a medullated nerve-fibre, as mentioned on a former page. The ganglion cells of the posterior horns have no axis cylinder process, all processes being branched and connected with the ground nerve network in the same way as the branched processes of the ganglion cells of the anterior horns.

Anastomoses between the processes of the ganglion cells of the anterior horns have been observed in a few instances (Carrière).

182. The ganglion cells of the anterior horns and those forming Clarke's column—*i.e.*, the ganglion cells with axis cylinder process—are considered as *motor*, the others as *sensory ganglion cells*; that is to say, the former are connected with a motor nerve-fibre, the latter with a sensory fibre; but it would be quite incorrect to say that all motor fibres are connected with the former, all sensory fibres with the latter.

183. The white and grey matter is supplied with a large number of blood-vessels, the capillaries being more abundant and forming a more uniform network in the grey than in the white matter; in the latter, most of them have a course parallel with the long axis. The blood-vessels are ensheathed in lymph spaces (*perivascular spaces* of His), and the ganglion cells are each surrounded by a lymph space (*pericellular space*).

CHAPTER XVII.

THE MEDULLA OBLONGATA.

184. As the cervical portion of the cord passes into the medulla oblongata, its parts alter position, arrangement, and name in the following manner :—

(*a*) The anterior median fissure is continued as far as the medulla extends. The posterior fissure of the cord is also continued on the medulla, but in the upper portion is lost, owing to the fact that the central canal, which in the cord is situated in about the middle, shifts in the medulla towards the posterior surface, and soon altogether opens into the fourth ventricle.

185. (*b*) The tracts of white matter bordering the anterior median fissure of the medulla, and separated from the other tracts by a distinct fissure on the surface, are the *pyramidal tracts*. As was mentioned on a former page, the median portion of the anterior columns of white matter of the cord—*i.e.*, the uncrossed or direct anterior fasciculus—is a direct prolongation of the pyramidal tract, and can be followed in this upwards into the pyramids—*i.e.*, the oblong prominences in the upper part of the medulla next to the anterior median fissure—and from there to the pons Varolii and farther into the crus cerebri. A major portion of the pyramidal tract crosses in the lower portion of the medulla, in the anterior median fissure—this forms the *pyramidal decussation* (Fig. 88). These crossed bundles enter the posterolateral column of the cord, that part of it which has been mentioned above as the fasciculus of Türk. The crossed portion of the pyramidal tract passes into

the pyramids and farther on into the pons Varolii and crus cerebri.

186. The greater portion of the anterior column

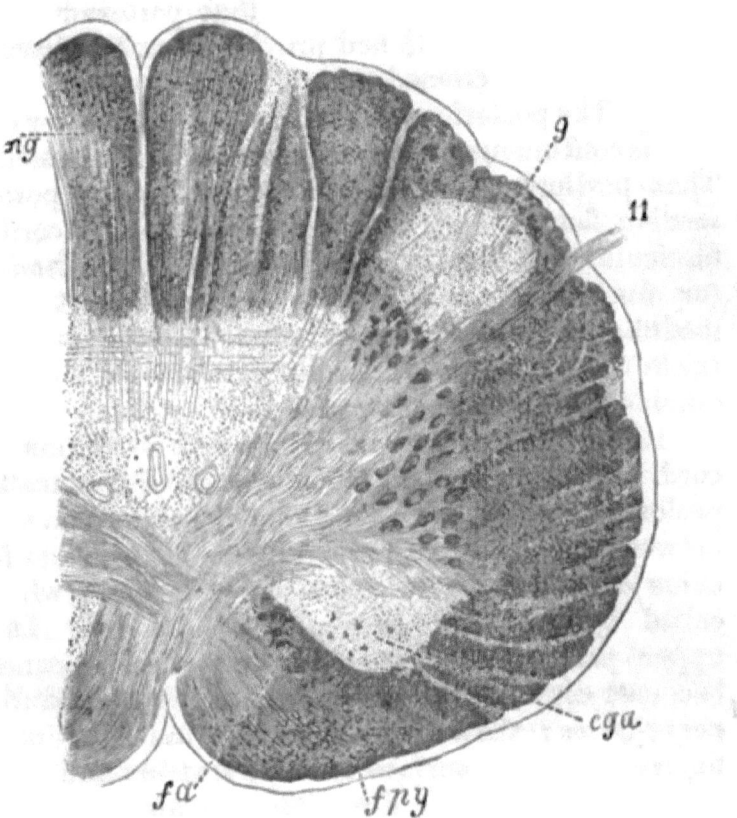

Fig. 88.—Transverse Section through the **Medulla Oblongata** in the Region of the Pyramidal Decussation.

fpy, Anterior pyramidal tract; *cga*, lateral nucleus of grey matter; *fa*, part of anterior column not decussating; *ng*, nucleus gracilis; *g*, gelatinous nucleus of posterior horn; 11, spinal accessory nerve. (Henle.)

of white matter of the cord is situated deeper in the medulla than the pyramidal tracts.

(c) The lateral column of white matter of the cord can be traced into the medulla as the *lateral tract*. In the upper part of the medulla it becomes hidden from view by the olivary bodies and the transversely arranged white tracts. The lateral tract of the medulla comprises all parts of the lateral column of the cord including that portion which was mentioned as the anterior radicular zone and the direct cerebellar fasciculus, but not that posterior division of it which was mentioned previously as the fasciculus of Türk, or the crossed pyramidal fasciculus.

(*d*) The posterior column of the white matter of the cord is continuous with the same column of the medulla. That portion of it which lies next to the posterior median fissure, and which is called in the cord the fasciculus of Goll, is in the medulla called the *fasciculus* (or funiculus) *gracilis*. In the upper part of the medulla, as the central canal opens into the fourth ventricle, the fasciculus gracilis turns in an oblique manner outside, and forms the lateral boundary of the ventricle.

187. (*e*) The lateral part of the posterior column of the cord, which was mentioned as the fasciculus cuneatus, is prolonged into the medulla under the same name. But between the two—*i.e.*, the fasciculus gracilis and fasciculus cuneatus—there exists another tract, which is called by Schwalbe *funiculus of Rolando*. In the upper part of the medulla the fasciculus cuneatus becomes covered by transverse bundles of medullated nerve-fibres; these pass from the anterior median fissure across the surface of the pyramids and olivary body in a transverse direction towards the posterior fissure, but before reaching this take an upward direction. These bundles are the *external arcuate fibres*. In the upper part of the medulla the external arcuate fibres, part of the funiculus cuneatus and funiculus of Rolando, as well as the direct cerebellar fasciculus of the lateral column, all join to form a prominent tract of white matter—*the corpus restiforme*—which enters the white matter of the cerebellar hemisphere on the same side; this is the *pedunculus cerebelli ad medullam oblongatam*, or the lower cerebellar peduncle.

188. (*f*) In the region of the pyramidal decussation—*i.e.*, the lower part of the medulla immediately following the cervical portion of the cord—the grey matter of the cord is changed in its disposition by the fasciculus of Türk or the crossed pyramidal fasciculus passing *en masse* from the lateral column of white matter *through* the anterior horns of the grey matter. Hereby the anterior portion of the grey matter of the cord is shut off from the rest of the grey matter, and is found lying near the surface of the lateral column of the lower portion of the medulla as the *lateral nucleus of grey matter* (Fig. 88). The main part of the anterior horn, however, is represented by the *reticular formation* of grey matter. This contains in its lateral portion, at any rate, the same large multipolar motor ganglion cells with axis cylinder processes of Deiters and nervous ground-network as the anterior horn of the cord; but in addition there are the numerous bundles of medullated nerve-fibres passing through it in transverse, oblique, and longitudinal directions. Some of these fibres belong to the continuation of the anterior columns of white matter of the cord, others join the fasciculus gracilis and cuneatus, and a third kind pass out from the middle line of the medulla.

189. (*g*) The grey matter of the posterior horns of the cord undergoes a change of disposition when passing into the medulla. Its hindmost portion is gradually shifted outwards by the development of the reticular formation of grey matter, and in about the middle of the medulla it is found lying near the surface of the lateral column as the *tubercle of Rolando*. The rest of the posterior horn remains at first collected around the central canal; but as this gradually approaches the posterior fissure, in order to open as the fourth ventricle above, the grey matter gradually expands laterally into the funiculus gracilis and cuneatus of

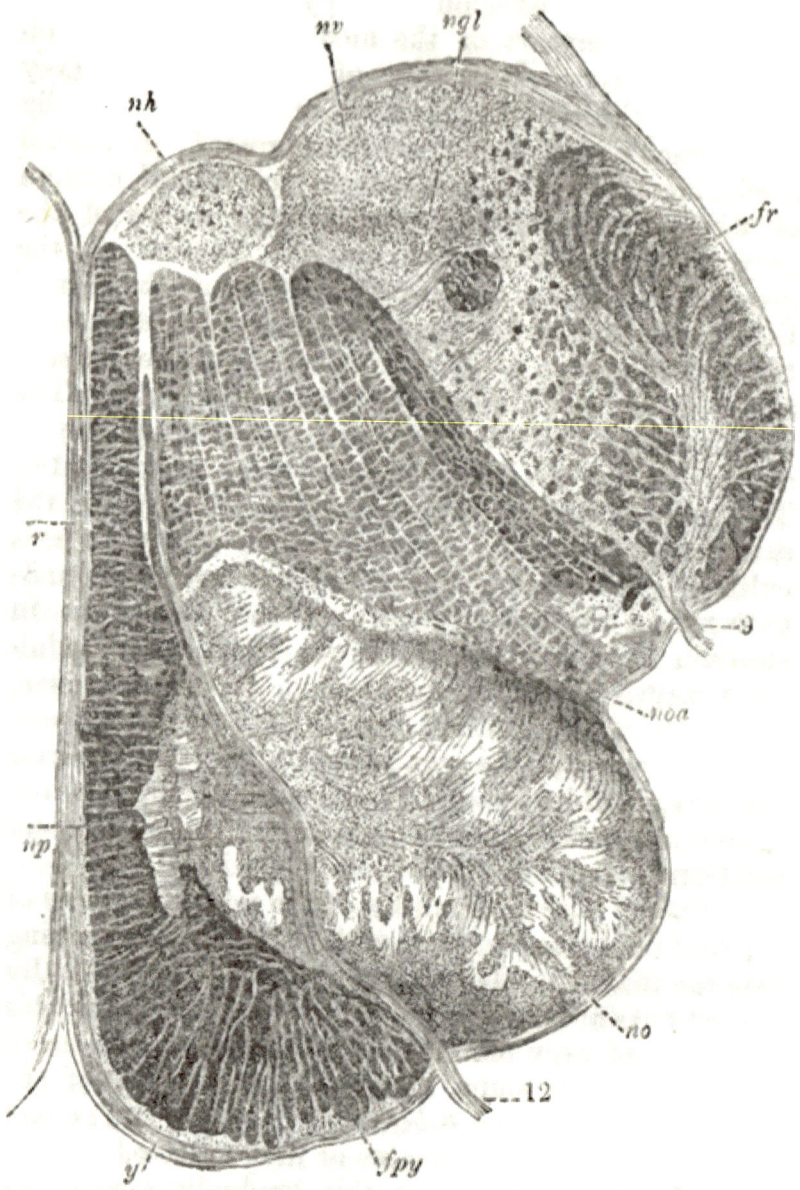

Fig. 89.—Transverse Section through the Medulla Oblongata in the Region of the Fourth Ventricle.

fpy, Anterior pyramidal tract; *fr*, restiform bodies; *np*, nucleus of pyramids; *no*, olivary nucleus; *noa*, accessory olivary nucleus; *nh*, nucleus of hypoglossal nerve; *nv*, nucleus of vagus; *ngl*, nucleus of glosso-pharyngeus; *r*, raphe; 9, glossopharyngeal nerve; 12, hypoglossal nerve; *y*, horizontal fibres. (Henle.)

the white matter, and forms one distinct accumulation of grey matter in each of these funiculi; they are respectively the *nucleus gracilis* and the *nucleus cuneatus*. The former is the grey matter, in the axis cylinder process of whose ganglion cells the nerve-fibres of the funiculus gracilis originate; but in the latter only a portion of the nerve-fibres of the funiculus cuneatus take their origin; since another part of it joins the restiform body, and with this passes into the cerebellum.

190. In the upper part of the medulla—*i.e.*, in the region of the fourth ventricle—the grey matter forms a continuous mass, the *floor of the fourth ventricle* (Fig. 89). In this region there is a distinct median septum, by which the medulla is divided into two halves; this is the *raphe*. It represents a thin membrane of nerve substance extending from the anterior longitudinal fissure to near the middle line of the floor of the fourth ventricle. This membrane consists of white matter in the shape of bundles of medullated nerve-fibres passing longitudinally, transversely, and obliquely; and of small masses of grey matter interspersed between the nerve-bundles, and especially at the side of the raphe, where nerve-fibre bundles pass out of it. The grey matter contains multipolar ganglion cells.

191. In a transverse section through the upper part of the medulla, we find at the side of the pyramid, and a little behind it, but covered on the outer surface by white matter—*i.e.*, the bundles of nerve-fibres constituting the fibræ arcuatæ externæ— a plicated lamina of grey matter which constitutes the *olivary nucleus*, or nucleus dentatus of the olivary body. It extends with its posterior portion into the reticular formation. Continuous with the olivary nucleus, but situated nearer the raphe, is a small lamina of similar grey matter; this is the accessory

olivary nucleus. In both nuclei are found numerous multipolar ganglion cells, each with an axis cylinder process.

192. The **grey matter at the floor of the fourth ventricle** is that from which the nerve roots of the cerebral nerves (facial auditory, glossopharyngeal, pneumogastric, accessory, and hypoglossal) originate. The ganglion cells in it are of various sizes, and are aggregated into groups which represent the "*nuclei*"—*i.e.*, the origin of the above nerves. The thin layer of grey matter forming the floor of the ventricle, in the strict sense, is neuroglia only, a continuation of the central grey nucleus of the cord.

The nerve cells in the hypoglossal nucleus are the largest; they are as large as the large cells of the anterior horns of the cord. The cells of the glossopharyngeal nerves are considerably smaller. The motor nerve-fibres (*e.g.*, those of the hypoglossal and pneumogastric) originate, as the axis cylinder process of the multipolar ganglion cells, in exactly the same manner as was mentioned in the cord, but the sensory nerve-fibres of these nerves originate from the nerve ground network, into which the processes of the ganglion cells of these nuclei break up.

193. In the lower part of the medulla, as long as there is still a closed central canal, we find next to this the last outrunners of the groups of ganglion cells representing the nucleus of the spinal accessory and hypoglossal.

As we pass upwards, and as the central canal opens as the fourth ventricle, the groups of ganglion cells below the floor of the ventricle are so arranged that we find near the median line the group representing the hypoglossal nucleus; then, farther outwards, several groups representing several sub-divisions of the pneumogastric nucleus; still farther upwards, but more in the anterior part of the medulla, the nucleus of the glossopharyngeal nerve; and, lastly, but more outwards and upwards, several divisions of the nucleus of the auditory nerve. The nerve-fibres, originating in these nuclei, pass in bundles through the substance of the medulla oblongata, so as to appear on the antero-lateral surface. Of course these nerves, the nuclei of which are situated nearer to the middle line— *e.g.*, the hypoglossal and spinal accessory—have to pass through the reticular formation, whereas those whose nuclei are situated more laterally pass only through the lateral part of the medulla.

CHAPTER XVIII.

THE CEREBRUM AND CEREBELLUM.

194. THE structure of the dura mater, arachnoidea, and pia mater of the brain is similar to that of the same membranes of the cord.

As has been shown by Boehm, Key and Retzius, and others, the deeper part of the dura contains peculiar ampullated dilatations connected with the capillary blood-vessels, and representing in fact the roots of the veins.

The *glandulæ Pacchioni*, or arachnoidal villi of Luschka, are composed of a spongy connective tissue, prolonged from the sub-arachnoidal tissue and covered with the arachnoidal membrane. These prolongations are pear-shaped or spindle-shaped, with a thin stalk. They are pushed through holes of the inner part of the dura mater into the venous sinuses of this latter, but are covered with endothelium. Injection matter passes from the sub-arachnoidal spaces through these stalks into the villi. The spaces of their spongy substance become thereby filled and enlarged, and finally the injection matter enters the venous sinus itself. The pia cerebralis is very rich in blood-vessels, like that of the cord, which pass to and from the brain substance. The capillaries of the pia mater possess an outer endothelial sheath. The plexus choroideus is covered with a layer of polyhedral epithelial cells, which are ciliated in the embryo and young.

195. As was mentioned of the cord, so also in the **brain the** sub-dural lymph space does not communicate with the sub-arachnoidal spaces or with the **ventricles** (Luschka, Key and Retzius). Nor is there a communication between the sub-arachnoidal space and the epicerebral space—*i.e.*, a space described by His to exist between pia mater and brain surface, but doubted by others. The relations between the cerebral nerves and the membranes of the brain, and the lymph-spaces of both, **are the** same as those described of the cord and the spinal nerves on a previous page.

196. **The pia mater passes** with the larger blood-vessels into **the brain substance** by the sulci of the cerebrum and cerebellum.

In the white and grey matter of the brain we find the same kind of supporting tissue that we described in the cord as neuroglia. It is also in the brain composed of a homogeneous *matrix*, of a network of *neuroglia fibrils*, and of branched, flattened neuroglia cells, called *Deiters' cells*.

In the white matter of the brain the neuroglia contains between the bundles of the **nerve-fibres** rows of small nucleated cells; these form special accumulations in the bulbi olfactorii, and in the cerebellum. Lymph corpuscles may be met with in the neuroglia, especially around the blood-vessels and ganglion cells.

All the ventricles, including the aqueductus Sylvii, are lined with a layer of neuroglia, being a direct continuation of the one lining the fourth ventricle, and this again being a direct continuation of the central grey nucleus of the cord. Like the central canal of the cord, also, the ventricles are lined with a layer of ciliated columnar, or short columnar epithelial cells.

197. The blood-vessels form a denser capillary network in the grey than in the white matter; in the latter the network is pre-eminently of a longitudinal arrangement, *i.e.*, parallel to the long axis of the bundles of the nerve-fibres. In the grey cortex of the hemispheres of the cerebrum and cerebellum, many of the capillary blood-vessels have an arrangement vertical to the surface, but are connected with one another by numerous transverse branches.

The blood-vessels of the brain are situated in spaces, *perivascular lymph-spaces*, traversed by fibres passing between the adventitia of the vessels, and the neuroglia forming the boundary of the space. There are no real lymphatic vessels in the grey or white substance.

198. The **white matter** consists of medullated nerve-fibres, which like those of the cord possess no neurilemma or nuclei of nerve corpuscles, and no constrictions of Ranvier. The nerve-fibres are of very various sizes, according to the locality. Divisions occur very often. When isolated they show the varicosities mentioned in the cord.

The **grey matter** consists, like that of the cord and medulla, besides the neuroglia, of a very fine network of elementary nerve-fibrils (Rindfleisch, Gerlach), into which pass, on the one hand, nerve-fibres, and, on the other, the branched processes of ganglion cells.

With regard to the structure of the ganglion cells of the brain and medulla, what has been mentioned of the ganglion cells of the cord holds good as to them. Like the former, those of the medulla and brain are situated in lymph spaces or the pericellular spaces (Obersteiner).

199. We now follow the above description of the structure of the medulla with that of the cerebellum and pons Varolii.

I. The **cerebellum** is composed of laminæ, folds, or convolutions, composed of secondary folds, each of which consists of a central tract of white matter covered with grey matter. The tracts of white matter of neighbouring convolutions of one lobe or division join, and thus form the principal tracts of white matter.

The white matter of the cerebellar hemisphere is connected (a) with the medulla oblongata by the corpus restiforme, this forming the inferior peduncle of the cerebellum; (b) with the cerebrum by the pedunculus cerebelli ad cerebrum, this forming the superior pedunculus; and (c) with the other cerebellar hemisphere by the commissure passing through the pons Varolii; this is the pedunculus cerebelli ad pontem, or the middle pedunculus.

200. On a vertical section, through a lamina of the cerebellum (Fig. 90), the following layers are seen: (a) the pia mater covering the general surface, and penetrating with the larger blood-vessels into the peripheral substance of the lamina; (b) a thick layer of cortical grey matter; (c) the layer of Purkinje's ganglion cells; (d) the nuclear layer, and (e) the central white matter.

201. The layer of ganglion cells of Purkinje is the most interesting layer; it consists of a single row of large multipolar ganglion cells, each with a large vesicular nucleus. Each possesses also a thin axis cylinder process, directed towards the depth, the cell sending out in the opposite direction—*i.e.*, towards the surface—a thick process which soon branches like the antlers of a deer, the processes being all very

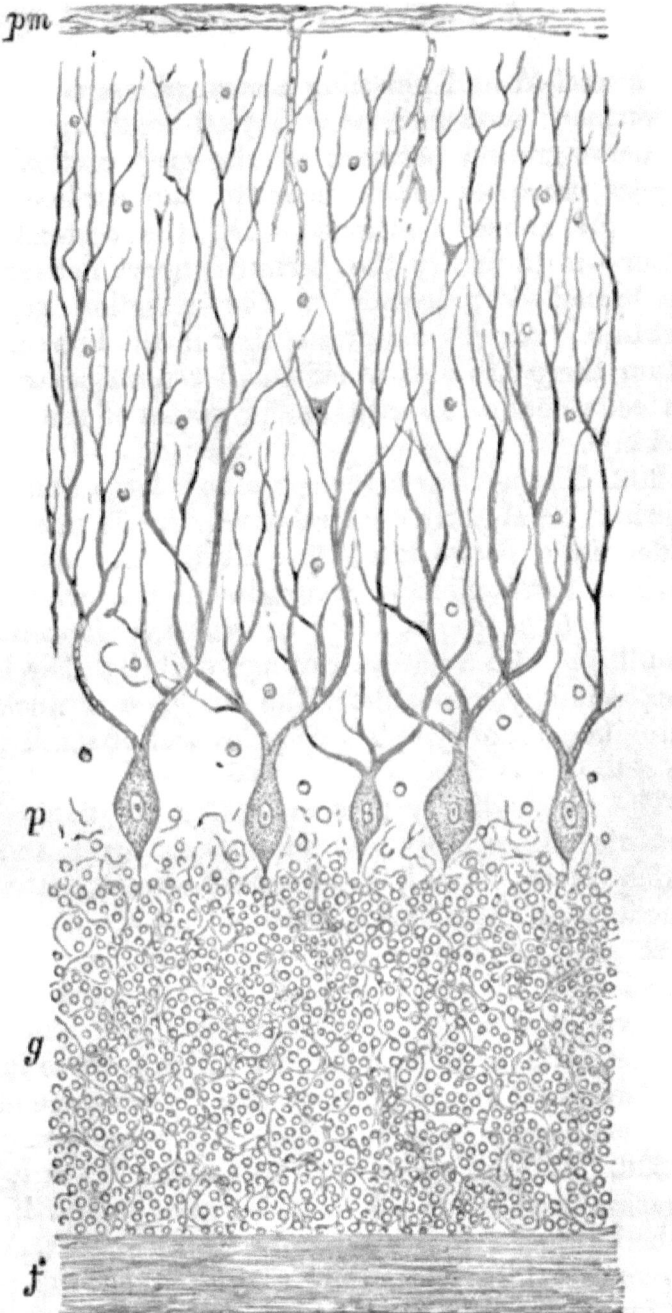

Fig. 90.—From a Vertical Section through the Grey Matter of the Cerebellum of the Dog.

pm, Pia mater; *p*, the ganglion cells of Purkinje; *g*, the nuclear layer; *f*, the layer of nerve-fibres (white matter). (Atlas.)

long-branched and pursuing a vertical course towards the surface; sooner or later they all break up into the fine nerve-ground network of the grey cortex. The longest processes reach near to the surface. The layer (*b*) above mentioned—*i.e.*, the cortical grey matter—is in reality the terminal nerve network for the branched processes of the ganglion cells of Purkinje. Sankey maintains that in the human cerebellum there are also other smaller multipolar ganglion cells connected with the processes of the cells of Purkinje.

202. The nuclear layer contains a large number of spherical or slightly oval relatively small nuclei embedded in a network of fine fibrils, the nature of which is not definitely ascertained—*i.e.*, whether it consists of neuroglia only, or whether it contains, in addition, also a network of nerve fibrils. The latter is exceedingly probable. The nuclei are nuclei of neuroglia cells, of lymph corpuscles and of small ganglion cells.

The axis cylinder process of the ganglion cell of Purkinje passes through the nuclear layer, and becoming invested with a medullary sheath, enters as a medullated nerve-fibre the central white matter. There are, however, medullated nerve-fibres of the central white matter, which are not connected with an axis cylinder process of a Purkinje's cell, but enter the nuclear layer and probably terminate there in the nerve network, or pass through it and terminate in the nerve network of the grey matter of the cortex.

203. II. The **pons Varolii** (Fig. 91) is a prolongation partly of the medulla and partly of the cerebellum. Of the latter only white matter passes transversely into the anterior portion of the pons, and forms there the *transverse bundles of nerve-fibres*, which give to the pons the horizontal striation.

As we pass upwards—*i.e.*, farther away from the

medulla, this part of the pons—*i.e.*, that composed of horizontal fibres—increases in thickness.

204. Of the medulla there is a greater portion

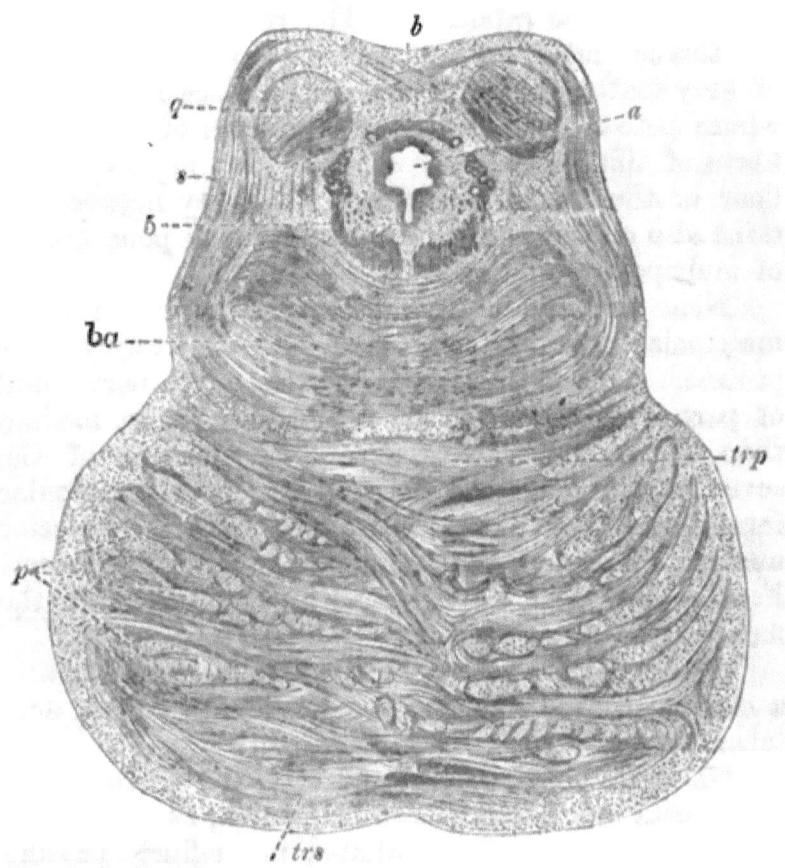

Fig. 91.—Transverse Section through the Lower Corpus Quadrigeminum and the Pons Varolii.

a, Aqueductus sylvii; *b*, crossing of the brachia of the lower corp. quadrig.; *q*, ganglion of the lower corp. quadrig.; *s*, pedunculus of the lower corp. quadrig.; *ba*, tegmentum; *5*, the descending root of the fifth; *p*, bundles of the anterior pyramidal tracts in cross section; *trp*, deep transverse bundles of the pons; *trs*, superficial transverse bundles of the pons. (Meynert, in Stricker's Manual.)

continued into the pons than of the cerebellum. (1) There are the pyramidal tracts; they do not lie on the surface as in the medulla, but are hidden by some

—the most anterior bundles—of the transverse fibres. The bundles of the pyramidal tract pass as longitudinal fibres merely through the anterior half of the pons, and enter the crura cerebri where they form the crusta. (2) The raphe. (3) The reticular formation; but this is limited to the posterior part. Small masses of grey matter and ganglion cells are scattered everywhere between the transverse bundles of the nerve-fibres of this formation. (4) The grey matter at the floor of the fourth ventricle. This grey matter contains also on the posterior surface of the pons groups of multipolar ganglion cells.

Near the middle line there is a group of large multipolar ganglion cells, each with an axis cylinder process. This is the nucleus for the sixth nerve, and of part of the seventh, the former lying more median than the latter. There is another nucleus of the seventh situated more deeply—*i.e.*, in the reticular formation. More outwards we meet with the superior nucleus of one of the roots of the auditory nerve. Farther upwards we meet with the nucleus of the motor roots of the fifth.

(5) In the lower part of the pons there exists also a continuation of the grey matter of the corpus dentatum of the olivary body.

205. The pons is connected with the cerebrum by the crusta of the crus cerebri, which, as mentioned above, are bundles of medullated nerve-fibres passing merely through the pons but being continuations of the anterior pyramidal tracts of the medulla.

206. III. The **hemispheres of the cerebrum.**—On a vertical section each convolution shows a *white centre* and a *grey cortex*. The former is composed of medullated nerve-fibres. The white matter of the convolutions of the cerebral hemispheres is arranged as (*a*) the *centrum ovale*—*i.e.*, the central mass of white matter from which the lamina of white matter

for each convolution branches off, and, (*b*), the commissure of white matter between the two hemispheres, *i.e.*, the *corpus callosum* and *anterior commissure*. The centrum ovale again consists of tracts of medullated nerve-fibre, which connect (*a*) the convolutions of the same hemisphere with one another, and (*b*) such as pass between the convolutions on the one hand, and the thalamus opticus, the pons, and medulla on the other. These tracts pass by the internal capsule (see below) to the thalamus opticus, and to the crus cerebri.

The *grey cortex* consists, according to Meynert, of the following layers (Fig. 92):—(1) a superficial layer of grey

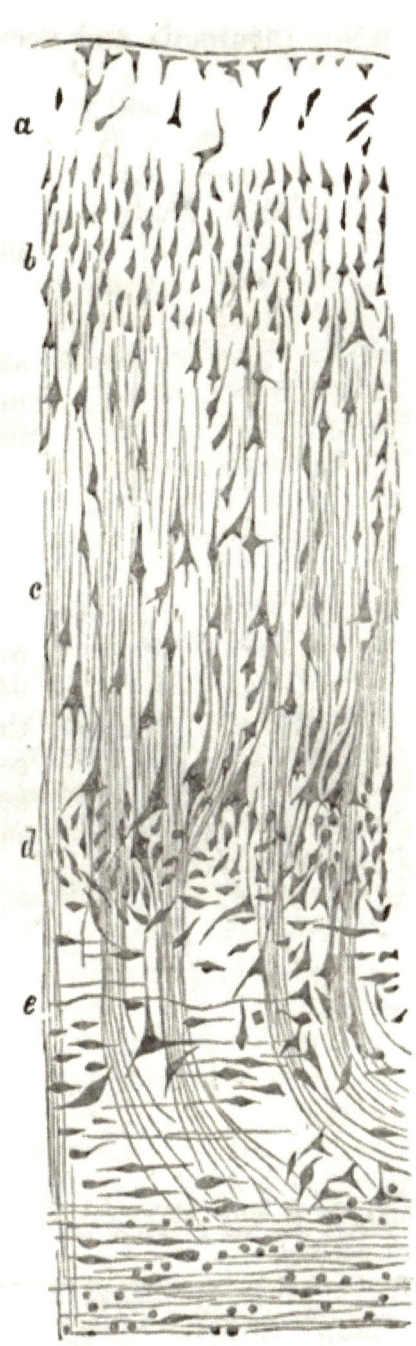

Fig. 92.— Vertical Section through the Grey Matter of a Cerebral Convolution.

a, Superficial layer; *b*, closely-packed small ganglion cells; *c*, the layer of the cornu Ammonis, this being the principal layer; *d*, the "granular formation," small multipolar ganglion cells; *e*, the layer of spindle-shaped ganglion cells. (Meynert, in Stricker's Manual.)

matter (neuroglia and nerve ground network), with few and small ganglion cells. (2) A layer of small more or less pyramidal ganglion cells densely aggregated. (3) The formation of the cornu Ammonis. This is the principal or broadest stratum of the cortex; it is composed of several layers of large, pyramidal ganglion cells, increasing in size as a deeper layer is reached.

The pyramidal cells of this third and of the previous second stratum consist of a pyramidal body including an oval vesicular nucleus (Fig. 93). From the body pass out the following principal processes:—(*a*) the process of the apex, directed towards the surface of the convolution; it can be traced for a longer or shorter distance. (*b*) The lateral basilar processes, and finally the median basilar process. This latter is fine, remains unbranched, and is an axis cylinder process, *i.e.*, becoming invested with a medullary sheath is a nervefibre of the central

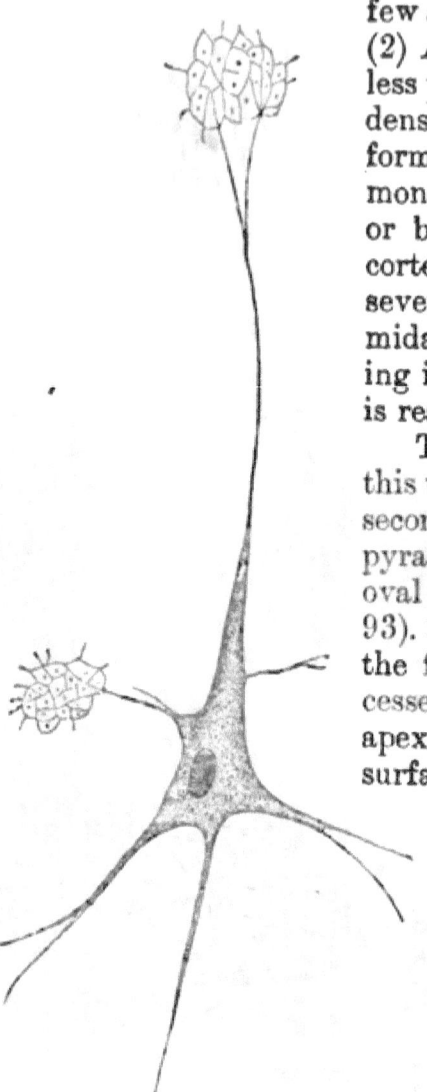

Fig. 93.—A large Pyramidal Ganglion Cell of the Grey Cortex of the Human Cerebrum.

The process of the apex and the other processes branch and break up into the fine nerve network. The median process of the base of the pyramid remains unbranched, and becomes an axis cylinder of a nerve-fibre. (Atlas.)

white matter. The other processes, sooner or later in their course, branch and break up finally into the nerve ground network of the grey matter. (4) A thin stratum of small irregular branched ganglion cells, the *granular formation* of Meynert. (5) A last stratum of spindle-shaped and branched ganglion cells, extending parallel to the surface.

207. According to Meynert, the grey cortex of the posterior portion of the occipital lobe about the sulcus hippocampi consists of eight layers, the granular formation being the principal one. In the grey cortex of the cornu Ammonis, on the other hand, the third layer is the principal layer, the fourth being wanting. In the claustrum (part of the wall of the fossa Sylvii), the spindle-shaped cells of the fourth layer form the principal stratum.

208. The **bulbus olfactorius** contains in most mammals, but not in man, a small central cavity lined with columnar ciliated epithelial cells. The substance of the bulb around this cavity consists of an upper part, which is white matter, and is a continuation of the tractus olfactorius. The lower part is grey matter, and contains the following layers, counting from below upwards: (1) a layer of non-medullated nerve-fibres, each with a neurilemma; this layer forms farther on the olfactory nerve going to the olfactory organ; (2) the stratum glomerulosum, composed of a number of glomeruli or convolutions, each of which consists of an olfactory nerve-fibre, and in addition to it numerous small neuroglia cells; (3) stratum gelatinosum of Lockhart Clarke, composed of a fine nerve network, and embedded in it multipolar ganglion cells; (4) a last and thickest layer of nuclei embedded in a network of fibrils, and similar in structure with the "nuclear layer."

209. IV. The **mesencephalon.**—The fourth ventricle above the upper part of the pons Varolii

closes again into a small canal—the *aqueductus Sylvii*—which having passed in front out of the region of the corpora quadrigemina opens out again as the third ventricle. The parts around the aqueductus Sylvii represent the mesencephalon (Fig. 91), developed from the middle brain vesicle in the embryo. They include the wall of the aqueductus Sylvii, the corpora quadrigemina, and the crura cerebri.

The *aqueductus Sylvii* is lined with epithelium and a layer of neuroglia continued from the fourth ventricle. The raphe of the medulla and of the pons are continued into the lower wall of the aqueductus. The lining layer of neuroglia is on its frontal aspect surrounded by a layer of grey substance continued from the grey substance of the floor of the fourth ventricle. It contains in a nerve network numerous multipolar ganglion cells grouped into nerve nuclei, connected with the third, fourth, and part of the fifth pair of nerves. In front of this layer is one of considerable thickness representing the *tegmentum*, which is the dorsal or posterior portion of the crus cerebri.

210. The **corpora quadrigemina.**—Each of the two inferior prominences consists of a superficial layer of white matter, and a deep grey portion, containing multipolar ganglion cells of various sizes embedded in a fine nerve network. Between this and the grey substance of the wall of the aqueductus Sylvii are tracts of white matter, forming part of the *fillet*. In each of the two superior prominences there is also a superficial layer of white matter, beneath which is a layer of grey matter (stratum cinereum); underneath this is the main portion—the *stratum opticum*—consisting of longitudinal tracts of nerve-fibres, between which are small masses of grey substance. Between this stratum opticum and the grey matter forming the wall of the aqueductus Sylvii is a layer of white matter, part of the *fillet*.

211. The **crus cerebri** of each side consists of an anterior, middle, and posterior portion. The anterior or ventral portion is the crusta, or pes; the posterior or dorsal portion is the tegmentum. Between the two is the substantia nigra. The **crusta** is composed of longitudinal tracts of medullated nerve-fibres passing from the margin of the pons Varolii to the internal capsule of the thalamencephalon, and farther into the white matter of the hemisphere.

212. The **tegmentum** has been mentioned above as being situated in front of the grey matter forming the anterior wall of the aqueductus Sylvii. The tegmentum is a prolongation of the reticular formation of the pons Varolii and medulla (see above), *i.e.*, small masses of grey substance separated by tracts of nerve-fibres, most of which run in a longitudinal or transverse direction. The longitudinal bundles include a continuation of the white matter of the cerebellum, mentioned in a previous page as the superior peduncle of the cerebellum, or the pedunculus cerebelli ad cerebrum. These undergo total decussation in the upper part of the mesencephalon, and ultimately enter the thalamus opticus.

213. The **substantia nigra** is grey matter situated between the two above; it has received its name from the numerous dark pigment granules lodged in the substance of its ganglion cells. These are small and multipolar.

214. V. The **thalamencephalon** and **corpus striatum**.—The former comprises the parts of the brain situated round the third ventricle, the most important being the thalamus opticus, the pineal gland, the corpora albicantia, the infundibulum and tuber cinereum, and the hypophysis cerebri. The *corpus striatum* is the ganglion of the cerebral hemisphere, with which it originates from the same part—*i.e.*, the frontal part of the first cerebral vesicle of the embryo.

L

215. The **thalamus opticus** consists of a superficial layer of white, and a centre of grey matter. In this numerous multipolar ganglion cells are noticed. The white matter in the outer portion is very considerable, and of great importance from its connections. From it radiate tracts of medullated nerve-fibres, which join the tracts of the internal capsule on their way to and from the different parts of the cerebral hemisphere.

The superior pedunculus cerebelli, after its decussation with that of the opposite side, passes into the white matter of the thalamus. The tractus opticus is connected with the outer white matter of the posterior portion of the thalamus—*i.e.*, the *pulvinar*.

216. The **corpus striatum**, as stated before, is considered as the ganglion of the cerebral hemisphere. It consists of the *nucleus caudatus* and the *nucleus lenticularis*. The former projects into the lateral ventricle, the latter is the outer portion of the corpus striatum. The nucleus lenticularis is separated from the nucleus caudatus and from the anterior portion of the thalamus opticus by tracts of medullated nerve-fibres, known as the *internal capsule*. On the outer surface of the nucleus lenticularis is a thin lamina of white matter which is the *external capsule*. This is separated from the white matter of the cerebral convolutions at this part—*i.e.*, the *island of Reil*—by a thin lamina of grey matter, called the *claustrum*. The nucleus caudatus and lenticularis consist of grey matter with larger and smaller groups of multipolar ganglion cells, permeated by tracts of medullated nerve-fibres, which originate in the grey matter. These tracts of white matter pass transversely and obliquely into the internal capsule, and are to be traced on the one hand to the white matter of the convolutions of the cerebral hemisphere,

which is, however, doubted by many observers, and on the other into the crusta of the crus cerebri.

217. The **internal capsule** is one of the most important masses of white matter; it contains the tracts of medullated nerve-fibres which pass between the white matter of the cerebral hemisphere and the crus cerebri—*i.e.*, the corona radiata of Reil; further, it contains tracts of medullated nerve-fibres passing between the thalamus opticus and the white matter of the cerebral hemispheres; and, finally, it contains tracts of nerve-fibres passing between the corpus striatum and the crus cerebri.

218. The **pineal gland,** or conarium, and the anterior lobe of the hypophisis cerebri, are epithelial in structure and origin, and will be described in a future chapter. The pineal gland contains a large amount of calcareous matter—brainsand.

The *corpora albicantia* are masses of white matter — *i.e.*, medullated nerve-fibres; each corpus albicans includes a centre of grey substance.

The *infundibulum* and *tuber cinereum* at the floor of the third ventricle are composed of grey matter; the latter extends between the corpora albicantia to the optic commissure, while the former is connected with the posterior or minor lobe of the hypophisis.

CHAPTER XIX.

THE CEREBRO-SPINAL GANGLIA.

219. THE ganglia connected with the posterior roots of the spinal nerves, and with some of the roots of the cerebral nerves—*e.g.*, Gasserian, otic, geniculate, ciliary, Meckel's ganglion, the ganglia of

the branches of the acoustic nerve, the submaxillary ganglion, &c.—possess a capsule of fibrous connective tissue continuous with the epineurium of the afferent and efferent nerve trunks. The interior of the ganglion is subdivided into smaller or larger divisions, containing nerve bundles with their perineurium, or larger and smaller groups of ganglion cells. In the spinal ganglia these latter are generally disposed about the cortical part, whereas the centre of the ganglia is chiefly occupied by bundles of nerve-fibres.

220. The ganglion cells differ very greatly in size, some being as big, and bigger, than a large multipolar ganglion cell of the anterior horn of the cord, others much smaller. Each cell has a large oval nucleus, including a network, with one or two large nucleoli. Its substance shows a distinct fibrillation. Each cell of the ganglia in man and mammals is *unipolar* (Fig. 94), flask or pear-shaped, and invested in a hyaline *capsule*, lined with a more or less continuous layer of *nucleated endothelial cell-plates*. The single process of the ganglion cell is finely and longitudinally striated, and is an axis cylinder process. Immediately after leaving the cell body it is much convoluted (Retzius); then it gets invested in a medullary sheath—*i.e.*, it becomes a medullated nerve-

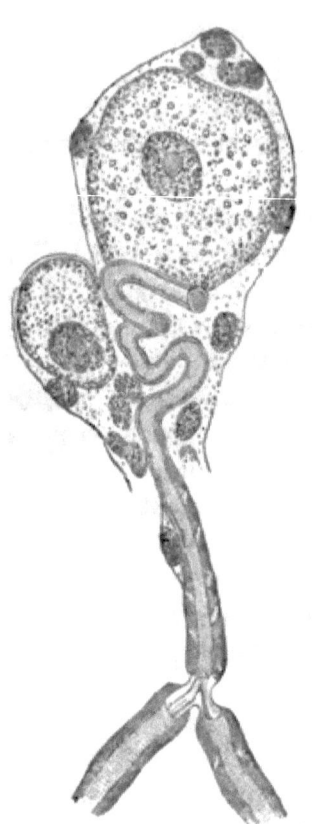

Fig. 94.—A large and small Ganglion Cell of the Ganglion Gasseri of the Rabbit.

The axis cylinder, after leaving the cell, becomes convoluted and transformed into a medullated nerve-fibre, which divides into two medullated fibres. (Key and Retzius.)

fibre. The capsule of the ganglion cell is continued on the axis cylinder process, and, farther on, on the medullated nerve-fibre, as the neurilemma; while the endothelial plates of the capsule pass into the nerve corpuscles lining the neurilemma, their number greatly diminishing (Fig. 95).

221. In the rabbit this medullated nerve-fibre at its first node of Ranvier, which is not at a great distance from the ganglion cell, divides into two medullated nerve-fibres in the shape of a T; one branch is supposed by Ranvier to pass to the cord, the other to the periphery. In man, this T-shaped division has also been observed by Retzius, but it cannot be said with certainty that in rabbit or man every axis cylinder process shows this T-shaped division. Retzius observed this T-shaped division also in the Gasserian, geniculate, and vagus ganglia in man.

The ganglion cells are not unipolar in all cerebral ganglia; in the ciliary and otic ganglia there are a good many ganglion cells which are multipolar.

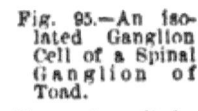

Fig. 95.—An isolated Ganglion Cell of a Spinal Ganglion of Toad.

The axis cylinder process becomes transformed into a medullated nerve-fibre. The capsule of the cell is prolonged as the neurilemma of the nerve-fibre. (Key and Retzius.)

222. Numerous ganglia of microscopic sizes are to be found in the sub-maxillary (salivary) gland; they are of different sizes, and are in reality ganglionic enlargements of larger or smaller nerve-bundles. Each ganglion is invested in connective tissue continuous with the perineurium, and the ganglion cells are unipolar, and of the same nature as those described above, each being possessed of an axis cylinder process, which becomes

soon connected with a nerve-fibre. At the back of the tongue there are similar small microscopic ganglia.

CHAPTER XX.

THE SYMPATHETIC SYSTEM.

223. THE sympathetic nerve-branches are of exactly the same nature in their connective tissue investments (epi- peri- and endo-neurium), and in the arrangement

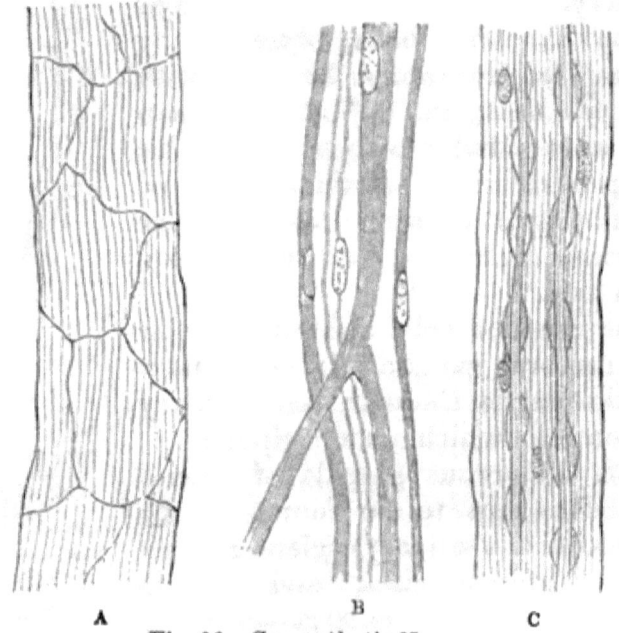

Fig. 96.—Sympathetic Nerves.

A, A small bundle invested in an endothelial sheath perineurium; B, one medullated and three non-medullated nerve-fibres of various sizes; the largest shows division; C, two varicose nerve-fibres. (Atlas.)

of the fibres in bundles (Fig. 96, A), as the cerebro-spinal nerves. Most of the nerve-fibres in the bundles are non-medullated or Remak's fibres (Fig. 96, B), each being an axis cylinder invested in a neurilemma, with oblong nuclei indicative of nerve corpuscles. But there are some medullated nerve-fibres to be met with in

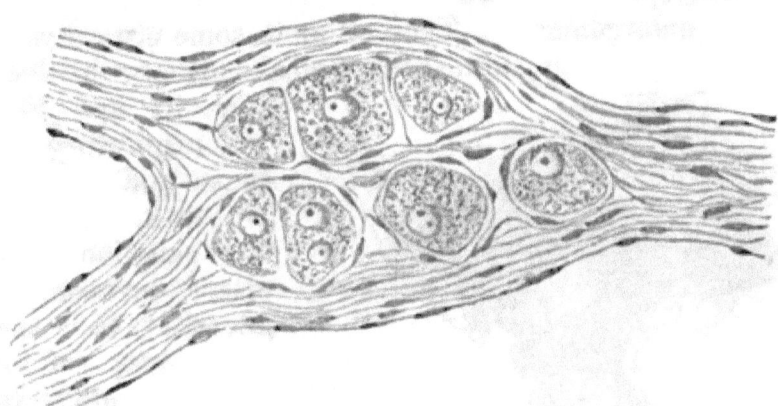

Fig. 97.—A group of Ganglion Cells interposed in a bundle of Sympathetic Nerve-fibres; from the Bladder of a Rabbit. (Handbook)

each bundle, at least, of the larger nerve-trunks. These in some cases show the medullary sheath more or less discontinuous, and with a varicose outline

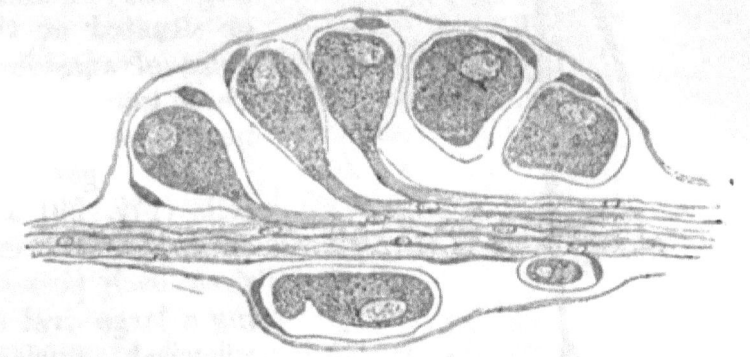

Fig. 98.—A small collection of Ganglion Cells along a small Bundle of Sympathetic Nerve-fibres in the Bladder of the Rabbit.
Each ganglion cell possesses a capsule. The substance of the ganglion cell is prolonged as the axis cylinder of a nerve-fibre. (Atlas.)

(Fig. 96, c), owing to a uniform local accumulation of fluid between it and the axis cylinder. The small or microscopic bundles of nerve-fibres have an endothelial (perineural) sheath. The small and large branches form always *rich plexuses*.

224. In connection with the macroscopic and microscopic sympathetic nerve-branches, are ganglionic enlargements. They occur in some organs very numerously—*e.g.*, alimentary canal, urinary bladder (Fig. 97 and Fig. 98), respiratory organs—and are of all sizes, from a few ganglion cells placed between, or laterally to, the nerve-fibres of a small bundle, to huge oval, spherical, or irregularly-shaped masses of ganglion cells placed in the course of a large nerve-bundle, or situated at the point of anastomosis of two or more nerve-branches.

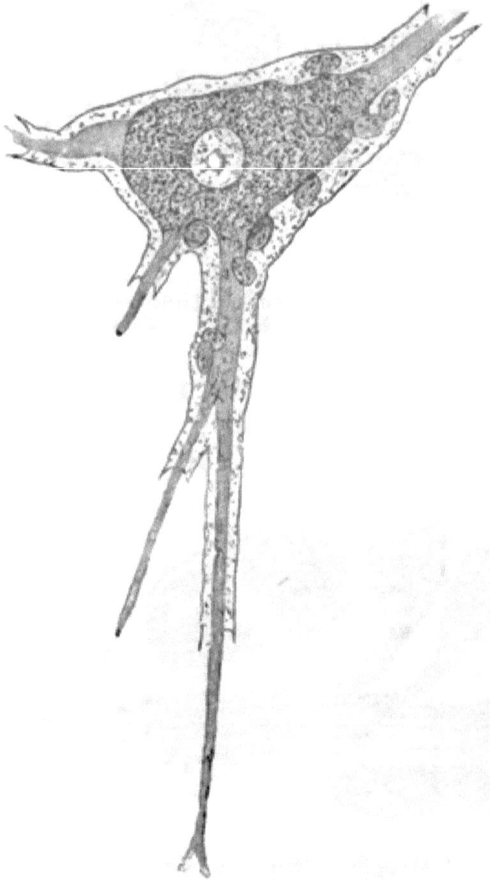

The ganglion cells (Fig. 99) are of very different sizes, each possessing a large oval or spherical nucleus with one or two nucleoli. Their shape is spherical or oval, flask-shaped, club-shaped, or pear-shaped; they possess either one, two, or more processes, being uni-, bi-,

Fig. 99.—Sympathetic Ganglion Cell of Man. The ganglion cell is multipolar; each process receiving a neurilemma from the capsule of the cell becomes a non-medullated nerve-fibre.

or multi-polar. The cell is invested in a capsule lined with nucleated cells, both being continued on the processes as neurilemma and nerve corpuscles respectively.

The processes of the ganglion cells are all axis cylinder processes, and invested in the neurilemma, thus representing non-medullated nerve-fibres. They do not become as a rule medullated.

225. In the frog (Beale, Arnold) and also in some few instances in the mammal, the sympathetic ganglion cell gives off one straight *axis cylinder process*, into which the substance of the ganglion cell is prolonged. This is entwined by a thin *spiral fibre* (Fig. 100), taking its origin by two or more rootlets from the ganglion cell substance, and circling round the (thicker) straight axis cylinder process. A single neurilemma ensheathes them both. Soon the spiral fibre leaves the axis cylinder process, becoming invested with a medullary sheath and its own neurilemma, thus forming a medullated fibre, whereas the straight axis cylinder continues its course as a non-medullated nerve-fibre (Key and Retzius).

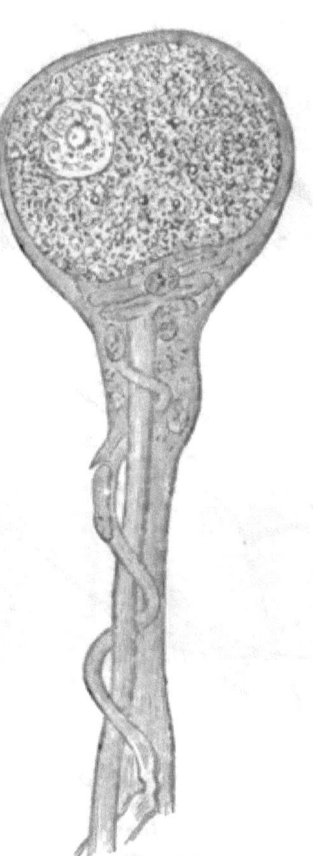

Fig. 100.—A Sympathetic Ganglion Cell of the Frog, showing the straight process and the spiral fibre; the latter becomes a medullated fibre. (Key and Retzius.)

226. The ganglia in connection with the plexuses of nerve branches of the heart, the ganglia in the plexus of non-medullated nerve-fibres existing between the longitudinal and circular coat

of the external muscular coat in the alimentary canal, known as the plexus myentericus of Auerbach, the ganglia in the plexus of nerve branches of the sub-

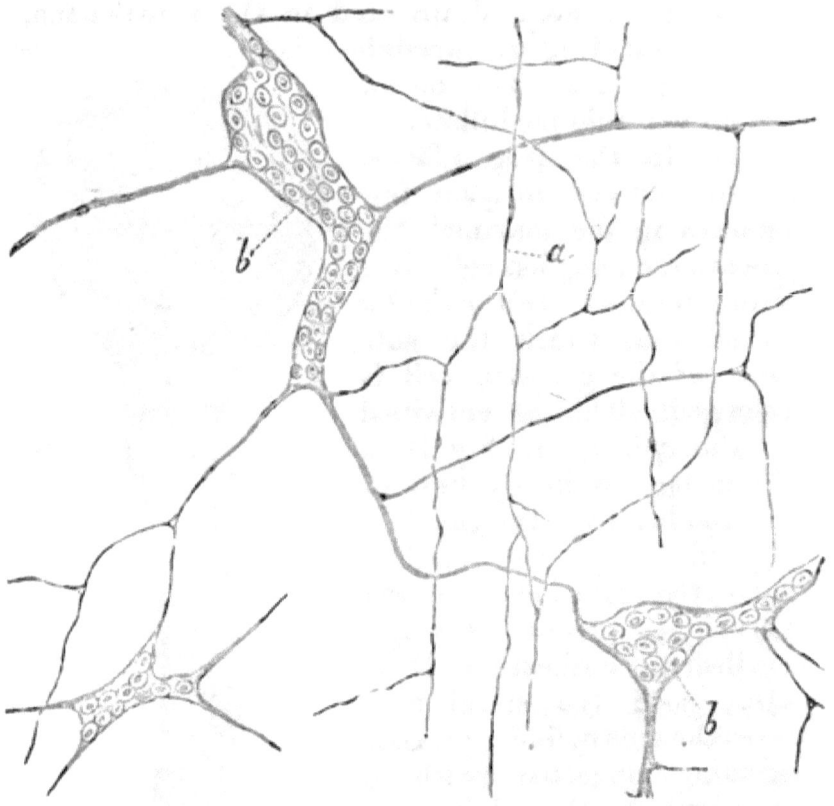

Fig. 101.—Plexus of fine Sympathetic Nerve-fibres, with Ganglionic Enlargements in the Nodal Points. From the Plexus of Meissner in the Submucous Tissue of the Intestine.

a, Fine nerve-fibres; b, groups of ganglion cells interposed between the nerve-fibres.

mucous tissue in the alimentary canal, and known as Meissner's plexus (Fig. 101), the ganglia in the plexuses of nerve branches in the outer wall of the bladder, in the bronchial wall, in the trachea, the ganglia in connection with the nerves supplying the ciliary muscle of the eye, &c., belong to the sympathetic system.

CHAPTER XXI.

TEETH.

227. A HUMAN tooth, adult and milk-tooth, consists (Fig. 102) of (*a*) the *enamel* covering the crown, (*b*) the dentine forming the real matrix of the whole tooth, and surrounding the pulp cavity both of the crown and fangs, and (*c*) the *cement*, or crusta petrosa, or substantia osteoidea. This cement covers the outside of the dentine of the fang or fangs, in the same way as the enamel covers the dentine of the crown. The crusta petrosa is covered on its outside by a dense fibrous tissue acting as a *periosteum* to it, and is fixed by it to the inner surface of the bone forming the wall of the alveolar cavity.

228. The **enamel** (Fig. 103) consists of thin microscopic prismatic elements, the *enamel prisms* placed closely, and extending in

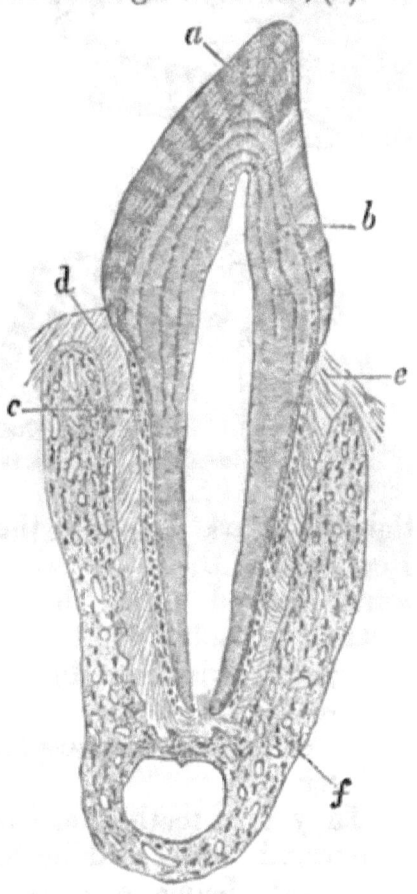

Fig. 102.—Longitudinal Section through the Præmolar Tooth of Cat.

a, Enamel; *b*, dentine; *c*, crusta petrosa; *e*, periosteum; *f*, bone of alveolus. (Waldeyer, in Stricker's Manual.)

a vertical direction from the surface to the dentine. When viewed in transverse section, the enamel prisms appear of an hexagonal outline, and separated by a very fine *interstitial cement substance*. The outline of the enamel prisms is not straight, but wavy, so that the prisms appear varicose. The prisms are aggregated into bundles, which are not quite parallel, but more or less slightly cover one another. On a longitudinal section through a tooth, the appearance of alternate

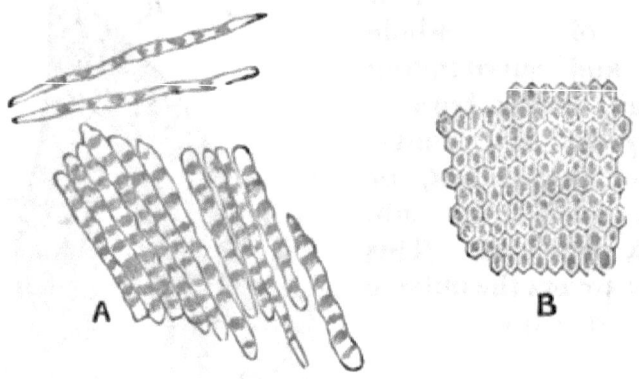

Fig. 103.—Enamel Prisms.
A, In longitudinal view; B, in cross section. (Kölliker.)

light and dark stripes in the enamel is thus produced. Besides this there are seen in the enamel dark horizontal curved lines, the brown parallel stripes of Retzius, probably due to inequalities in the density of the enamel prisms produced by the successive formation of layers of the enamel. The enamel consists chiefly of lime-salts: phosphate, carbonate, and fluorate of calcium.

In young teeth the free surface of the enamel is covered with a delicate cuticle (the cuticle of Nasmyth), being a single layer of non-nucleated scales. In adult teeth this cuticle is wanting, having been rubbed off.

229. The **dentine** is the essential part of the hard

substances of the tooth. It forms a complete investment of the pulp cavity of the crown and fang, being slightly thicker in the former than in the latter region. The dentine is composed of (Fig. 104): (1) a homogeneous *matrix;* this is a reticular tissue of fine fibrils impregnated with lime-salts, and thus resembles the matrix of bone; (2) long fine canals, the *dentinal canals or tubes* passing in a more or less spiral manner, and vertically from the inner to the outer surface of the dentine. These tubes are branched; they open in the pulp cavity with their broadest part, and become finer as they approach the outer surface of the dentine. Each canal is lined with a delicate sheath—the *dentinal sheath.* Inside the tube is a fibre, the *dentinal fibre,* a solid elastic fibre originating with its thickest part at the pulp side of the dentine from cells lining the outer surface of the pulp, and called *odontoblasts.*

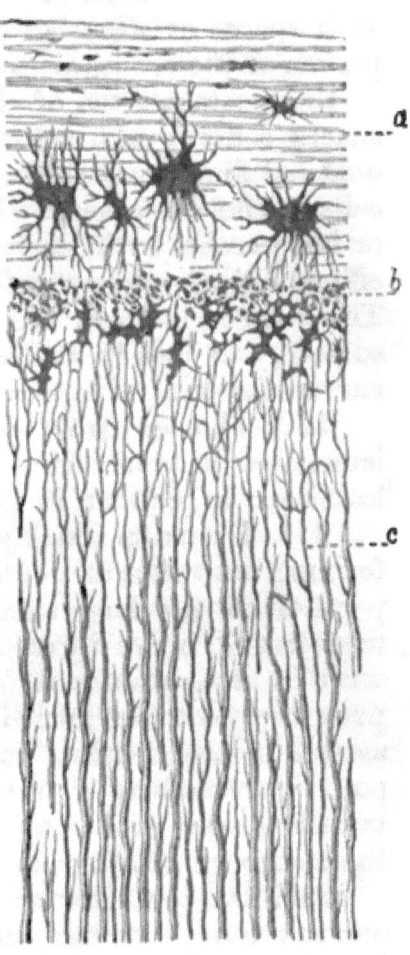

Fig. 104.—From a Section through a Canine Tooth of Man.

a, Crusta petrosa, with large bone corpuscles; *b,* interglobular substance; *c,* dentinal tubules. (Waldeyer, in Stricker's Manual.)

On the outer surface of the dentine, both in the region of the enamel and crusta petrosa, the dentinal tubes pass into a layer of intercommunicating irregular branched spaces, the *interglobular spaces* of Czermak, or the *granular layer* of Purkinje. These communicate with spaces existing between the bundles of enamel prisms of the crown, as well as with the bone laminæ of the crusta petrosa of the fang. The interglobular spaces contain each a branched nucleated cell. The dentinal fibres anastomose with the processes of these cells. The *incremental lines of Salter* are lines more or less parallel to the surface, owing to imperfectly calcified dentine—the *interglobular substance of Czermak*. The *lines of Schreger* are curved lines parallel to the surface, and due to the optical effect of simultaneous curvatures of dentinal fibres.

230. The **cement** is osseous substance, being lamellated bone matrix with bone corpuscles. These latter are larger than in ordinary bone.

231. The **pulp** is richly supplied with blood-vessels, forming networks, and extending chiefly in a direction parallel to the long axis of the tooth. Numerous medullated nerve-fibres forming plexuses are met with in the pulp tissue; on the outer surface of the pulp they become non-medullated fibres, and probably ascend in the dentinal tubes. The matrix of the pulp is formed by a transparent network of richly branched cells, similar to the network of cells forming the matrix of gelatinous connective tissue.

232. On the outer surface of the pulp—*i.e.*, the one in contact with the inner surface of the dentine—is a layer of nucleated cells, which are elongated, more or less columnar. These are the *odontoblasts proper*. Between them are wedged in more or less *spindle-shaped nucleated cells*, the outer or distal process of which passes into a dentinal fibre. The odontoblasts proper are concerned in the production of the dentinal matrix, according to some by a continuous growth of the distal or outer part of the cell and

a petrification of this increment, according to others by a secretion by the cell of the dentinal matrix. Waldeyer, Tomes, and others, consider the odontoblasts proper concerned in the production both of the dentinal matrix and dentinal fibres. The odontoblasts proper and the spindle-shaped cells are continuous with the branched cells of the pulp matrix.

233. **Development of teeth.**—The first rudiment of a tooth in the embryo appears as early as the second month. It is a solid cylindrical prolongation of the stratified epithelium of the surface into the depth of the embryonal mucous membrane. Along the border of the jaws the epithelium appears thickened, and the subjacent mucous membrane forms there a depression—the *primitive dental groove*. Into this groove the solid cylindrical prolongation of the surface epithelium takes place. This prolongation represents the rudiment of the *enamel organ*. While continuing to grow in the depth, it soon broadens at its deepest part, and the surrounding vascular mucous membrane condenses at the bottom of the prolongation as the rudiment of the tooth papilla. While the distal part of the enamel organ continues to grow towards the depth, it gradually embraces the tooth papilla in the shape of a cap—the *enamel cap*. During this time the connection between the surface epithelium and the enamel cap becomes greatly thinned out and pushed to one side, owing to the growth of the enamel cap and papilla taking place chiefly to one side of the original dental groove.

234. The enamel cap (Fig. 105) is composed of three strata—an inner, middle, and outer stratum. The inner stratum is the layer of the *enamel cells;* it is a layer of beautiful columnar epithelial cells; they were originally continuous with the deep layer or the columnar cells of the surface epithelium. The middle stratum is the thickest and is of great transparency,

owing to a transformation of the middle layer of the epithelial cells into a spongy gelatinous tissue, due to accumulation of fluid between the epithelial cells of this layer, and to a reduction of its substance to thin nucleated plates, apparently branched. The

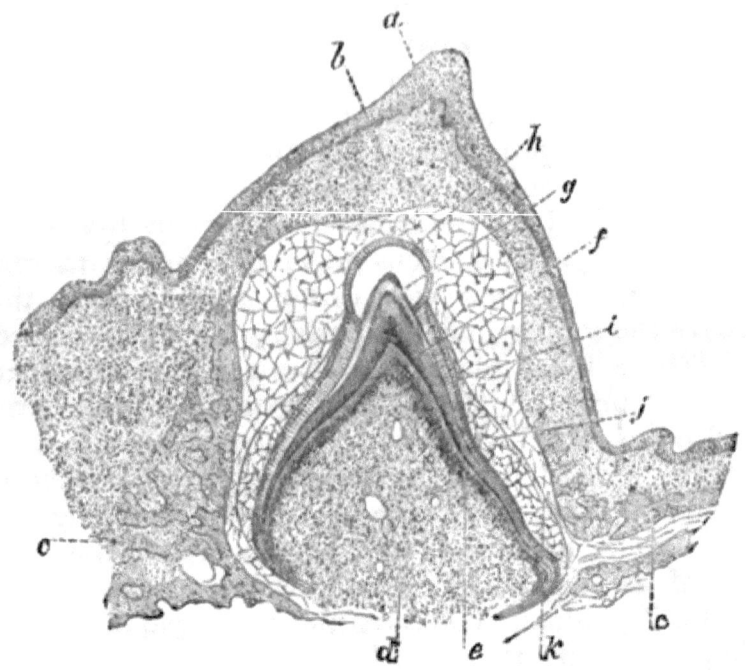

Fig. 105.—From a Section through the Tooth and Lower Jaw of Fœtal Kitten.

a, Epithelium of the free surface of the gum; *b*, the mucous membrane of same; *c*, spongy bone of jaw; *d*, papilla of tooth; *e*, odontoblasts; *f*, dentine; *g*, enamel; *h*, membrane of Nasmyth; *i*, enamel cells; *j*, middle layer of enamel organ; *k*, outer layer of enamel organ.

outer stratum consists of one or more layers of polyhedral cells, continuous with the deep layers of cells of the epithelium of the surface of the gum. Outside the enamel cap is the gelatinous vascular tissue of the mucous membrane of the gum.

235. The fœtal tooth papilla is a vascular embryonal or gelatinous tissue; on its outer surface a con-

densation of its cells is soon noticeable into a more or less continuous stratum of elongated or columnar cells, the odontoblasts.

236. The dentine is formed in connection with the odontoblasts (Fig. 105); on its outside appears the enamel formed by the *enamel cells*, *i.e.*, the inner layer of the enamel organ. The dentine and enamel are deposited gradually, and in layers. At first they are soft tissues, showing a vertical differentiation corresponding to the individual cells of the enamel cells and odontoblasts respectively. Soon lime salts are deposited in it, at first imperfectly, but afterwards a perfect petrifaction takes place. The layer of most recently formed enamel and dentine is more or less distinctly marked off from the more advanced layer, the most recently formed layer of the enamel being situated next to the enamel cells, that of the dentine next to the odontoblasts.

The milk tooth remains buried in the mucous membrane of the gum. When it breaks through, the enamel remains covered with—*i.e.*, carries with it— the inner stratum of the enamel organ only, *i.e.*, the enamel cells (Fig. 105, h); these at the same time as the surface of the enamel increases become much flattened, and, finally losing their nuclei, are converted into a layer of transparent scales, the *membrane* or *cuticle of Nasmyth*.

237. Long before the milk tooth breaks through the gum, there appears a solid cylindrical mass of epithelial cells extending into the depth from the connection between the enamel organ and the epithelium of the surface of the gum mentioned above. This epithelial outgrowth represents the germ for the enamel organ of the permanent tooth; but it remains stationary in its growth till the time arrives for the milk tooth to be supplanted by a permanent tooth. Then that rudiment undergoes exactly the

same changes of growth as the enamel organ of the milk tooth did in the first period of fœtal life. A new tooth is thus formed in the depth of the alveolar cavity of a milk tooth, and the growth of the former in size and towards the surface gradually lifts the latter out of its socket.

CHAPTER XXII.

THE SALIVARY GLANDS.

238. THE salivary glands, according to their structure and secretion, are of the following kinds :—
(1) *True salivary* (Fig. 106), *serous*, or albuminous

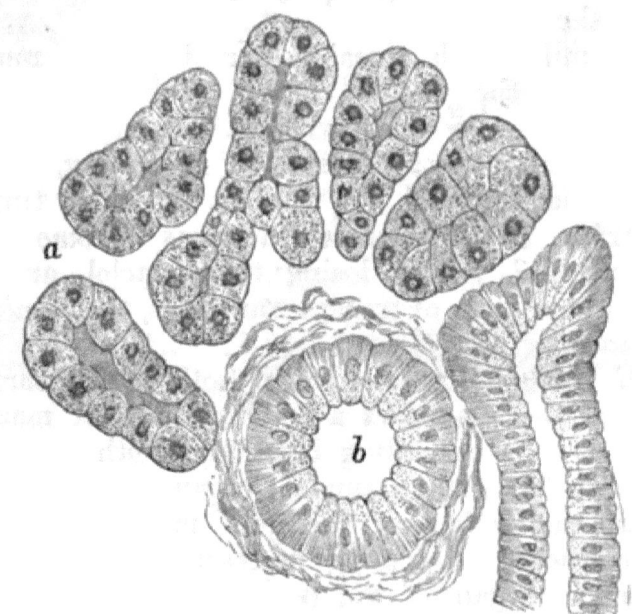

Fig. 106.—From a Section through a Serous or True Salivary Gland; part of the Human Sub-maxillary.

a, The gland alveoli, lined with the albuminous "salivary cells;" b, intralobular duct cut transversely. (Atlas.)

glands, such as the parotid of man and mammals, the sub-maxillary and orbital of rabbit, the sub-maxillary of the guinea-pig. They secrete true, thin, watery saliva.

(2) *Mucous glands*, such as the sub-maxillary and orbital of cat and dog (Fig. 107), the sub-lingual

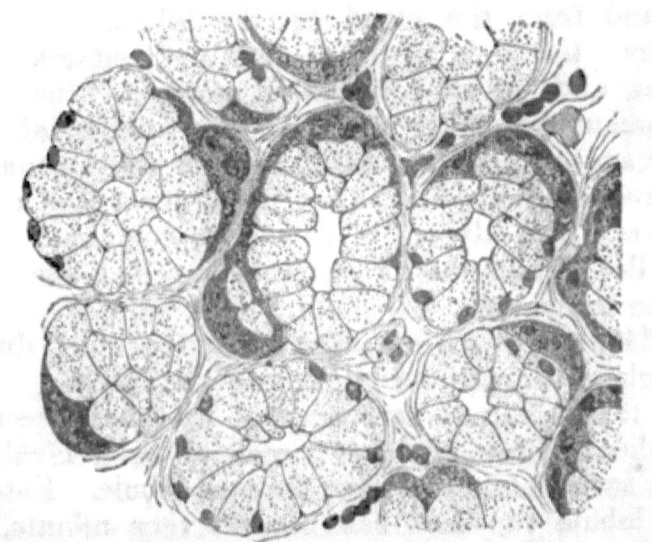

Fig. 107.—From a Section through the Orbital (mucous) Gland of Dog. Quiescent state.
The alveoli are lined with transparent "mucous cells," and outside these are the demilunes of Heidenhain. (Heidenhain.)

of cat, dog, rabbit, and guinea-pig. They secrete thickish, less watery mucus.

(3) *Mixed salivary*, or *muco-salivary glands*, such as the sub-maxillary and sub-lingual of man and ape.

In addition to the three salivary glands—parotid, sub-maxillary, and sub-lingual—there are in some cases, as in the rabbit and the guinea-pig, two minute additional glands, one intimately annexed to the parotid and the other to the sub-maxillary, and of the nature of a mucous gland. These are the superior and inferior *admaxillary glands*.

239. **The framework.**—Each salivary gland is enveloped in a fibrous connective tissue capsule, in connection with which are fibrous trabeculæ and septa in the interior of the gland, by which the substance of the latter is subdivided into *lobes*, these again into *lobules*, and these finally into the alveoli or acini. The duct, large vessels and nerves pass to and from the gland by the hilum. The connective tissue is of loose texture, contains elastic fibres, and, in some instances more, in others less, numerous lymphoid cells. In the sub-lingual gland they are so numerous that they form continuous rows between the alveoli. The connective tissue matrix between the alveoli is chiefly represented by fine bundles of fibrous tissue, and branched connective tissue corpuscles.

240. **The ducts.**—Following the chief duct of the gland through the hilum into the interior, we see that it divides into several great branches, according to the number of lobes; each of these breaks off into several branches, one for each lobule. Entering the lobule the duct has become very minute, and passing along it gives off laterally several minute ducts, all within the lobule being the *intralobular ducts*, or the *salivary tubes* of Pflüger, the bigger ducts being the *interlobular*, and, further, the *interlobar ducts*. Each of the latter consists of a limiting membrana propria, strengthened, according to the size of the duct, by thicker or thinner trabeculæ of connective tissue. In the chief branches there is present in addition non-striped muscular tissue. The interior of the duct is a cavity lined with a layer of columnar epithelial cells. In the largest branches there is, outside this layer and inside the membrana propria, a layer of small polyhedral cells.

241. The **intralobular ducts,** or the salivary tubes of Pflüger, consist of a limiting membrana

propria, with a single layer of columnar epithelial cells. Each of these has a spherical nucleus in about the middle; the outer half of the cell substance shows very marked longitudinal striation, due to more or less coarse fibrillæ (see Fig. 106). The inner half, *i.e.*, the one bordering the lumen of the duct, is only very faintly striated. The outline of these salivary tubes is never smooth, but irregular, hence the diameter of the tube varies from place to place.

Not in all salivary glands do the epithelial cells of the intralobular ducts show this coarse fibrillation in the outer part of their substance; *e.g.*, it is not present in the sub-lingual gland of the dog and guinea-pig.

242. The ends of the branches of the salivary tubes are connected with the secreting parts of the lobule, *i.e.*, the acini or alveoli. These always very conspicuously differ in structure from the salivary tubes, and, as a rule, are larger in diameter. That part of the duct which is in immediate connection with the alveoli is the *intermediary part*, this being interposed, as it were, between the alveoli and the salivary tube with fibrillated epithelium. The intermediary part is much narrower than the salivary tube, and is lined with a single layer of very flattened epithelial cells, each with a single oval nucleus; the boundary is formed by the membrana propria, continued from the salivary tube. The lumen of the intermediary part is much smaller than that of the salivary tube, and is generally lined with a fine hyaline membrane, with here and there an oblong nucleus in it.

At the point of transition of the salivary tube into the intermediary part there is generally a sudden diminution in size of the former, and the columnar cells of the salivary tube are replaced by polyhedral cells; this is the neck of the intermediary part. In some salivary glands, especially in the mucous, this neck is the only portion of the intermediary part present, e.g., in the sub-maxillary and orbital glands of dog and cat, and in the sub-lingual of the rabbit. In others, especially in the serous salivary glands, as in the parotid of man and mammals, sub-maxillary of rabbit and guinea-pig, and in the mixed salivary— e.g., sub-maxillary and sub-lingual of man and ape— there exists after the neck a long intermediary part, which gives off several shorter or longer branches of the same kind, all of which terminate in alveoli.

243. The **alveoli** or **acini** are the essential or secreting part of the gland; they are flask-shaped, club-shaped, shorter or longer cylindrical tubes, more or less wavy, or if long, more or less convoluted; many of them are branched. Generally several open into the same intermediary part of a salivary tube. The alveoli are much larger in diameter than the intermediary part, and slightly larger, or about as large, as the intralobular ducts. But there is a difference in this respect between the alveoli of a serous and a mucous salivary gland; in the former the alveoli are smaller than in the latter.

The membrana propria of the intermediary duct is continued as the membrana propria of the alveoli. This is a reticulated structure, being in reality a basket-shaped network of hyaline branched nucleated cells (Boll). The lumen of the alveoli is very minute in the serous, but is considerably larger in the mucous glands; it is in both glands smaller during secretion than during rest.

244. The epithelial cells lining the alveoli are called the *salivary cells* — they are of different characters in the different salivary glands, and chiefly determine the nature of the gland. The cells are separated from one another by a fluid albuminous cement substance. (1) In the serous or true salivary glands, as parotid of man and mammals, sub-maxillary

of rabbit and guinea-pig, the salivary cells form a single layer of shorter or longer columnar or pyramidal *albuminous cells*, composed of a densely reticulated protoplasm, and containing a spherical nucleus in the outer part of the cell. (2) In the mucous glands, such as the sub-lingual of the guinea-pig, or the admaxillary of the same animal, the cells lining the alveoli form a single layer of goblet-shaped *mucous cells*, such as have been described in par. 25. Each cell consists of an inner principal part, composed of a transparent mucoid substance (contained in a wide-meshed reticulum of the protoplasm), and an outer small, more opaque part, containing a compressed and flattened nucleus. This part is drawn out in a fine extremity, which, being curved in a direction parallel to the surface of the alveolus, is imbricated on its neighbours.

245. In the case of the sub-maxillary and orbital glands of the dog, the sub-lingual of rabbit, there exist, in addition to, and outside of the *mucous cells* lining the alveoli, but within the membrana propria, from place to place crescentic masses, being the *demilunes of Heidenhain*, or the *crescents of Gianuzzi* (see Fig. 107). Each is composed of several polyhedral granular-looking cells, each with a spherical nucleus; the cells at the margin of the crescent are of course thinner than those forming the middle. Heidenhain and his pupils, Lavdovski and others, have shown that, during prolonged exhausting stimulation of the sub-maxillary and orbital of the dog, all the lining cylindrical *mucous* cells become replaced by small polyhedral cells, similar to those constituting the crescents, while at the same time the alveoli become smaller (Fig. 108). These observers maintain that this change is due to a total destruction of the mucous cells, and a replacement of them by new ones, derived by multiplication

from the crescent cells. This is improbable, since, during ordinary conditions of secretions, there is no disappearance of the mucous cells as such; they change in size, becoming larger during secretion, and their contents are converted into perfect mucus. It is probable that, on prolonged exhaustive stimulation, the mucous cells collapse into the small cells, seen by Heidenhain and his pupils.

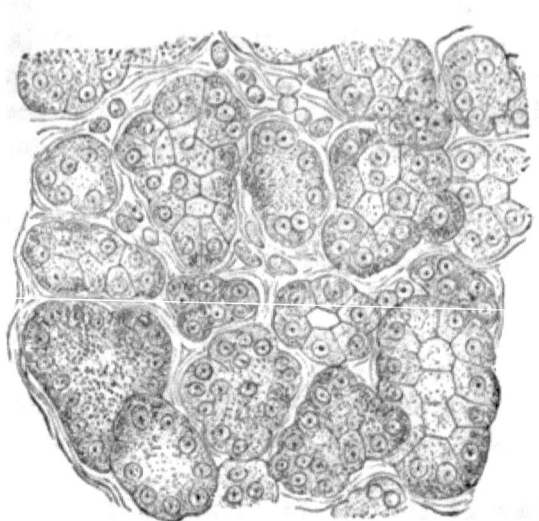

Fig. 103.—From a Section through the Orbital Gland of Dog, after prolonged electrical stimulation.
The alveoli are lined with small granular cells.
(Lavdovski.)

246. The alveoli of the sub-lingual of the dog are again different in structure both from those of the sub-maxillary of the dog and of the sub-lingual of guinea-pig, for the alveoli are there lined either with mucous cells or with columnar albuminous cells, or the two kinds of cells follow one another *in the same alveolus.*

This gland is a sort of intermediate form between the sub-lingual of man and the sub-maxillary of man and ape, *i.e.*, the mixed or muco-salivary glands. In these the great number of alveoli are serous, *i.e.*, small, with small lumen, and lined with albuminous cells, whereas there are always present a few alveoli exactly like those of a mucous gland. The two kinds of alveoli are in direct continuity with one another. In some conditions there are only very few mucous alveoli to be met with within the lobule, so few sometimes that they seem to be altogether absent; in others they are numerous, but even under most favourable conditions form only a fraction of the number of the serous alveoli. In the sub-lingual of man they are much more frequent, and for this reason this gland possesses a great resemblance to the sub-lingual of the dog.

What appear to be *crescents* in the mucous alveoli

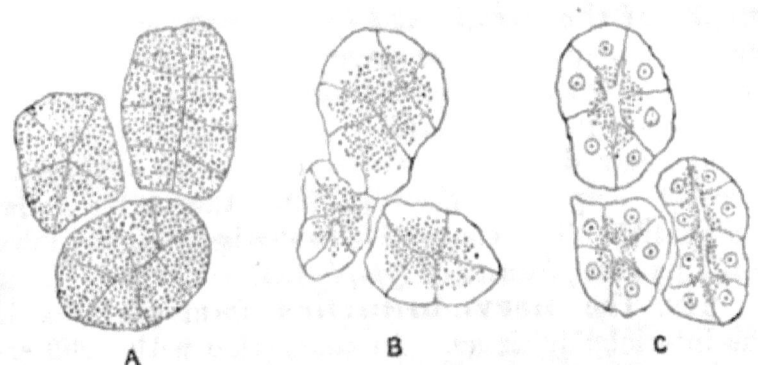

Fig. 109.—Alveoli of Serous Gland.

A, At rest; B, first stage of secretion; C, prolonged secretion. (Langley.)

of the human gland are an oblique view of albuminous cells lining the alveoli at the transition between the mucous and serous part of the same gland-tube.

247. The columnar salivary cells lining the alveoli of the sub-maxillary of the guinea-pig in some conditions show *two distinct portions*, an outer homogeneous or slightly and longitudinally striated substance, and an inner, more transparent, granular-looking part, and in this respect the cells resemble those of the pancreas. (See a future chapter.)

Langley has shown (Fig. 109) that during the period preparatory to secretion the cells lining the alveoli of the serous salivary glands become enlarged and filled with coarse granules; during secretion these granules become used up, so that the cell-substance grows more transparent, beginning from the outer part of the cell and gradually progressing towards the lumen of the alveolus.

248. **Blood-vessels and lymphatics.**—The lobules are richly supplied with blood-vessels. The arteries break up into numerous capillaries, which with their dense networks surround and entwine the alveoli. Between the interalveolar connective tissue carrying the capillary blood-vessels and the membrana propria of the alveoli exist *lymph spaces* surrounding the greater part of the circumference of the alveoli and forming an intercommunicating system of spaces. They open into *lymphatic vessels* accompanying the intralobular ducts, or at the margin of the lobule directly empty themselves into the interlobular lymphatics. The connective tissue between the lobes contains rich *plexuses of lymphatics.*

249. The **nerve-branches** form plexuses in the interlobular tissue. In connection with them are larger or smaller *ganglia*. They are very numerously met with in the sub-maxillary, but are absent in the parotid. Some ganglia are present in connection with the nerve-branches surrounding the chief duct of the sub-lingual gland.

Pflüger maintains that the ultimate nerve-fibres are connected with the salivary cells of the alveoli in man and mammals, but this remains to be proved.

CHAPTER XXIII.

THE MOUTH, PHARYNX, AND TONGUE.

250. **The glands.**—Into the cavity of the mouth and pharynx open very numerous minute glands, which, as regards structure and secretion, are either serous or mucous. The latter occur in the depth of the mucous membrane covering the lips of the mouth, in the buccal mucous membrane, in that of the hard palate, and especially in that of the soft palate and the uvula, in the depth of the mucous membrane of the tonsils, at the back of the tongue, and in the mucous membrane of the pharynx. The serous glands are found in the back of the tongue, in close proximity to the parts containing the special organs for the perception of taste—the taste goblets or buds (*see* below.) All glands are of very minute size, but when isolated they are perceptible to the unaided eye as minute whitish specks, as big as a pin's head, or bigger. The largest are in the lips, at the back of the tongue and soft palate, where there is something like a grouping of the alveoli around the small branches of the duct, so as to form little lobules.

251. The chief duct generally opens with a narrow mouth on the free surface of the oral cavity; it passes in a vertical or oblique direction through the superficial part of the mucous membrane. In the deeper, looser part (submucous tissue) it branches in two or more small ducts, which take up a number of alveoli. Of course, on the number of minute ducts and alveoli depends the size of the gland.

In man, all ducts are lined with a single layer of columnar epithelial cells, longer in the larger than in the smaller ducts; in mammals, the epithelium is a single layer of polyhedral cells. No fibrillation is noticeable in the epithelial cells. At the transition of the terminal ducts into the alveoli there is occasionally a slight enlargement, called the *infundibulum;* here the granular-looking epithelial cells of the duct change into the columnar transparent mucous cells lining the alveoli.

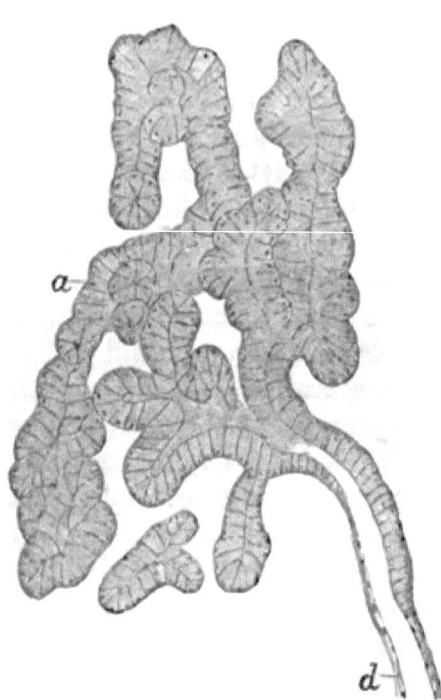

Fig. 110.—Part of a Lobule of a Mucous Gland in the Tongue of Dog.

a, Gland tubes (alveoli) viewed in various directions; they are lined with transparent "mucous cells;" *d,* duct, lined with small polyhedral cells. (Atlas.)

252. The alveoli of these glands are identical with those of the mucous glands described above (Fig. 110)—*e.g.*, the sublingual gland, as regards size, tubular branched nature, the lining epithelium, and lumen.

In some instances (as in the soft palate and tongue) the duct near the opening is lined with ciliated columnar epithelium. The stratified epithelium of the surface is generally continued a short distance into the mouth of the duct.

253. The **serous glands** at the root of the tongue (von Ebner) differ from the mucous chiefly in the size, epithelium, and lumen of the alveoli. These are of exactly the same nature and structure as those of the serous or true salivary glands.

Saliva obtained from the mouth contains numbers of epithelial scales, detached from the surface of the mucous membrane, groups of bacteria and micrococci, and lymph corpuscles. Some of these are in a state of disintegration, while others are swollen up by the water of the saliva. In these there are contained numbers of granules in rapid oscillation, called Brownian molecular movement.

254. The **mucous membrane** lining the cavity of the mouth is a thin membrane covered on its free surface with a thick stratified pavement epithelium, the most superficial cells being scales, more or less hornified.

Underneath the epithelium is a somewhat dense feltwork of fibrous connective tissue, with numerous elastic fibrils in networks. This part is the *mucosa*, and it projects into the epithelium in the shape of cylindrical or conical *papillæ*.

According to the thickness of the epithelium, the papillæ differ in length. The longest are found where the epithelium is thickest—*e.g.*, in the mucosa of the lips, soft palate, and uvula.

Numerous lymph corpuscles are found in the mucosa of the palate and uvula. Sometimes they amount to diffuse adenoid tissue. The deeper part of the mucous membrane is the *submucosa*. It is looser in its texture, but it is also composed of fibrous connective tissue with elastic fibrils. The glands are here embedded; fat tissue in the shape of groups of fat cells up to continuous lobules of fat cells are here to be met with. The large vascular and nervous trunks pass to and from the submucosa.

255. **Striped muscular tissue** is found in the submucosa. Of the lips, soft palate, uvula, and palatine arches, it forms a very conspicuous portion—*e.g.*, musculus sphincter orbicularis, with its outrunners into the mucous membrane of the lip, the muscles of

the palate, uvula (levator and tensor palati), and the arcus, palato-pharyngeus, and palato-glossus.

256. The last branches of the *arteries* break up in a dense *capillary network* on the surface of the mucosa, and from it loops ascend into the papillæ. Of course, fat tissue, glands, and muscular tissue receive their own supply. There is a very rich *plexus of veins* in the superficial part of the mucosa. They are conspicuous by their size and the thinness of their wall.

The *lymphatics* form networks in all layers of the mucosa, including the papillæ. The large efferent trunks are situated in the submucosa. The last outrunners of the *nerve branches* form a *plexus of non-medullated fibres* in the superficial layer of the mucosa, whence numerous *primitive fibrillæ* ascend into the epithelium to form networks. Meissner's tactile corpuscles have been found in the papillæ of the lips and in those of the tongue.

257. In the **pharynx** the relations remain the same, except in the upper or nasal part, where we find numerous places covered with columnar ciliated epithelium. As in the palatine tonsils so also here, the mucosa is infiltrated with diffuse adenoid tissue, and with lymph follicles in great numbers. This forms the *pharynx tonsil* of Luschka.

In the palatine tonsil and in the pharynx tonsil there are numerous crypts leading from the surface into the depth. This is due to the folding of the infiltrated mucosa. Such crypts are, in the pharynx, sometimes lined all through with ciliated epithelium, although the parts of the free surface around them are covered with stratified pavement epithelium.

258. **The tongue** is a fold of the mucous membrane. Its bulk is made up of striped muscular tissue (genio-, hyo-, and stylo-glossus ; according to direction : longitudinalis superior and inferior, and transversus linguæ). The lower surface is covered

with a delicate mucous membrane, identical in structure with that lining the rest of the oral cavity, whereas the upper part is covered with a membrane, of which the mucosa projects over the free surface as exceedingly numerous fine and short hair-like processes, the *papillæ filiformes*, or as less numerous isolated somewhat longer and broader mushroom-shaped *papillæ fungiformes*. The papillæ, as well as the pits between them, are covered with stratified pavement epithelium. Each has numbers of minute secondary papillæ. Their substance, like the mucous membrane of the tongue, is made up of fibrous connective tissue. This is firmly and intimately connected with the fibrous tissue forming the septa between the muscular bundles of the deeper tissue. The mucous membrane is on the whole thin. It contains large vascular trunks, amongst which the plexus of veins is very conspicuous. On the surface of the mucosa is a rich network of capillary blood-vessels, extending as complex loops into the papillæ. Lymphatics form rich plexuses in the mucosa and in the deep muscular tissue. Fat tissue is common between the muscular bundles, especially at the back of the tongue.

259. There are two varieties of glands present in the tongue, the mucous and serous. The latter occur only at the back, and in the immediate neighbourhood of the taste organs; the mucous glands are chiefly present at the back; but in the human tongue there are small mucous glands (glands of Nuhn) in the tip. All the glands at the back are embedded between the bundles of striped muscular tissue, and thus the movements of the tongue have the effect of squeezing out the secretion of the glands. Near and about the glands numerous nerve bundles are found connected with minute ganglia.

At the root of the tongue the mucous membrane is much thicker, and contains in its mucosa numerous lymph follicles and diffuse adenoid tissue. Thus

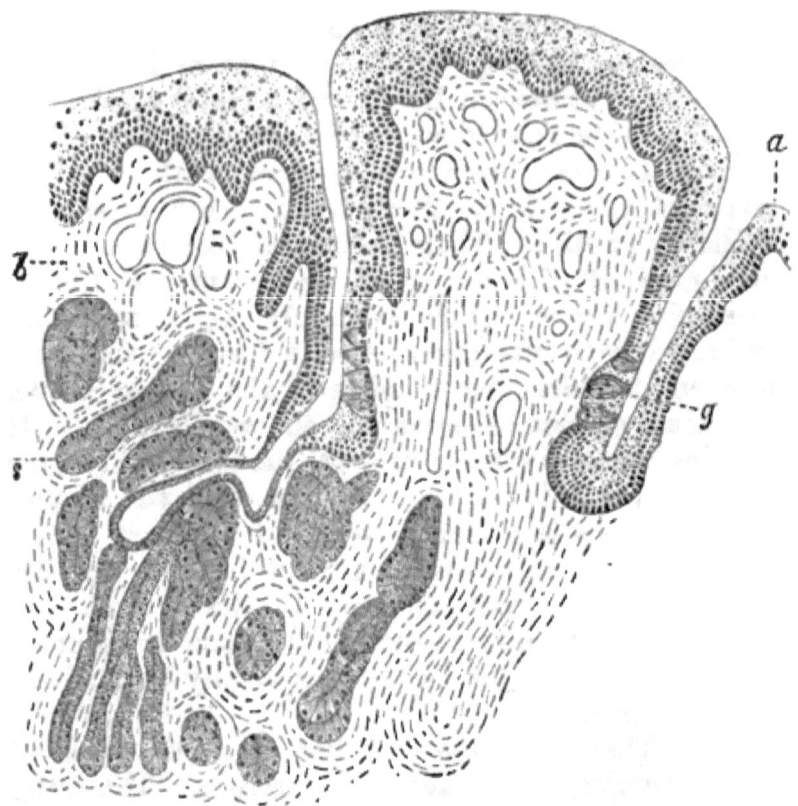

Fig. 111.—From a Vertical Section through a Circumvallate Papilla of the Tongue of a Child.

a, The stratified pavement epithelium covering the fold around the papilla; *b*, the mucous membrane; *s*, the serous glands; *g*, the pit between the fold and the papilla; in the epithelium of this latter are seen the "taste goblets." (Atlas.)

numerous knob-like or fold-like prominences of the mucosa are produced. There are also minute pits or crypts leading into the depth of these prominences.

260. The **papillæ circumvallatæ** (Fig. 111) are large papillæ fungiformes, each surrounded by a fold of the mucosa. They contain taste goblets or buds—

Chap. XXIII.] MOUTH, PHARYNX, AND TONGUE. 193

i.e., the terminal taste organs. At the margin of the tongue, in the region of the circumvallate papillæ, there are always a few permanent folds, which also contain taste goblets. In some domestic animals these folds assume a definite organisation—*e.g.*, in the rabbit there is an oval or circular organ composed of **numbers** of parallel and permanent folds, *papillæ foliatæ*. The papillæ fungiformes of the rest of the tongue also contain in some places a taste goblet. But most of the taste goblets are found on the papillæ circumvallatæ and foliatæ. In both kinds of structures **the** taste goblets are placed in several **rows close** round **the** bottom **of** the pit, separating, **in the** papillæ circumvallatæ, the papillæ fungiformes **from** the fold **of the** mucosa **surrounding** it : **in the** papillæ **foliatæ** the pits are represented by groves separating the individual folds from one another.

261. The **taste goblets** or **taste buds** are barrel-shaped structures (Fig. 112), extending in a vertical direction through the epithelium, from the free surface to the mucosa. Each is composed of a layer of flattened epithelial cells, elongated in the direction of the goblet, and forming its cover; these are the *tegmental cells*. The interior of the goblet is made up of a bundle of spindle-shaped or staff-shaped *taste cells*. Each includes an oval nucleus, and is drawn out into an outer and an inner fine

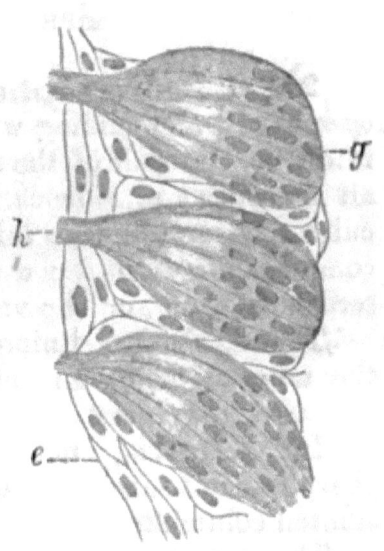

Fig. 112.—Three Taste Goblets, highly magnified.

g, The base of the goblet next the mucosa; *h*, the free surface; *e*, the epithelium of the surface. (Atlas.)

extremity. The former extends to the free surface, projecting just through an opening of the goblet, and resembles a fine hair; the latter is generally branched, and passes towards the mucosa; there, probably, it becomes connected with a nerve-fibre, the mucosa of these parts containing rich plexuses of nerve-fibres.

Into the pits surrounded by taste goblets open the ducts of the serous glands only (von Ebner).

CHAPTER XXIV.

THE ŒSOPHAGUS AND STOMACH.

262. I. THE **œsophagus.**—Beginning with the œsophagus, and ending with the rectum of the large intestine, the wall of the alimentary canal consists of an inner coat or mucous membrane, an outer or muscular coat, and outside this a thin fibrous coat, which, commencing with the cardia of the stomach, is the serous covering, or the visceral peritoneum.

The epithelium lining the inner or free surface of the mucous membrane of the œsophagus is a thick, stratified, pavement epithelium.

In Batrachia, not only the oral cavity and pharynx, but also the œsophagus, are lined with ciliated columnar epithelium.

The mucous membrane is a fibrous connective tissue membrane, the superficial part of which is dense—the *mucosa;* this projects, in the shape of small papillæ, into the epithelium.

The deeper, looser portion of the mucous membrane is the *submucosa;* in it lie small mucous glands, the ducts of which pass in a vertical or oblique direction through the mucosa, in order to open on the free surface. In man, these glands are relatively scarce; in carnivorous animals (dog, cat) they form an almost continuous layer (Fig. 113).

263. Between the mucosa and submucosa are longitudinal bundles of non-striped muscular tissue.

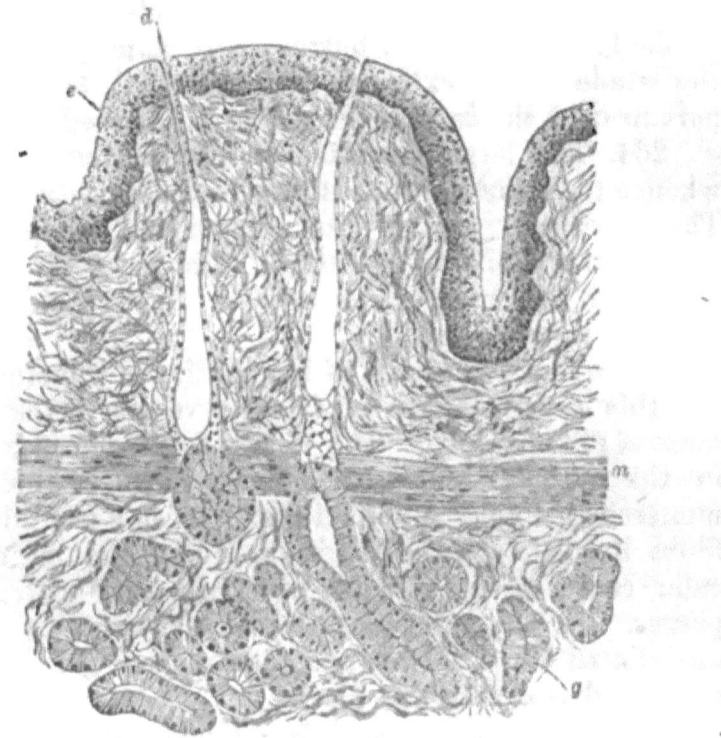

Fig. 113.—From a Longitudinal Section through the Mucous Membrane of the Œsophagus of Dog.

e, The stratified pavement epithelium of the surface; *m*, the muscularis mucosæ; between the two is the mucosa; *g*, the mucous glands; *d*, ducts of same. (Atlas.)

At the beginning of the œsophagus they are absent, but soon make their appearance—at first as small bundles separated from one another by masses of connective tissue; but lower down, about the middle, they form a continuous stratum of longitudinal bundles. This is the *muscularis mucosæ*.

Outside the submucosa is the muscularis externa. This consists of a thicker inner circular and an outer thinner longitudinal coat. And outside this is the outer, or limiting, fibrous coat of the œsophagus. In man, the outer muscular coat consists of non-striped muscular tissue, except at the beginning (about the upper third, **or** less) of **the œsophagus, which is** composed of the striped **variety**; but in many mammals almost the **w**hole of **the** external muscular coat, **except the part** nearest **the** cardia, is made up of striped **fibres.**

264. The large vessels pass into the submucosa, **whence** their finer branches pass to the surface parts. **The** superficial part of the mucosa and the papillæ contain the capillary networks. The outer muscular coat and the muscularis mucosæ have their own vascular supply.

There is a rich plexus of lymphatics in the mucosa, and this leads to a plexus of larger vessels in the submucosa (Teichmann). The nerves form rich plexuses in the outer fibrous **coat**; these plexuses include numerous ganglia. A second **plexus of** non-medullated **fibres** lies between the longitudinal and circular muscular coat; **a** few ganglia **are** connected with this plexus. In the submucosa are also plexuses of non-medullated fibres. Now and then a small ganglion is **connected** also with this plexus.

265. II. **The stomach.**—Beginning with the cardia, the mucous membrane of the stomach is covered with a single layer of beautiful thin columnar epithelial cells, most of which are mucus-secreting goblet cells. On the surface of the mucous membrane of **the** stomach open numerous fine ducts of glands, **placed** very closely side by side. These extend, more or less vertically, as minute tubes, into the depth of the mucous membrane. In the pyloric end, where the mucous membrane presents a pale aspect, the glands are called the *pyloric glands;* in the rest of the stomach, whose mucous

membrane presents a reddish or red-brown appearance, they are called the *peptic glands*. Owing to the very numerous fine ducts opening on the surface of the mucous membrane, the tissue of this latter appears on a vertical section to be made up of thinner or thicker folds, or villi—plicæ villosæ. But they are not real villi.

The part of the mucous membrane containing the glands is the *mucosa*; outside this is a loose connective tissue containing the large vessels—this is the *submucosa*. Between the two, but belonging to the mucosa, is the *muscularis mucosæ*, a thick stratum of bundles of non-striped muscular tissue, arranged in most parts of the stomach as an inner circular and an outer longitudinal layer. The tissue of the mucosa is dense, owing to its containing, placed closely side by side, the gland tubes. Between them is a delicate connective tissue, in which the minute capillary blood-vessels pass in a direction vertical to the surface. Numerous small bundles of non-striped muscular fibres pass from the muscularis mucosæ towards the surface—up to near the epithelium of the surface—forming longitudinal muscular sheaths, as it were, around the gland tubes.

The plicæ villosæ of the superficial part of the mucosa contain fibrous connective tissue and numerous lymphoid cells.

266. The **peptic glands** (Fig. 114) are more or less wavy tubes, extending down to the muscularis mucosæ. The deep part is broader than the rest, and is more or less curved, seldom branched. This is the fundus of the gland; near the surface of the mucosa is the thinnest part of the tube; this is the *neck*. Two or three neighbouring glands join and open into the short cylindrical *duct* mentioned above. The duct is lined with a layer of columnar epithelial cells, continuous and identical with those of the

free surface of the mucous membrane. The cavity of the duct is continued as a very fine canal into the neck and through the rest of the gland tube. Next

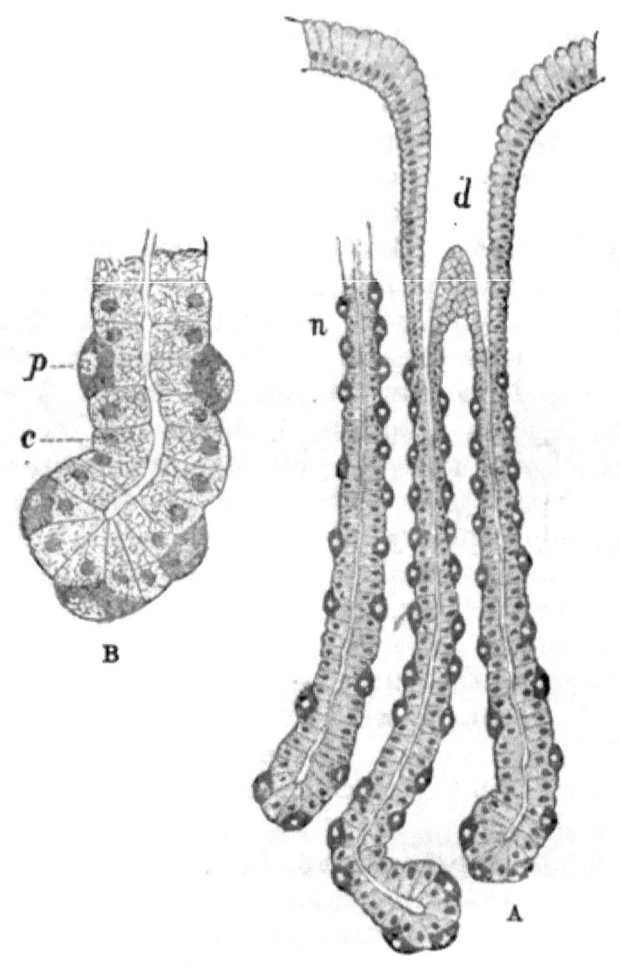

Fig. 114.—Peptic Glands.

A, Under a low power; d, duct; n, neck; B, part of the fundus of a gland tube under a high power; p, parietal cells; c, chief cells.

to the lumen is a continuous single layer of more or less transparent, granular-looking, epithelial cells, each with a reticulated protoplasm and a spherical or slightly oval nucleus. In the neck these cells are polyhedral, but farther downwards increase to cylindrical cells, and in the fundus of the gland tube they are long columnar. This layer of cells bordering the lumen is the layer of *chief cells* (Heidenhain.) Outside them is the limiting membrana propria of the gland tube. But from place to place, between the membrana propria and the chief cells, are *single* oval spherical or angular large granular and opaque-looking cells, called the *parietal cells* (Heidenhain). These are more numerous in the neck than in any other part of the gland; at the fundus they are few and far between, whereas at the neck they form almost a continuous layer. Their protoplasm is densely reticulated.

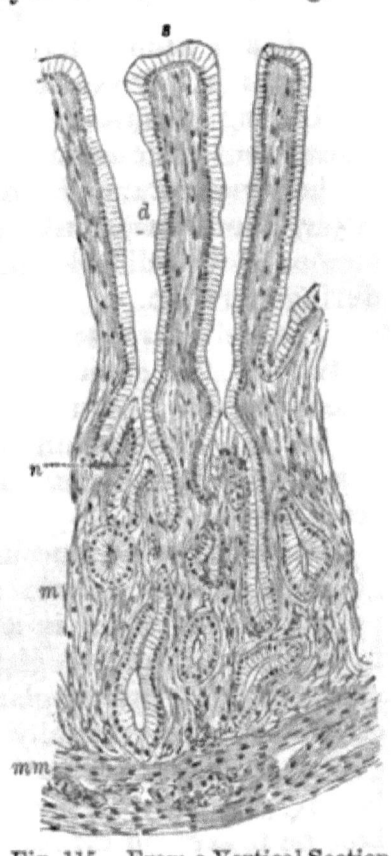

Fig. 115.—From a Vertical Section through the Mucous Membrane at the Pyloric End of the Stomach.

s, Free surface; *d*, ducts of pyloric glands; *n*, neck of same; *m*, the gland alveoli; *mm*, muscularis mucosæ. (Atlas.)

267. The **pyloric glands** (Fig. 115).—The *duct* of each pyloric gland is several times longer than in the peptic. The duct of the former occupies in some places as much as half of the thickness of the mucosa, whereas that of the latter does not exceed in the fundus of the stomach or in the cardia, more than one-fourth or one-fifth of the thickness.

The epithelium lining the duct of the pyloric glands is the same as in that of the peptic. Each duct takes up two or three tubes by their short, narrow, thin neck. The main part of each gland tube is convoluted and slightly branched. The neck is lined with a layer of polyhedral cells, whereas the gland tube has a lining of columnar transparent cells, and its lumen is very conspicuous.

During exhaustion these cells are smaller and less transparent than during secretion. Their protoplasm in the former state is a denser reticulum than in the latter, the transparent interstitial substance in the meshes of the cell reticulum being increased in amount during secretion.

The cells are serous, not mucous, and the secretion of the glands cannot therefore be mucous. According to Ebstein the secretion is pepsin, and so he and Heidenhain consider the pyloric glands as simple peptic glands. But this view is not generally accepted.

Between the mucous membrane with peptic glands and the **pyloric** end of the stomach with pyloric glands there is a narrow *intermediary zone*, in which the peptic glands appear by degrees to merge into the pyloric glands. That is, the short duct of the former gradually elongates, the gland tubes get shorter in proportion and convoluted, their lumen gradually enlarges, and the parietal cells become fewer and ultimately disappear.

268. The mucosa contains isolated *lymph follicles*, glandulæ lenticulares, and in the pyloric part also groups of these—glandulæ agminatæ.

The submucosa is of very loose texture, and enables the mucosa to become easily folded in all directions.

The *muscular coat* is very thick, and consists of an outer longitudinal and an inner thicker circular stratum of non-striped muscular tissue. Numerous oblique bundles are found in the inner section of the circular stratum.

The gland tubes are ensheathed in a longitudinal network of *capillary blood-vessels* derived from arteries of the submucosa. This network forms on the surface a special dense horizontal layer, from which the venous branches are derived. The outer muscular coat and the muscularis mucosæ possess their own vascular supply.

269. The *lymphatics* form **a network** in the mucosa near the fundus of the glands. Into this plexus lead lymphatics running longitudinally between the glands anastomosing with one another freely, and extending to near the surface (Lovèn). Another plexus belongs **to** the submucosa.

Between the longitudinal and circular **stratum of the** outer **muscular coat, and** extending parallel **to the** surface, is a plexus **of** non-medullated nerve-**branches** with a few ganglia in its nodes. This corresponds **to the** *plexus of Auerbach* **of the** intestine, and is destined for the outer muscular coat. A second plexus of non-medullated nerve-branches with ganglia also extending parallel to the surface lies in the submucosa. This corresponds to the *plexus of Meissner* of the intestine, and is destined for the muscularis mucosæ and the mucosa.

According to Rabe, the gastric gland tubes in the horse are surrounded by a rich plexus of nerve-fibres, terminating **in peculiar** spindle-shaped cells.

CHAPTER XXV.

THE SMALL AND LARGE INTESTINE.

270. THE epithelium covering the inner or free surface of the mucous membrane of the small and large intestine is a single layer of columnar cells, their protoplasm more or less distinctly longitudinally fibrillated ; their free surface appears covered with a

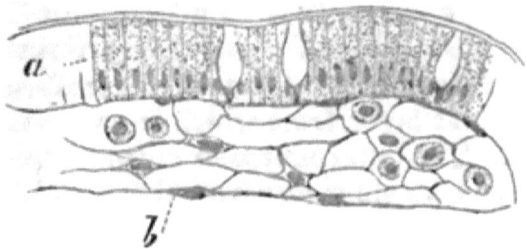

Fig. 116.—From a Longitudinal Section through a Villus of the Small Intestine.

a, The epithelium of the surface; *b*, non-striped muscular fibres. Immediately underneath the epithelium is a basement membrane with oblong nuclei; the tissue of the villus is made up of a reticulum of cells; in its meshes are lymph corpuscles.

vertically and finely *striated basilar border*. Many cells are goblet cells. Underneath the epithelium is a basement membrane, the sub-epithelial endothelium of Debove (see par. 39).

As in the stomach, so also in the small and large intestine, the *mucosa* is connected with the outer muscular coat by a loose-textured fibrous *submucosa*, in which lie the large vascular trunks, and in many places larger or smaller groups of fat cells and lymph corpuscles. Between the mucosa and submucosa, but belonging to the former, is a layer of non-striped muscular tissue, the *muscularis mucosæ*. This is in many places composed of inner circular and outer longitudinal bundles, but there are a good many places, especially in the small intestine, where only a layer of longitudinal bundles can be made out.

The tissue of the mucosa is similar in structure to adenoid tissue (Fig. 116), consisting of a reticular matrix with flat-

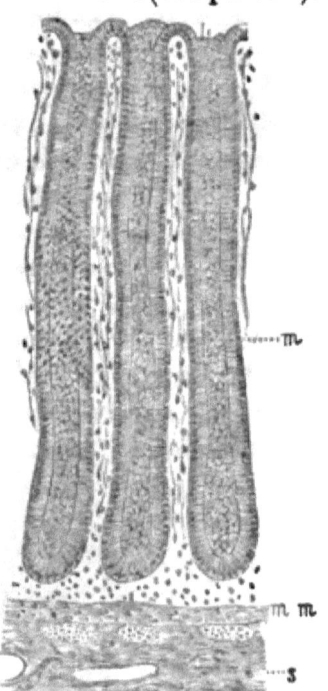

Fig. 117.—From a Vertical Section through the Mucous Membrane of the Large Intestine of Dog.

m, The mucosa containing the crypts of Lieberkühn, closely placed side by side; each crypt is lined with a layer of columnar epithelium; *mm*, muscularis mucosæ; *s*, submucosa. (Atlas.)

tened large nucleated endotheloid cells and numerous smaller lymph corpuscles. The mucosa of the small and large intestine contains simple gland tubes, the *crypts or follicles of Lieberkühn* (Fig. 117); they are placed vertically and closely side by side, extend-

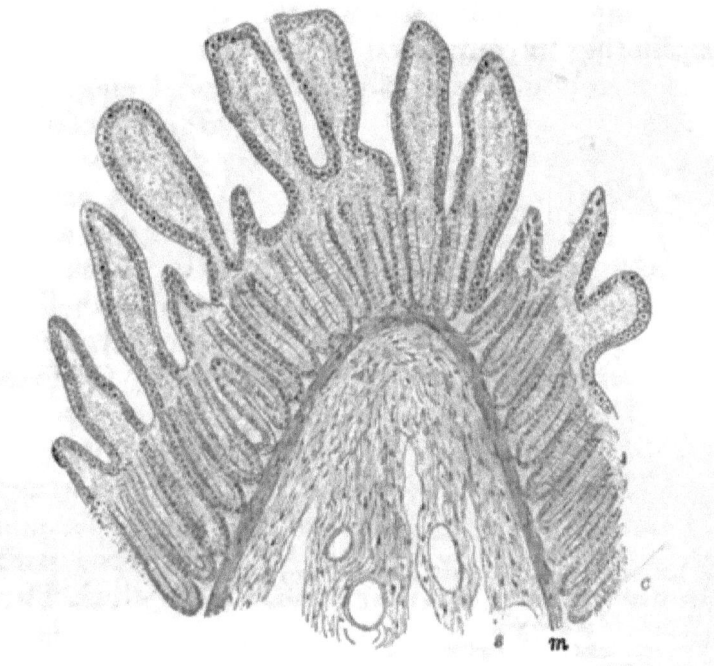

Fig. 118.—From a Vertical Section through a Fold of the Mucous Membrane of the Jejunum of Dog.

c, The mucosa, containing the crypts of Lieberkühn, and projecting as the villi; m, muscularis mucosæ; s, submucosa. (Atlas.)

ing from the free surface, where they open, to the muscularis mucosæ. These glands possess a large lumen, and are lined with a single layer of columnar epithelial cells, many of them goblet cells.

271. In the small intestine the tissue of the mucosa projects beyond the general surface in the shape of very numerous fine, longer or shorter, cylindrical, conical or leaf-shaped *villi* (Fig. 118). These are, of course, covered with the columnar epithelium

of the general **surface,** and their tissue is the same as that of the mucosa—*i.e.*, adenoid tissue—with the addition of: (*a*) One or two central wide chyle (lymph) vessels (see Fig. 120), their wall being a single layer of endothelial plates. (*b*) Along these chyle vessels are longitudinal bundles of non-striped **muscular** tissue, extending **from the** base to the apex of **the villus,** terminating in connection with the cells of **the base**ment **membrane**—*i.e.*, the subepithelial endothelium.

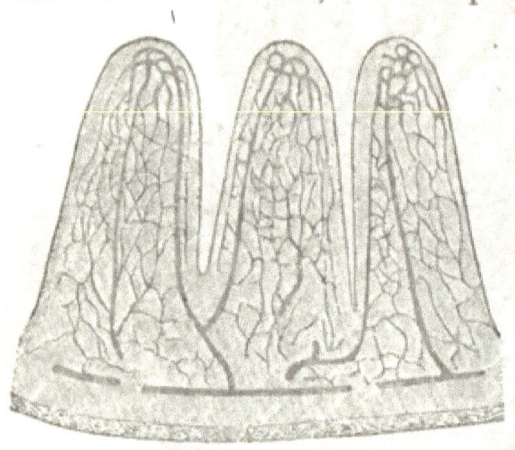

Fig. 119.—From a Vertical Section through the Small Intestine of Mouse; the Blood-vessels are injected.

The networks of the capillaries of the villi are well shown. (Atlas.)

(*c*) A network of capillary blood-vessels extending over the whole of the villus close to the epithelium of the surface (Fig. 119). This capillary network derives its blood from an artery in about the middle or upper part of the villus. Two venous vessels carry away the blood from the villus.

The Lieberkühn's crypts open between the bases of the villi.

At the sides of the villi of the small intestine, and at the sides of the plicæ villosæ of the stomach (see a former chapter), there exist amongst the epithelium of the surface peculiar goblet-shaped groups of epithelial cells, which, as Watney has shown, are due to local multiplication of the epithelial cells.

272. Lymph follicles occur singly in the submucosa, and extend with their inner **part or** summit through the muscularis mucosæ into **the mucosa to near** the

internal free surface of the latter (Fig. 120). These are the *solitary lymph follicles* of the small and large intestine; in the latter they are larger than in the former.

Agminated glands, or *Peyer's glands*, are larger or smaller groups of lymph follicles, more or less fused

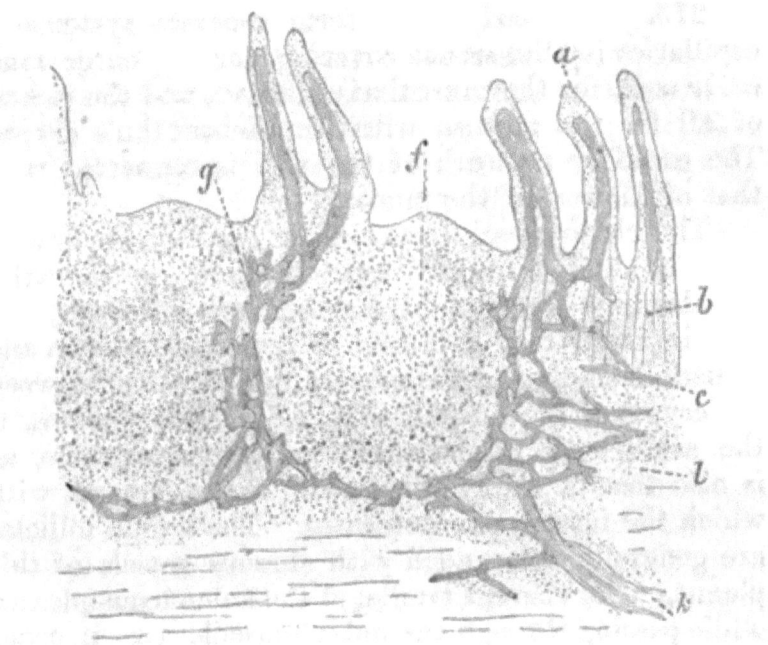

Fig. 120.—From a Section through a **part of a** Human Peyer's Patch, showing the distribution of the **Lymphatic Vessels** in the Mucosa and Submucosa.

a, Villi, with central chyle vessel ; *b*, **Lieberkühn's crypts** ; *c*, region of muscularis mucosæ; *f*, lymph follicle ; *g*, **network of lymphatics** around the lymph follicle ; *l*, lymphatic network **of the submucosa** ; *k*, an efferent lymphatic trunk. (Frey.)

with one another, and situated with their main part in the submucosa, but extending with their summit to the epithelium of the free surface of the mucosa. In the lower part of the ileum these Peyer's glands are very numerous. The epithelium covering the summits of these lymph follicles is invaded by, and more or less replaced by, the lymph corpuscles of the adenoid tissue of the follicles (Watney), similar to what is the case in the tonsils (see par. 124).

The outer muscular coat consists of an inner thicker circular and an outer thinner longitudinal stratum of non-striped muscular tissue.

In the large intestine, in the "ligamenta," only the longitudinal layer is present, and is much thickened.

273. The blood-vessels form separate systems of capillaries for the serous covering, for the outer muscular coat, for the muscularis mucosæ, and the richest of all for the mucosa with its Lieberkühn's crypts. The capillary network of the villi is connected with that of the rest of the mucosæ.

The chyle vessel, or vessels of the villi, commence with a blind extremity near the apex of the villi. At the base the chyle vessel becomes narrower, and empties itself into a plexus of lymphatic vessels and sinuses belonging to the mucosa, and situated between the crypts of Lieberkühn (Fig. 120). This network is the same both in the small and large intestine, as is also that of the lymphatics of the submucosa with which the former communicates. The lymph follicles are generally surrounded with sinuous vessels of this plexus. The efferent trunks of the submucous plexus, while passing through the outer muscular coat in order to reach the mesentery, take up the efferent vessels of the plexus of the lymphatics of the muscular coat.

The **chyle**, composed of granules and globules of different but minute sizes, passes from the inner free surface of the mucous membrane of the small intestine through the epithelium (probably through its fluid interstitial cement substance) into the reticulum of the villus matrix, and from thence the central chyle vessel, and farther into the plexus of vessels of the mucosa and submucosa.

Owing to the peripheral disposition of the capillaries in the villi, and owing to the greater filling with blood of the capillaries during digestion, the villi are thrown into a state of turgescence during this period, in consequence of which the central chyle vessels are kept distended. Absorption is thus greatly supported. The contraction of the muscular tissue of the villi and of the muscular coat of the in-

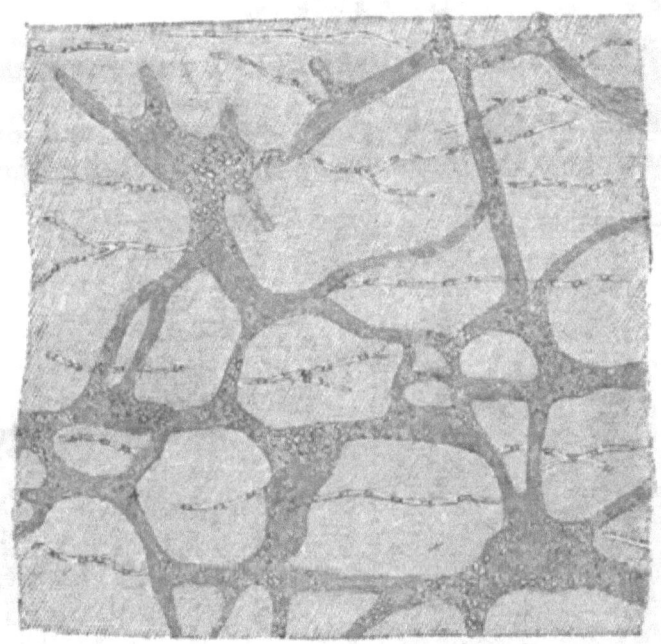

Fig. 120A.—Plexus **Myentericus** of Auerbach of the Small Intestine of a new-born Child.
The minute circles and ovals indicate ganglion cells. (Atlas.)

testine greatly facilitates the absorption and discharge of the chyle.

274. The non-medullated nerves form a rich plexus, called the the *plexus myentericus of Auerbach* (Fig. 120A), with groups of ganglion cells in the nodes; this plexus lies between the longitudinal and circular muscular coat. Another plexus connected with the former lies in the submucous tissue; this is the *plexus of Meissner*, with ganglia. In both

plexuses the branches are of a very variable thickness; they are groups of simple axis cylinders, held together by a delicate endothelial sheath.

CHAPTER XXVI.

THE GLANDS OF BRUNNER, AND THE PANCREAS.

275. **AT** the passage of the pyloric end of the stomach into the duodenum (Fig. 121), and in the first

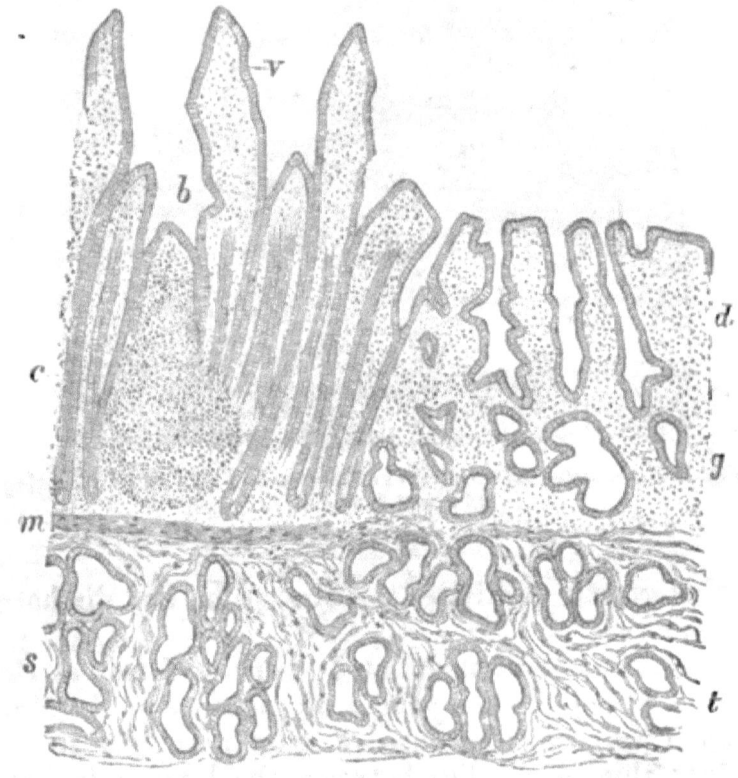

Fig. 121.—Vertical Section through the Mucous Membrane of the end of Stomach and commencement of Duodenum.

v, Villi of duodenum; *b*, a lymph follicle; *c*, Lieberkühn's crypts; *d*, mucosa of pyloric end of stomach; *g*, the alveoli of the pyloric glands; *t*, the same in the submucosa; they are continued into the duodenum as—*s*, the Brunner's glands; *m*, the muscularis mucosæ. (Atlas.)

part of the latter, is a continuous layer of gland tissue in the submucosa, composed of convoluted, more or less branched tubes grouped into lobules, and permeated by bundles of non-striped muscular tissue, outrunners of the muscularis mucosæ. These are the *glands of Brunner*. Numerous thin ducts lined with a single layer of columnar epithelial cells pass through the mucosa, and open into the crypts of Lieberkühn between the bases of the villi. The gland tubes of Brunner's glands are *identical in structure with the pyloric glands, with which they form a direct anatomical continuity*.

276. The **pancreas** (Fig. 122) is in most respects identical in structure with a serous or true salivary gland. The connective tissue framework, the distribution of the blood-vessels and lymphatics, and of the gland tissue in lobes and lobules, with the córresponding inter- and intra-lobular ducts, is similar in both cases. The epithelium lining the latter ducts is only faintly striated, not by any means so distinctly as in the salivary tubes. The alveoli or acini are club-shaped, flask-shaped, shorter or longer cylindrical, and convoluted.

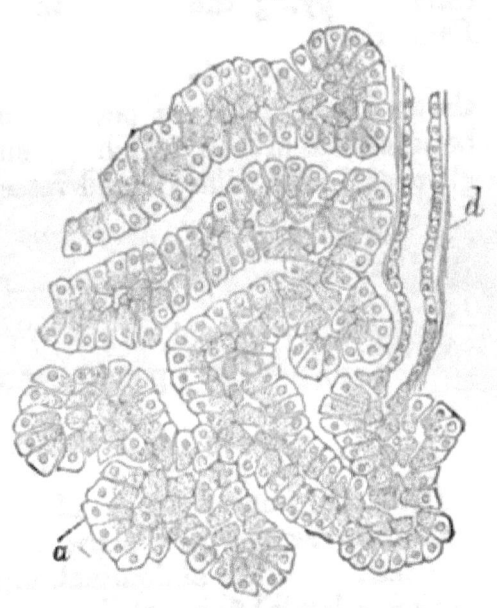

Fig. 122.—From a Section through the Pancreas of Dog.

a, The alveoli (tubes) of the gland; the lining cells show an outer homogeneous and an inner granular-looking portion; *d*, a minute duct. (Atlas.)

277. The intermediary part of the duct and its passage into the alveoli is the same as in the salivary glands. The cells lining the alveoli are columnar or pyramidal, and show an outer homogeneous, or faintly and longitudinally striated zone (Langerhans, Heidenhain), and an inner more transparent granular-looking zone. The nucleus of the cell is spherical, and lies in about the middle. According to the state of secretion the two zones vary in amount, one at the expense of the other.

The lumen of the alveoli is very minute, and in the beginning of the alveoli, *i.e.*, next to the intermediary part of the duct, are seen spindle-shaped cells occupying the lumen, the *centroacinous cells* of Langerhans.

In the rabbit's pancreas Kühne and Lea have shown that there are peculiar accumulations of cells between the alveoli, which are supplied with veritable glomeruli of capillary blood-vessels.

CHAPTER XXVII.

THE LIVER.

278. The outer surface of the liver is covered with a delicate *serous membrane*, the peritoneum, which, like that of other abdominal organs, has on its free surface a layer of endothelium. It consists chiefly of fibrous connective tissue.

At the hilum or porta hepatis this connective tissue is continued into the interior, and becomes one with the connective tissue of the *Glisson's capsule*, or the *interlobular connective tissue* (connective tissue of the portal canals). This tissue is fibrous, and more or less lamellated; by it the substance of the liver is subdivided into numerous, more or less polyhedral, solid *lobules* or *acini* (Fig. 123), each about $\frac{1}{20}$th of an inch in diameter. According to whether the interlobular tissue forms complete boundaries or not, the acini appear well defined from one another (pig, ice-bear), or more or less fused (man and carnivorous animals and rodents).

Within each acinus there is only very scanty connective tissue, in the shape of extremely delicate bundles and flattened connective-tissue cells. Occasionally, especially in the young liver, migratory cells are to be met with in the acini and in the tissue between them.

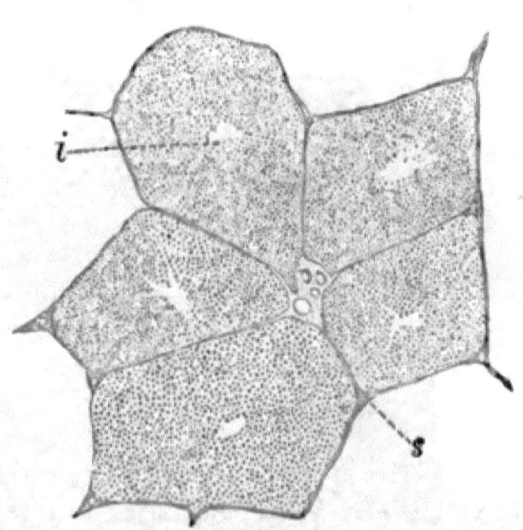

Fig. 123.—From a Section through the Liver of Pig. Five lobules are shown. They are well separated from one another by the interlobular tissue.

s, Interlobular connective tissue, containing the interlobular blood-vessels, *i.e.*, the branches of the hepatic artery and portal vein, and the interlobular bile ducts; *i*, intralobular or central vein. (Atlas.)

279. The vena portæ having entered the hilum gives off rapidly numerous branches, which follow the interlobular tissue in which they are situated, and they form rich *plexuses around each acinus;* these are the *interlobular veins* (Fig. 124). Numerous capillary blood-vessels are derived from these veins. These capillaries pass in a radiating direction to the centre of the acinus, at the same time anastomosing with one another by numerous transverse branches. In the

centre of the acinus the capillaries become confluent into one large vein, the *central* or *intralobular* vein. The intralobular veins of several neighbouring acini join so as to form the *sublobular veins*, and these lead into the efferent veins of the liver, or

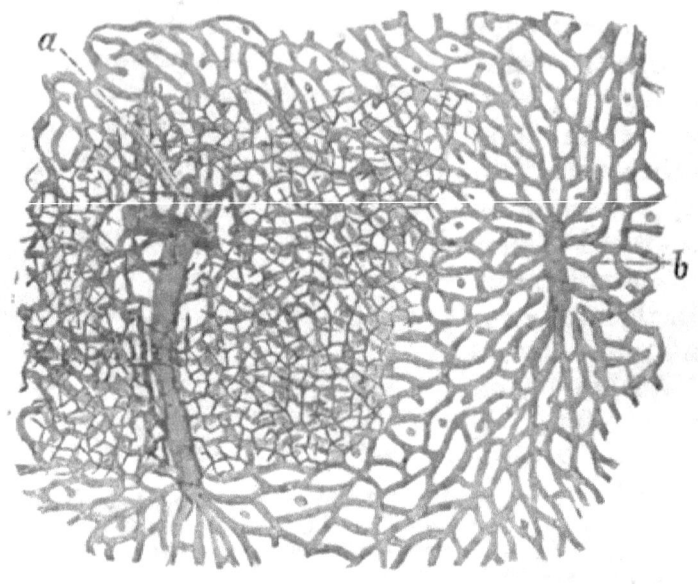

Fig. 124.—From a Vertical Section through the Liver of Rabbit; the Blood-vessels and Bile-vessels injected.

a, Interlobular veins surrounded by interlobular bile-ducts; these latter take up the network of fine intralobular bile capillaries; the meshes of this network correspond to the liver cells; b, the intralobular or central vein. (Atlas.)

the *hepatic veins*, which finally pass into the vena cava inferior.

280. The substance of each acinus—*i.e.*, the tissue between the capillary blood-vessels—is composed of uniform polygonal protoplasmic epithelial cells, of about $\frac{1}{1000}$th of an inch in diameter; these are the *liver cells*. Owing to the peculiar, more or less radiating, arrangement of the capillaries, the liver cells appear to form columns or cylinders, also more or less radiating from the periphery towards the

centre of the acinus. Sometimes the liver cells contain minute pigment granules.

Each liver cell shows a more or less fibrillated and reticulated protoplasm (Kupfer), and in the centre a spherical nucleus with its reticulum, generally with one or more nucleoli.

During activity the liver cells are larger and look more granular than after action.

The liver cells are joined with one another by an albuminous cement substance, in which are left fine channels; these are the *bile capillaries*, or *intralobular bile vessels* (Fig. 125). In a successfully injected preparation, the liver cells appear separated everywhere from one another by a bile capillary, and *these form for the whole acinus a continuous intercommunicating network* of minute channels. Where the liver cells are in contact with a capillary blood-vessel, there, of course, are no bile capillaries, since these exist *only between liver cells*.

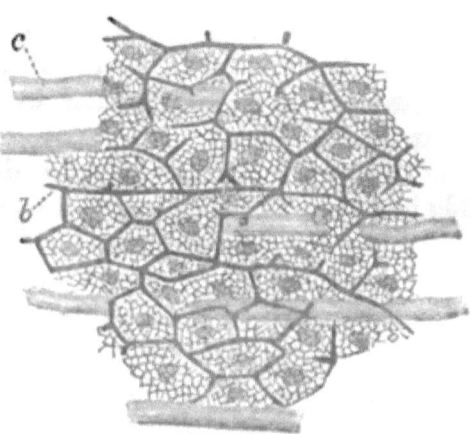

Fig. 125.—From a Lobule of the Liver of Rabbit, in which Blood and Bile Vessels had been injected, more highly magnified than in Fig. 124.

b, Bile capillaries between the liver cells; these are well shown as nucleated polygonal cells, each with a distinct reticulum; *c*, capillary blood-vessels. (Atlas.)

281. At the margin of the acinus the bile capillaries are connected with the lumen of minute tubes; these possess a membrana propria and a lumen lined with a single layer of transparent polyhedral epithelial cells. These are the *small interlobular bile ducts* (Fig. 124). Their epithelial cells are in reality continuous with the liver cells. They join so as to form

the *larger interlobular bile ducts*, lined with more or less columnar epithelium. The first part of the bile duct lined with polyhedral cells corresponds to the intermediary part of the ducts of the salivary glands. The interlobular bile ducts form networks in the interlobular tissue. Towards the hilum they become of great diameter, and their wall is made up of fibrous tissue, with non-striped muscular tissue. Small mucus-secreting glands are in their wall, and open into their lumen.

The wall of the hepatic duct, and of the gall bladder, are merely exaggerations of this structure.

282. The hepatic artery follows in its ramification the interlobular veins. The arterial branches form plexuses in the interlobular tissue, and they supply the capillary blood-vessels of the interlobular connective tissue, and especially of the bile ducts. The capillary blood-vessels of the bile ducts join so as to form small veins which finally empty themselves into the interlobular veins. The anastomoses of the capillary blood-vessels, derived from the arterial branches, directly with the capillary blood-vessels of the acini, are insignificant (Cohnheim and Litten). The serous covering of the liver contains special arterial branches—rami capsulares. Networks of lymphatics—*deep lymphatics*—are present in the interlobular connective tissue, forming plexuses around the interlobular blood-vessels and bile ducts, and occasionally forming a perivascular lymphatic around a branch of the hepatic vein. Within the acinus, the lymphatics are represented only by spaces and clefts existing between the liver cells and capillary blood-vessels; these are the *intralobular* lymphatics (Macgillivry, Frey, and others). They anastomose at the margin of the acinus with the interlobular lymphatics.

In the capsule of the liver is a special network

of lymphatics, called the *superficial lymphatics*. Numerous branches pass between this network and those of the interlobular lymphatics.

CHAPTER XXVIII.

THE ORGANS OF RESPIRATION.

283. I. **The larynx.**—The supporting framework of the larynx is formed by cartilage. In the epiglottis the cartilage is elastic and reticulated, *i.e.*, the cartilage plate is perforated by numerous smaller and larger holes. The cartilages of Santorini and Wrisbergii, the former attached to the top of the arytenoid cartilage, the latter enclosed in the aryteno-epiglottidean fold, are also elastic. The thyroid, cricoid, and arytenoid cartilages are hyaline. All these are covered with the usual perichondrium.

A small nodule of elastic cartilage is enclosed in the front part of the true vocal cord. This is the cartilage of Luschka.

The mucous membrane lining the cavity of the larynx (Fig. 126) has the following structure :—

The epithelium covering the **free** surface is *stratified columnar ciliated*, *i.e.*, the most superficial layer is made up of conical cells with cilia on their surface; then between the extremities of these cells are wedged in spindle-shaped and inverted conical cells. Numerous goblet cells are found amongst the superficial cells. The two surfaces of the epiglottis and the true vocal cord are covered with *stratified pavement epithelium*.

Underneath the epithelium is a basement membrane separating the former from the mucous membrane proper.

284. The mucous membrane is delicate connective tissue with numerous lymph corpuscles. In the posterior surface of the epiglottis, in the false vocal cord,

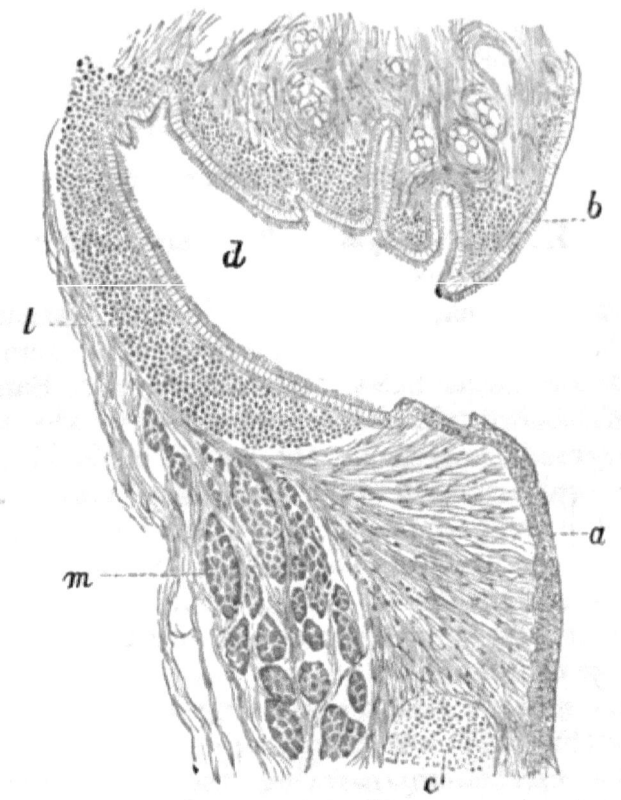

Fig. 126.—From a Longitudinal Section through the Ventricle of the Larynx of a Child.

a, True vocal cord; *b*, false vocal cord; *c*, a nodule of elastic cartilage (cartilage of Luschka); *d*, ventricle; *l*, lymphatic tissue; *m*, bundles of the thyro-arytenoid muscle in transverse section. (Atlas.)

and in the lower parts of the larynx, but especially in the ventricle, this infiltration amounts to diffuse adenoid tissue, and even to the localisation of this as lymph follicles. In both surfaces of the epiglottis, and in the true vocal cords, the mucosa extends into the stratified pavement epithelium in the shape of minute papillæ.

In the lower part of the larynx the mucous membrane contains bundles of elastic fibres connected into **networks, and** running in a longitudinal direction These elastic fibres are found chiefly in the superficial parts of the mucous membrane. In **the** true vocal cord the mucosa is entirely made up of elastic fibres extending in the direction of the vocal cord.

285. The deeper part of the mucous membrane is of loose texture, and corresponds to the sub-mucosa; in it are embedded numerous *mucous glands*, the ducts of which pass through the mucosa and open on the free surface. The alveoli of the glands are of the nature of mucous alveoli, *i.e.*, a considerable lumen lined with a layer **of mucous goblet** cells. **There are,** however, **also alveoli lined with** columnar **albuminous** cells, and such as have both **side by** side, **as in the** case of the sub-lingual gland **of the dog.** The ciliated epithelium of the surface in some places extends **also** for a short distance into the duct. The true vocal cords have no mucous glands.

The blood-vessels terminate as the capillary network in the superficial—*i.e.*, sub-epithelial—layer of the mucosa; where there are papillæ—*i.e.*, in the epiglottis and true vocal cord—these receive a loop of capillary blood-vessels. The lymphatics form superficial networks **of** fine vessels, and **deep** submucous networks of large **vessels. These are of enormous width and size in the** membrane **of the anterior surface of the** epiglottis. The finer nerves form superficial plexuses of **non-medullated** fibres. Here, according to Luschka and Boldyrew, there are end bulbs. Taste-buds have been found in the posterior surface of the epiglottis (Verson, Schofield, Davis), and also in the deeper parts of the larynx (Davis).

286. II. **The trachea.**—The trachea is very similar in structure to the lower part of the larynx, from which it differs merely in possessing the rings

218 ELEMENTS OF HISTOLOGY. [Chap. XXVIII.

of hyaline cartilage, and in possessing, in the posterior or membranous portion, *circular bundles of non-*

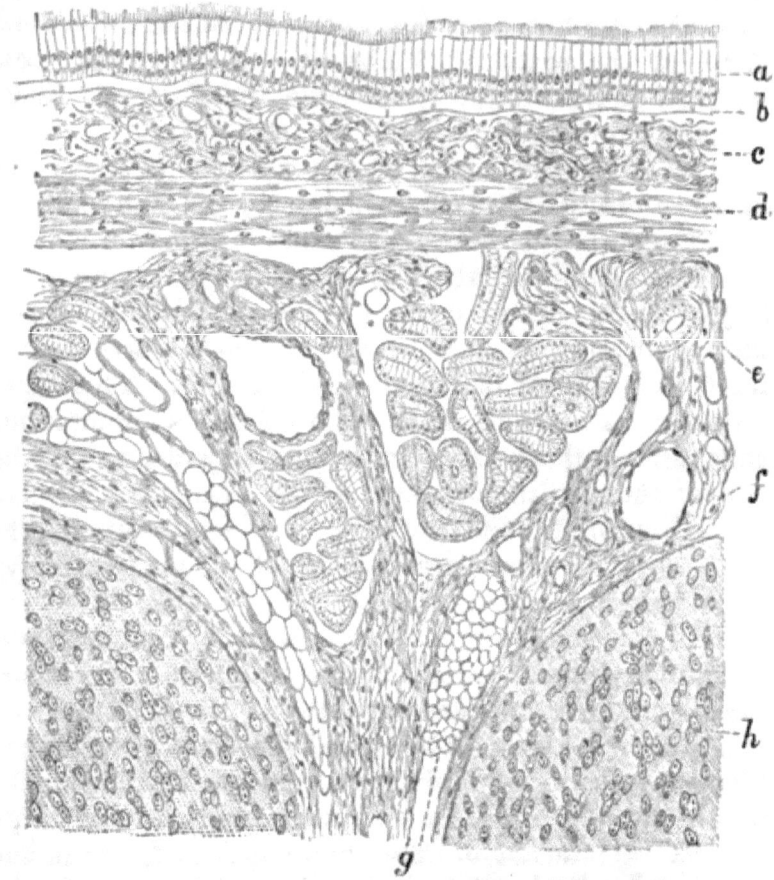

Fig. 127.—From a Longitudinal Section through the Trachea of a Child.

a, The stratified columnar ciliated epithelium of the internal free surface; b, the basement membrane; c, the mucosa; d, the networks of longitudinal elastic fibres; the oval nuclei between them indicate connective tissue corpuscles; e, the submucous tissue containing mucous glands; f, large blood-vessels; g, fat cells; h, hyaline cartilage of the tracheal rings. (Atlas.)

striped muscular tissue, extending, as it were, between the ends of the rings. Its component parts are (Fig. 127):—

(a) A stratified columnar ciliated epithelium.
(b) A basement membrane.

(c) A mucosa, with the terminal networks of capillary blood-vessels, and infiltrated with adenoid tissue.

(d) A layer of longitudinal elastic fibres.

(e) A loosely-textured submucous tissue, containing the large vessels and nerves and small mucous glands. Occasionally the gland or its duct is embedded in a lymph follicle.

287. III. **The bronchi and the lung.**—The bronchi branch within the lung dendritically into finer and finer tubes. The finest branches are the terminal bronchi. In the bronchi we find, instead of rings of hyaline cartilage, as in the trachea, larger and smaller oblong or irregularly-shaped plates of hyaline cartilage distributed more or less uniformly in the circumference of the wall. Towards the small microscopic bronchi, these cartilage plates gradually diminish in size and number. The epithelium, the basement membrane, the sub-epithelial mucosa, and the layer of longitudinal elastic fibres, remain the same as in the trachea. The submucous tissue contains small mucous glands.

288. Between the sub-epithelial mucosa and submucosa is a *continuous layer of circular non-striped muscular tissue.* In the smaller microscopic bronchi this layer is one of the most conspicuous. By the contraction of the circular muscular coat the mucosa is placed in longitudinal folds.

The state of contraction and distension of the small bronchi bears an important relation to the aspect of the epithelium, which appears as a single layer of columnar cells in the distended bronchiole, and stratified when the bronchiole is contracted.

The distribution of the blood-vessels is the same as in the trachea. Lymph follicles are met with in the submucous tissue of the bronchial wall in animals and man.

The lymphatic networks of the bronchial mucous membrane are very conspicuous. Those of the submucous tissue, *i.e.*, the peribronchial lymphatics, anastomose with those surrounding the pulmonary bloodvessels.

Pigment and small particles can be easily absorbed through the cement substance of the epithelium into the radicles of the superficial lymphatics, whence they pass readily into the (larger) peribronchial lymphatics.

In connection with the nerve branches in the bronchial wall are minute ganglia.

289. Each terminal bronchiole branches into several wider tubes called the *alveolar ducts*, or *infundibula*; each of these branches again into several similar ducts. All ducts, or infundibula, are closely beset in their whole extent with spherical, or, being pressed against one another, with polygonal, vesicles—the *air-cells* or *alveoli*—opening by a wide aperture into the alveolar duct or infundibulum, but not communicating with each other. The infundibula are much wider than the terminal bronchioles, and also wider than the alveoli.

290. All infundibula with their air-cells, belonging to one terminal bronchiole, represent a conical structure, the apex of which is formed by the terminal bronchus. Such a conical mass is a *lobule* of the lung, and the whole tissue of the lung is made up of such lobules closely aggregated, and arranged as *lobes*. The lobules are separated from one another by delicate fibrous connective tissue; this forms a continuity with the connective tissue accompanying the bronchial tubes and large vascular trunks, and with these is traceable to the hilum. On the other hand, the interlobular connective tissue of the superficial parts of the lung is continuous with the fibrous tissue of the surface called the pleura pulmonalis. This

membrane contains numerous elastic fibres, and on the free surface is covered with a layer of endothelium.

In some instances (guinea-pig) the pleura pulmonalis contains bundles of non-striped muscular tissue.

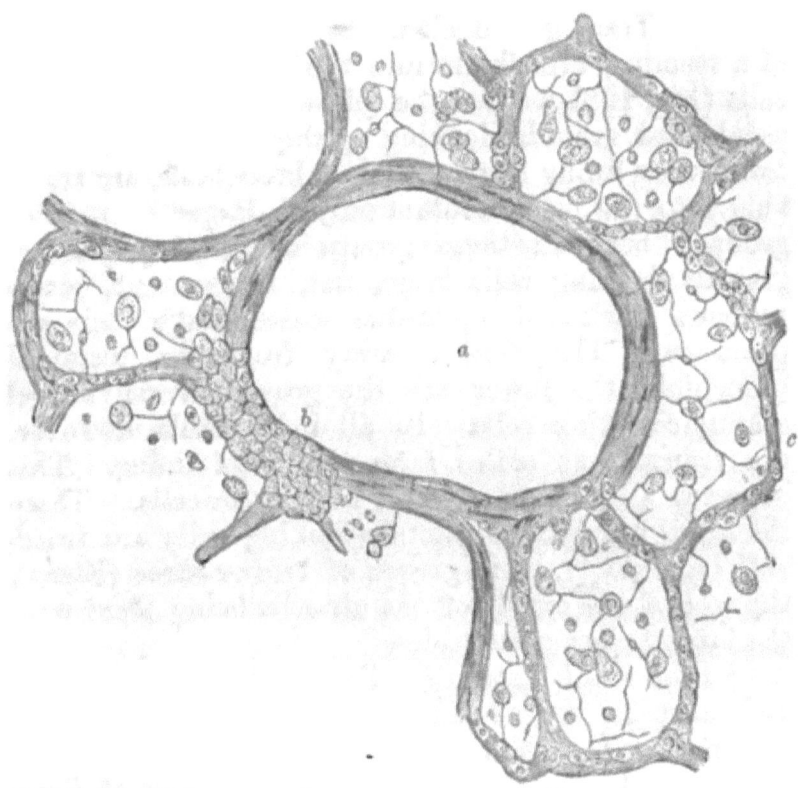

Fig. 128.—From a Section through the Lung of Cat, **stained with Nitrate of Silver.**

a, An infundibulum or alveolar duct in cross section ; *b*, groups of polyhedral cells lining one part of the infundibulum, the rest being lined with flattened transparent epithelial scales ; *c*, the alveoli lined with flattened epithelial scales ; here and there between them is seen a polyhedral granular epithelial cell. (Atlas.)

The lobes of the lung are separated from one another by large septa of connective tissue—the ligamenta pulmonis.

291. The **terminal bronchi** contain no cartilage

or mucous glands in their wall. This is made up of three coats: (a) a tiny epithelium—a single layer of small polyhedral granular-looking cells; (b) a circular muscular coat of non-striped muscular tissue; and (c) a fine adventitia of elastic fibres, arranged chiefly as longitudinal networks.

292. Tracing the elements constituting the wall of a terminal bronchiole into the infundibula and air-cells (Fig. 128), we find the following changes: (a) the polyhedral granular-looking epithelial cells forming a continuous lining in the terminal bronchiole, are traceable into the infundibulum only as larger or smaller groups; between these groups of small polyhedral granular-looking cells large, flat, transparent, homogeneous, nucleated, epithelial scales make their appearance. The farther away from the terminal bronchiole, the fewer are the groups of polyhedral granular-looking cells. In all infundibula, however, the transparent scales form the chief lining. This becomes still more marked in the air-cells. There the small polyhedral granular-looking cells are traceable only singly, or in groups of two or three (Elens), the rest of the cavity of the air-cells being lined with the large transparent scales.

In the fœtal state all cells lining the infundibula and air-cells are of the small polyhedral granular-looking variety (Küttner). With the expansion of the lungs during the first inspiration many of these cells change into the large transparent scales, in order to make up for the increment of surface. A lung expanded ad maximum shows much fewer or none of the small polyhedral cells; while a lung that is collapsed shows them in groups in the infundibula, and isolated or in twos or threes in the alveoli.

293. (b) The circular coat of non-striped muscular tissue of the terminal bronchiole is continued as a continuous circular coat—but slightly thinner—on the

alveolar ducts or infundibula, in their whole extent, but not beyond them, *i.e.*, not on the air-cells.

(c) The adventitia of elastic networks is continued on the infundibula, and thence on the air-cells, where they form an essential part of the wall of the alveoli, being its framework.

Amongst the network of elastic fibres forming the wall of the alveoli is a network of branched connective tissue cells, contained as usual in similarly shaped branched lacunæ, which are the radicles of the lymphatic vessels.

294. The blood-vessels and lymphatics.—The branches of the pulmonary artery and veins are contained within the connective tissue separating the lobes and lobules, whence they can be traced into their finer ramifications towards the infundibula and air-cells. Each of these latter is surrounded by a sort of basket-shaped **dense network of** capillary blood-vessels (Fig. 129). The capillary networks of adjacent alveoli are continuous

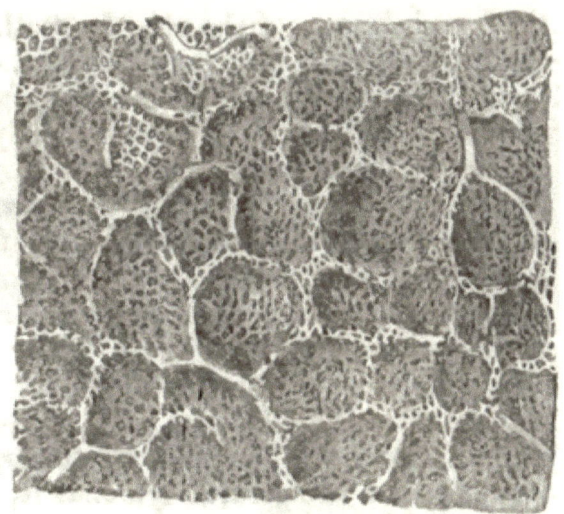

Fig. 129.—Networks of Capillary Blood-vessels surrounding the Alveoli of the Human Lung. (Kölliker.)

with one another, and stand in communication on the one hand with a branch of the pulmonary artery, and on the other with branches of the pulmonary vein. The branches of the bronchial artery belong to the bronchial walls, which are supplied by them with capillary networks.

The lacunæ and canaliculi in the wall of the alveoli, mentioned above, are the rootlets of lymphatic vessels, which accompany the pulmonary vessels, and form a network around them; these are the deep lymphatics, or the *perivascular lymphatics*. They are connected also with the networks of lymphatics surrounding the bronchi, *i.e.*, the *peribronchial lymphatics*. The rootlets of the superficial air-cells empty themselves into the *sub-pleural plexus of lymphatics*, a rich plexus of large lymphatics with valves. All these lymphatics lead by large trunks into the bronchial lymph glands.

295. Between the flattened transparent epithelial cells lining the alveoli are minute openings, *stomata* (Fig. 128), leading from the cavity of the air-cells into the lymph lacunæ of the alveolar wall. These stomata are more distinct during expansion, *i.e.*, inspiration, than in the collapsed state. Inspiration, by its expanding the lungs, and consequently also **the lymphatics**, favours greatly absorption. **Through these stomata,** and also through the interstitial **cement substance of** the lining epithelium, formed particles —**such as soot** particles **of** a smoky atmosphere, pigment artificially **inhaled,** cellular elements, such as mucus or pus corpuscles, that have been carried into the alveoli from the bronchi by natural inspiration, germ-particles, &c., find their way into the radicles of the lymphatics; thence into the perivascular and sub-pleural lymphatics, and finally into the bronchial glands.

CHAPTER XXIX.

THE SPLEEN.

296. THE **capsule** enveloping the spleen is a serous membrane—the peritoneum. It is a connective

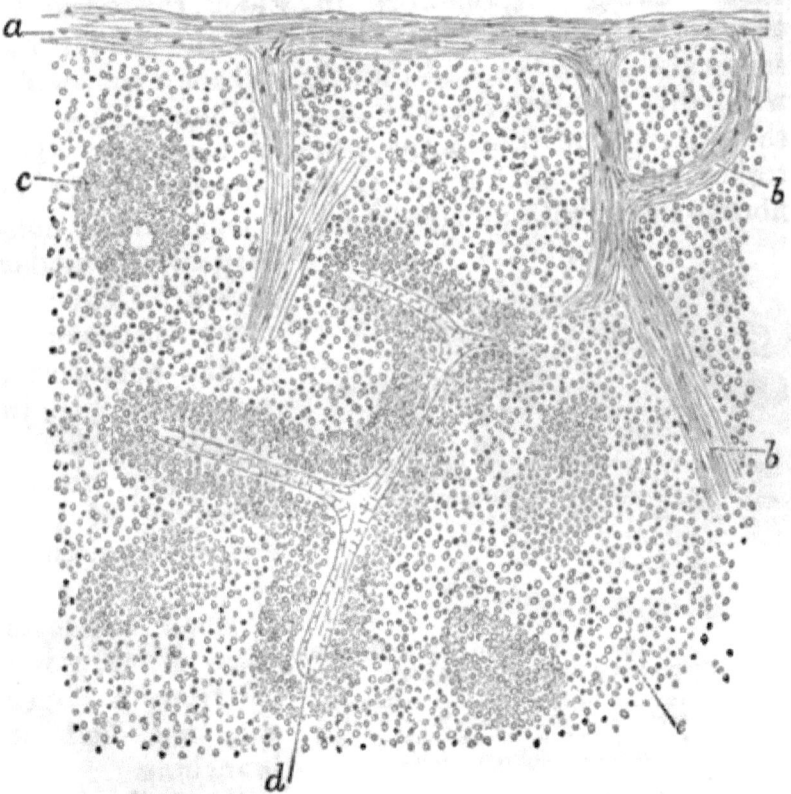

Fig. 130.—From a Vertical Section through the Spleen of Ape.

a, The capsule; *b*, the trabeculæ; *c*, Malpighian corpuscle; *d*, artery ensheathed in a Malpighian corpuscle; *e*, pulp tissue. (Atlas.)

tissue membrane with networks of elastic fibres, and covered on its free surface with an endothelium. The

P

deep part of the capsule contains *bundles of non-striped muscular tissue* forming plexuses. In man the bundles are relatively thin, but in some mammals—*e.g.*, dog, pig, horse—they are continuous masses arranged sometimes as a deep longitudinal and a superficial circular layer.

In connection with the capsule are the *trabeculæ* (Fig. 130). These are microscopical, thicker or thinner cylindrical bands branching and anastomosing, and thus making a framework in which the tissue of the spleen is contained. Towards the hilum the trabeculæ are larger, and they form there a continuity with the connective tissue of the hilum. They are the carriers of the large vascular branches. The trabeculæ in the human spleen consist chiefly of fibrous tissue with an admixture of longitudinal non-striped muscular tissue. This is more pronounced in the dog, horse, pig, guinea-pig, in which the trabeculæ are chiefly composed of non-striped muscular tissue. Following a small trabecula after it is given off from a larger one, we find it branching into still smaller ones, which ultimately lose themselves amongst the elements of that part of the spleen tissue called spleen pulp (Fig. 131).

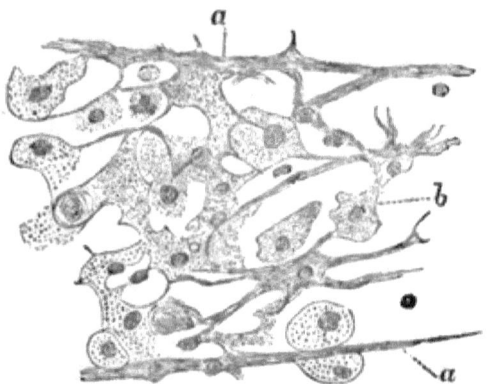

Fig. 131.—From a Section through the Pulp of the Spleen of the Pig.

a, Last outrunners of the muscular trabeculæ; *b*, the flattened cells forming the honeycombed matrix of the pulp; in the meshes of this matrix are contained lymphoid cells of various sizes. (Atlas.)

The meshes of the network of the trabeculæ are filled up with the parenchyma. This consists of two kinds of tissues : (*a*) the Malpighian corpuscles; and (*b*) the pulp tissue.

297. The **Malpighian corpuscles** are masses of adenoid tissue connected with the branches of the splenic artery. Following the chief arterial trunks as they pass in the big trabeculæ towards the interior of the spleen, they are seen to give off numerous smaller branches to the spleen parenchyma; these are ensheathed in masses of *adenoid tissue* which are either cylindrical or irregularly-shaped, and in some places form oval or spherical enlargements. These sheaths of adenoid tissue are traceable to the end of an arterial branch; and in the whole extent the adenoid tissue or Malpighian corpuscle is supplied by its artery with a network of capillary blood-vessels.

298. The rest of the spleen parenchyma is made up of the **pulp.** The matrix of this is a honeycombed, spongy network of fibres and septa, which are the processes and bodies of large, flattened, endotheloid cells, each with an oval nucleus. In some, especially young, animals, some of these cells are huge and multinucleated. The spaces of the honeycombed tissue are of different diameters, some not larger than a blood corpuscle, others large enough to hold several. All spaces form an intercommunicating system. The spaces contain nucleated lymph corpuscles, more or less connected with and derived from the cell plates of the matrix. But they do not fill the spaces, so that some room is left, large enough to allow blood corpuscles to pass.

The spaces of the honeycombed pulp matrix are in communication, on the one hand, with the ends of the capillary blood-vessels of the Malpighian corpuscles, and, on the other, they open into the venous radicles or *sinuses* (Fig. 132), which are oblong spaces lined with a layer of more or less polyhedral endothelial cells. These sinuses form networks, and lead into the large venous branches passing in the big trabeculæ to the hilum. The venous sinuses in man and ape possess a special adventitia formed of circular elastic fibrils.

Not all arterial branches are ensheathed in

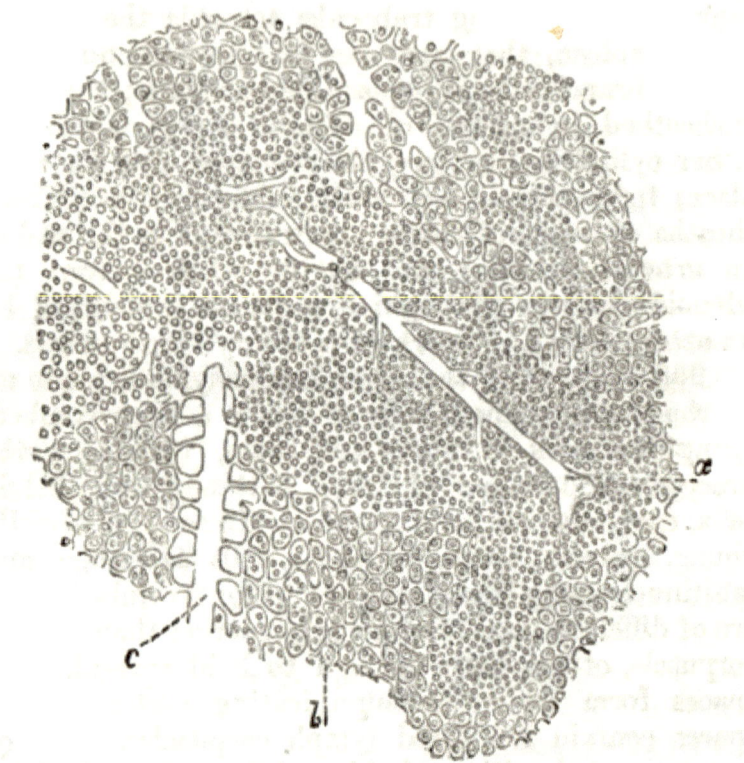

Fig. 132.—From a Section through the Spleen of a Guinea-pig; the Blood-vessels had been injected (not shown in the figure).

a, Artery of Malpighian corpuscle ; *b*, pulp ; between its cells are the minute blood-channels opening into *c*, the radicles of the veins. (Atlas.)

Malpighian corpuscles ; some few fine arterial branches open directly into the spaces of the pulp matrix, being invested in a peculiar reticular or concentrically arranged cellular tissue (not adenoid). These are the capillary sheaths of Schweigger Seidel.

299. The blood passes then from the arterial branches through the capillaries of the Malpighian

corpuscles, whence it travels into the labyrinth of minute spaces in the honeycombed pulp matrix; thence it passes into the venous sinuses, and finally into the venous trunks. The current of blood on its passage through the pulp tissue becomes, therefore, greatly retarded. Under these conditions **numerous** red blood-corpuscles appear to be taken up by the cells of the pulp, some of which contain several in their interior. In these corpuscles the blood-discs become gradually broken up, so that, finally, only granules and small clumps of blood-pigment are left in them. The presence of blood-pigment in the corpuscles of the pulp is explained in this **way**; and it is therefore **said** that the **pulp tissue is a** destroyer of red blood-corpuscles.

The pulp tissue is most probably the birthplace of colourless blood-corpuscles; and according to Bizzozero and Salvioli it is also the birthplace of red blood-corpuscles.

The *lymphatics* form plexuses in the capsule (Tomsa, Kyber). These are continuous with the plexus of lymphatics of the trabeculæ; and these again with the plexus of lymphatics in the adventitia of the arterial trunks.

Non-medullated nerve-fibres have been traced along the arterial branches.

CHAPTER XXX.

THE KIDNEY, URETER, AND BLADDER.

300. A. **The framework.**

The kidney possesses a thin investing capsule composed of fibrous tissue, more or less of a lamellar arrangement. Bundles of fibrous tissue pass with blood-vessels between the deeper part of the capsule and the parenchyma of the periphery. According to Eberth, a plexus of non-striped muscle cells is situated underneath the capsule.

The ureter entering the hilum enlarges into the pelvis of the kidney, and with its minor recesses or prolongations forms the calices. Both the pelvis and the calices are limited by a wall which is a direct continuation of the ureter. The internal free surface is lined with stratified transitional epithelium. Underneath the epithelium is a fibrous connective tissue membrane (the mucosa), containing the networks of capillary blood-vessels and fine nerve-fibres. Outside the mucosa and insensibly passing into it is the loose-textured submucosa, with groups of fat cells. There are present in the sub-mucosa bundles of non-striped muscular tissue, continued from the ureter, in the shape of longitudinal and circular bundles.

In the pelvis of the kidney of the horse small glands (simple or branched tubes), lined with a single layer of columnar epithelial cells, have been observed by Paladino, Sertoli, and Egli. The last-named mentions also that in the pelvis of the human kidney there are gland-tubes similar in structure to sebaceous follicles.

301. The large vascular trunks enter, or pass from the tissue of the calices into the parenchyma of the kidney between the cortex and medulla, and they are accompanied by bundles of fibrous connective tissue and here and there a few longitudinal bundles of non-striped muscular tissue.

In the parenchyma there is a very scanty fibrous connective tissue, chiefly around the Malpighian corpuscles and around the arterial vessels, especially in the young kidney. In the papillæ there is relatively a great amount of fibrous tissue. On the surface of the papillæ (facing the calices) there is a continuous layer of fibrous tissue, and this on its free surface is covered with stratified transitional epithelium.

The parenchyma of the kidney consists entirely of the urinary tubules and the intertubular blood-vessels, and there is an interstitial or intertubular connective tissue framework in the shape of honeycombed hyaline membranes with flattened nucleated branched or spindle-shaped cells. The meshes of the honeycomb are the spaces for the urinary tubules and blood-vessels.

302. B. **The parenchyma.**—I. The *urinary tubules* (Fig. 133).—In a transverse or longitudinal section through the kidney we notice the *cortex*, the *boundary layer* of Ludwig and the *papillary portions*, the last terminating in the conical *papillæ* in the cavity of the calices.

The boundary layer and the papillary portion form the medulla. A papilla with the papillary portion and boundary layer, continuous with it, constitute a *Malpighian pyramid*. The relative proportion of the thickness of the three parts is about 3·5 cortex, 2·5 boundary layer, and 4 papillary portion.

303. The cortex consists of vast numbers of convoluted tubules with their cæcal origin, the Malpighian corpuscle; this is the *labyrinth* separated into vertical divisions of equal breadth by regularly disposed vertical straight striæ originating a short distance from the outer capsule, and radiating towards the boundary layer which they enter. Each of these striæ is a bundle of straight tubules, and represents a *medullary ray*. The boundary layer shows a uniform longitudinal striation, in which opaque and transparent striæ alternate with one another. The opaque striæ are continuations of the medullary rays, the transparent striæ are bundles of blood-vessels.

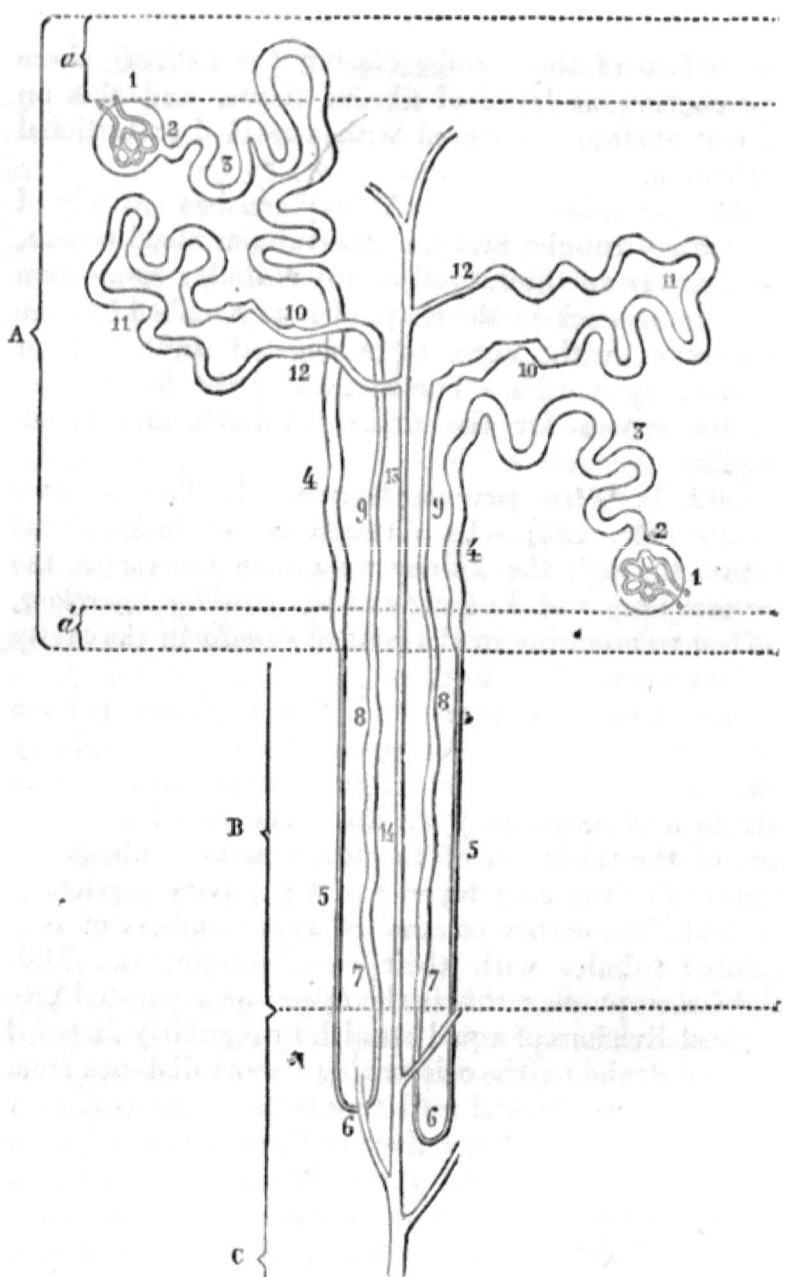

Fig. 133.—Diagram showing the course of the Uriniferous Tubules in the different parts of the Cortex and Medulla.

(For description of Fig. 133, see foot of next page.)

The papillary portion is uniformly and longitudinally striated.

Tracing a medullary ray from the boundary layer into the cortex, it is seen that its breadth gradually diminishes, and it altogether ceases at a short distance from the outer capsule. A medullary ray is, consequently, of a conical shape, its apex being situated at the periphery of the cortex, its base in the boundary layer. Such a pyramid is called a *pyramid of Ferrein*.

304. All urinary tubules commence as convoluted tubules in the part of the cortex named the labyrinth, but not in the medullary rays, with a cæcal enlargement called a *Malpighian corpuscle*, and terminate—having previously joined with many other tubules into larger and larger ducts—at one of the many minute openings or mouths at the apex of a papilla. On their way the tubes several times alter their size and nature.

From its start to its end there is a continuous fine *membrana propria* forming the boundary wall of the urinary tubule, and this membrana propria is lined with a *single layer of epithelial cells* differing in size, shape, and structure from place to place; in the centre of the tubule is a *lumen*, differing in size according to the size of the tubule.

305. (1) Each *Malpighian corpuscle* (Fig. 134) is composed of the capsule—the *capsule of Bowman*—and the *glomerulus*, or Malpighian tuft of capillary blood-vessels.

A, Cortex limited on its free surface by the capsule; a, subcapsular layer not containing Malpighian corpuscles; a', inner stratum of cortex without Malpighian corpuscles; B, boundary layer; C, papillary part next the boundary layer; 1, Bowman's capsule; 2, neck of capsule; 3, proximal convoluted tube; 4, spiral part; 5, descending limb of Henle's loop-tube; 6, the loop itself; 7, 8, and 9, the ascending limb of Henle's loop-tube; 10, the irregular tubule; 11, the distal convoluted tubule; 12, the first part of the collecting tube; 13 and 14, larger collecting tube; in the papilla itself, not represented here, the collecting tube joins others, and forms the duct. (Atlas.)

The *capsule* of Bowman is a hyaline membrana propria, supported, as mentioned above, by a small amount of connective tissue. On its inner surface there is a continuous layer of nucleated epithelial

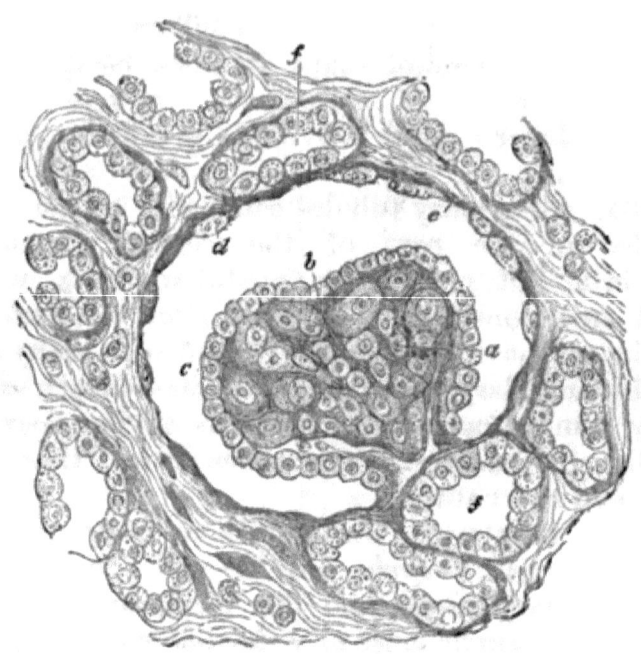

Fig. 134.—From a Section through the Cortical Substance of the Kidney of a Human Fœtus, showing a Malpighian corpuscle.

a, Glomerulus; *b*, tissue of the glomerulus; *c*, epithelium covering the glomerulus; *d*, flattened epithelium lining Bowman's capsule; *e*, the capsule itself; *f*, uriniferous tubules in cross section. (Handbook.)

cells, in the young state of polyhedral shape, in the adult state squamous.

The *glomerulus* is a network of convoluted capillary blood-vessels, separated from one another by scanty connective tissue, chiefly in the shape of a few connective tissue corpuscles. The capillaries are grouped together in two to five lobules. The whole surface of the glomerulus is lined with a delicate membrana propria, and a continuous layer of nucleated epithelial cells, polyhedral, or even columnar in the young, squamous in the adult state. The membrana propria and epithelium dip in, of course, between the lobules of the glomerulus, and represent in reality the visceral layer of the capsule of the Malpighian corpuscle, the capsule of Bowman being the parietal layer. The glomerulus is connected at one pole with an *afferent* and *efferent* arterial vessel, the former being the larger.

Between Bowman's capsule and the glomerulus there is a space, the size of which differs according to the state of secretion, being chiefly dependent on the amount of fluid present.

The Malpighian corpuscles are distributed in the labyrinth of the cortex only, with the exception of a thin peripheral layer near the outer capsule, and a still thinner layer near the boundary layer. The Malpighian corpuscles near the boundary layer are the largest, those near the periphery the smallest; in the human kidney their mean diameter is about $\frac{1}{120}$ of an inch.

306. On the side opposite to that where the afferent and efferent arterioles join the glomerulus, the capsule of Bowman passes through a narrow *neck* into the cylindrical urinary tubule in such a way, that the membrana propria and epithelium of the capsule are continued as the membrana propria and lining epithelium of the tubule respectively, and the space between the capsule of Bowman and the glomerulus becomes the cavity or lumen of the urinary tubule.

307. (2) After it has passed the neck, the urinary tubule becomes convoluted; this is the *proximal convoluted tubule* (Fig. 135). It is of considerable length and is situated in the labyrinth. It has a distinct lumen, and its epithelium is a single layer of polyhedral or short, columnar, angular, or club-shaped cells, each

with a spherical nucleus. These cells commence generally at the neck, but in some animals—*e.g.*, in the

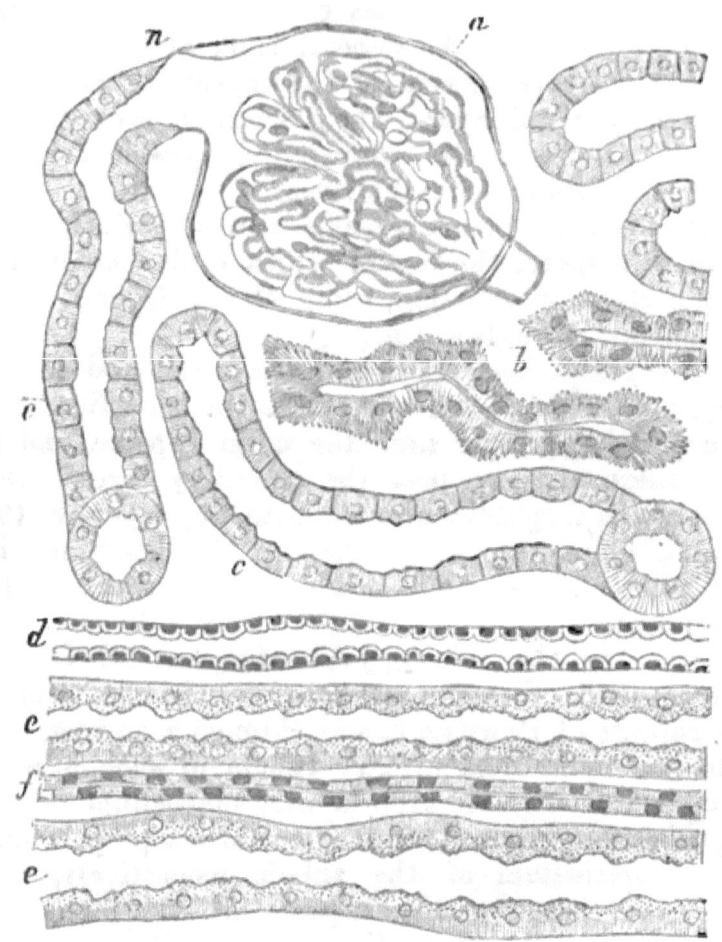

Fig. 135.—From a Vertical Section through the **Kidney** of Dog, showing part of the Labyrinth and the adjoining **Medullary** Ray.

a, The capsule of Bowman; the capillaries of the glomerulus are arranged in lobules; *n*, neck of capsule; *b*, irregular tubule; *c*, proximal convoluted tubules; *d*, a collecting tube; *e*, part of the spiral tubule; *f*, portion of the ascending limb of Henle's loop-tube; *d*, *e*, *f* form the medullary ray. (Atlas.)

mouse—they already have begun in the Malpighian corpuscle. The outer part of the cell protoplasm—*i.e.*, next the membrana propria—is distinctly striated, owing to the presence of rod-shaped fibrils (Heidenhain) vertically arranged. The inner part of the cell substance—*i.e.*, between the nucleus and the inner free margin—appears granular. Epithelial cells the protoplasm of which possesses the above rod-shaped fibrils, will in the following paragraphs be spoken of as fibrillated cells.

The proximal convoluted tube appears sometimes thicker than at other times; in the first case, its lumen is smaller, but its lining epithelial cells are distinctly more columnar. This state is probably connected with the state of secretion.

308. (3) The convoluted tube passes into the *spiral tubule* (Schachowa). This differs from the former in being situated not in the labyrinth, but in a medullary ray, in which it forms one conspicuous element, and in not being convoluted, but more or less straight, slightly wavy, and spiral. Its thickness and lumen are the same as in the former; its epithelium is a single layer of polyhedral cells, with distinct indication of fibrillation.

309. (4) Precisely at the line where the cortex joins the boundary layer, the spiral tube becomes suddenly greatly reduced in thickness; it becomes at the same time very transparent; its lumen is distinct; its membrana propria is now lined with a single layer of scales, each with an oval flattened nucleus. This altered tubule is the *descending loop-tube* of Henle, and it pursues its course in the boundary layer as a straight tubule, in the continuation of the medullary ray.

In aspect and size this part of the urinary tubule resembles a capillary blood-vessel, but differs from it inasmuch as, in addition to the lining layer of flattened epithelial cells, it possesses a membrana propria.

310. (5) The so formed descending Henle's loop-tube passes the line between the boundary layer and papillary portion, and having entered this latter pursues its course for a short distance, when it sharply bends backwards as the *loop* of Henle's tube; it now runs back towards the boundary layer, and precisely at the point of entering this becomes suddenly enlarged. Up to this point the structure and size of the loop are exactly the same as those of the descending limb.

311. (6) Having entered the boundary layer it pursues its course in this latter to the cortex in a more or less straight direction within the medullary ray as the *ascending loop-tube*. Besides being bigger than the descending limb and the loop, its lumen is comparatively smaller, and its lining epithelium is a layer of polyhedral, distinctly fibrillated epithelial cells. The tube is not quite of the same thickness all along the boundary layer, but is broader in the inner than in the outer half; besides, the tube is not quite straight, but slightly wavy or even spiral.

(7) Having reached the cortex, it enters this as the *cortical part of the ascending loop-tube*, forming one of the tubes of a medullary ray; it is at the same time narrower than in the boundary layer, and is more or less straight or wavy. Its lumen is very minute, its lining cells are flat polyhedral with a small flattened nucleus, and there is an indication of fibrillation (Fig. 135).

(8) Sooner or later on its way in the cortex in a medullary ray it leaves this latter to enter the labyrinth, where it winds between the convoluted tubes as an angular *irregular tubule* (Fig. 135). Its shape is very irregular, its size alters from place to place, its lumen is very minute, its epithelium a layer of polyhedral, pyramidal, or short columnar cells

—according to the thickness of the tube; each cell possesses a flattened oval nucleus next to the lumen, and a very coarsely and conspicuously fibrillated protoplasm.

312. (9) This irregular tubule passes into the *distal convoluted tubule* or intercalated tubule of Schweigger Seidel. This forms one of the convoluted tubes of the labyrinth, and in size, aspect, and structure, is identical with the proximal convoluted tubule.

(10) The distal convoluted tube passes into a short, thin, more or less *curved* or wavy *collecting tubule*, lined with a layer of transparent, flattened, polyhedral cells; this is still contained in the labyrinth.

(11) This leads into a somewhat larger *straight collecting tube*, lined with a layer of transparent polyhedral cells and with distinct lumen. This tube forms part of a medullary ray, and on its way to the boundary layer takes up from the labyrinth numerous curved collecting tubules.

(12) It then passes unaltered as a *straight collecting tube* through the boundary layer into the papillary portion.

313. (13) In this part these tubes join under acute angles, thereby gradually enlarging. They run in a straight direction towards the apex of the papilla, and the nearer to this the fewer and the bigger they become. These are the *ducts* or tubes of Bellini. They finally open on the apex into a calix. The lumen and the size of the lining epithelial cells—viz., whether more or less columnar—are in direct relation to the size of the collecting tube. The substance of the epithelial cells is a transparent protoplasm, and the nucleus is more or less oval.

314. In many places nucleated cells, spindle-shaped or branched, can be traced from the membrana

propria of the tubule between the lining epithelium; and, in some cases, even a delicate nucleated membrane can be seen lining the surface of the epithelium next the lumen. In the frog, the epithelium lining, the Malpighian corpuscles, and the exceedingly long neck of the urinary tubule, are possessed of long filamentous cilia, rapidly moving during life. In the neck of some of the urinary tubules in mammals there is also an indication of cilia to be noticed.

Heidenhain has shown that indigo-sulphate of sodium, injected into the circulating blood of the dog and rabbit, is excreted through certain parts of the urinary tubules only—viz., those which are lined with "fibrillated" epithelium. He maintains that this excretion is effected through the cell substance; but, in the case of carmine being used as pigment, I have not found the excretion to take place through the substance of the epithelial cells, but through the homogeneous interstitial or cement substance *between* the epithelial cells.

315. II. *The blood-vessels* (Fig. 136).

The large branches of the renal artery and vein are situated in the submucous tissue of the pelvis, and they enter, or pass out respectively from, the part of the parenchyma corresponding to the junction of the cortex and boundary layer, where they follow a more or less horizontal course, and give off, or take up respectively, smaller branches to or from the cortex and medulla.

(1) In the cortex the arterial trunks give off to the cortex small branches, which *singly enter the labyrinth* in a direction vertical to the surface of the kidney. These are the *interlobular arteries*. Each of these, on its way towards the external capsule of the kidney, gives off, on all sides of its circumference, shorter or longer lateral branches; these are the *afferent arterioles* for the Malpighian corpuscles, each

one entering a Malpighian corpuscle and breaking up into the capillaries of the glomerulus.

On its way towards the external capsule, the interlobular arteries become greatly reduced, and finally enter the capillary network of the most peripheral part of the cortex; but some of these arterioles may be also traced into the outer capsule, where they become connected with the capillary networks of this latter. The efferent vessel of a Malpighian glomerulus at once breaks up into a dense network of capillary blood-vessels, which entwine in all possible directions the urinary tubules of the labyrinth. This network is continued with that of capillaries of the medullary rays, the meshes being there elongated, and the capillary blood-vessels, for obvious reasons, more of a straight

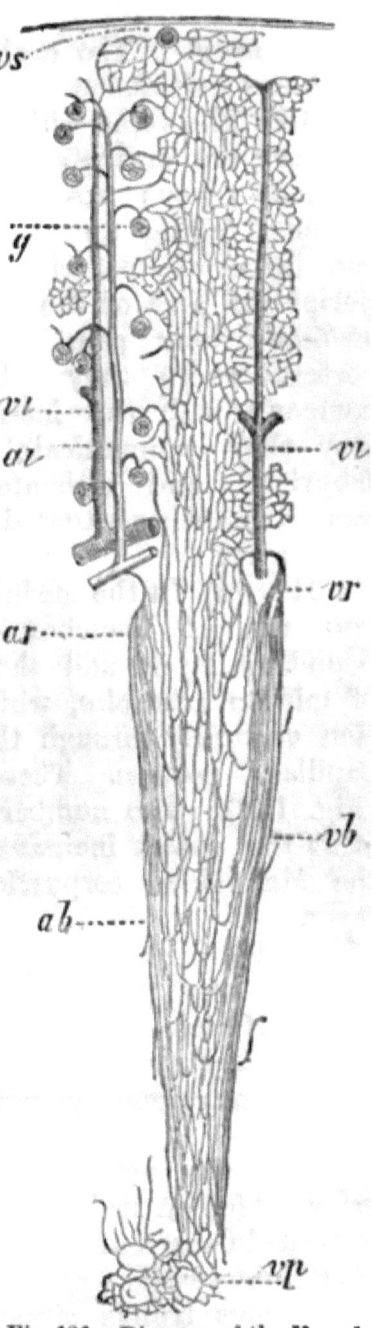

Fig. 136.—Diagram of the Vessels of the Kidney.

ai, Interlobular artery; *vi*, interlobular vein; *g*, glomerulus of Malpighian corpuscle; *vs*, vena stellata; *ar*, arteriæ rectæ; *vr*, venæ rectæ; *ab*, bundle of arteriæ rectæ; *vb*, bundle of venæ rectæ; *vp*, network of vessels around the mouth of the ducts at the apex of the papillæ. (Ludwig, in Stricker's Manual.)

Q

arrangement. The capillaries of the whole cortex form one continuous network.

316. The veins which take up the blood from this network are arranged in this manner :—There are formed venous vessels underneath the external capsule, taking up like rays on all sides, minute radicles connected with the capillaries of the most peripheral part of the cortex. These are the *venæ stellatæ;* they pass into the labyrinth of the cortex, where they follow a vertical course in company with the interlobular arteries. On the way they communicate with the capillaries of the labyrinth, and ultimately open into the large venous branches situated between cortex and boundary layer.

317. (2) In the medulla. From the large arterial trunks short branches come off, which enter the boundary layer, and there split up into a bundle of minute arterioles, which pass in a straight direction vertically through the boundary layer into the papillary portion. These are the *arteriæ rectæ* (Fig. 136). The number of vessels of each bundle is at the outset increased by the efferent vessel of the Malpighian corpuscles nearest to the boundary layer.

On their way through the boundary layer, and through the papillary portion of the medulla, these arterioles give off the capillary network for the urinary tubules of these parts, the network, for obvious reasons, possessing an elongated arrangement.

From this network originate everywhere minute veins, which on their way towards the cortical margin increase in size and number; they form also bundles of straight vessels—*venæ rectæ*—and ultimately enter the venous trunks situated between the boundary layer and cortex.

Chap. XXX.] KIDNEY, URETER, AND BLADDER.

The bundles of the arteriæ rectæ and [venæ] rectæ form severally, in the boundary lay[er, the] transparent striæ mentioned on a previous page as alternating with the opaque striæ, these latter being bundles of urinary tubules.

At the apex of each papilla there is a network of capillaries around the mouth of each duct.

318. The outer capsule of the kidney contains a network of capillary blood-vessels; the arterial branches leading into them are derived from two sources: (*a*) from the outrunners of the interlobular arteries of the cortex, and (*b*) from extrarenal arteries. The veins lead (*a*) into the venæ stellatæ, and (*b*) the extrarenal veins.

The *lymphatic* vessels form a plexus in the capsule of the kidney. They are connected with lymph spaces between the urinary tubes of the cortex. The large blood-vessels are surrounded by a plexus of lymphatics, which take up lymph spaces between the urinary tubules, both in the cortex and the boundary layer.

319. The **ureter** is lined with stratified transitional epithelium. Underneath this is the mucosa, a connective tissue membrane with capillary blood-vessels. The submucosa is a loose connective tissue Then follows a muscular coat composed of non-striped muscular tissue, arranged as an inner and outer longitudinal and a middle circular coat. Then follows an outer limiting thin fibrous coat or adventitia. In this last have been observed minute ganglia in connection with the nerve branches.

320. The **bladder** is similar in structure, but the mucous membrane and muscular coat are very much thicker. In the latter, which consists of non-striped fibres, are distinguished an inner circular, a middle oblique, and an outer longitudinal stratum. The last is best developed in the fundus.

Numerous sympathetic ganglia, of various sizes, are found in connection with the **nerve** branches underneath the adventitia (peritoneal covering), and in the muscular coat (F. Darwin). The epithelium lining the bladder is stratified transitional, and it greatly varies in the shape of its cells and their stratification, according to the state of expansion of the bladder.

CHAPTER XXXI.

THE MALE GENITAL ORGANS.

321. (1) THE **testis** of man and mammals is enveloped in a capsule of white fibrous tissue, the *tunica adnata*. This is the visceral layer of the tunica vaginalis. Like the parietal layer, it is a serous membrane, and is therefore covered with endothelium. Minute villi are occasionally seen projecting from this membrane into the cavity of the tunica vaginalis. These villi are generally covered with germinating endothelium (see par. 33). Inside the tunica adnata, and firmly attached to it, is the *tunica albuginea*, a fibrous connective tissue membrane of lamellar structure. Towards the posterior margin of the human testis its thickness increases, and forms there a special accumulation—in cross section more or less conical, with posterior basis—the mediastinum testis, or corpus Highmori.

Between the tunica adnata and tunica albuginea is a rich plexus of lymphatics, which, on the one hand, takes up the lymphatics of the interior, and on the other leads into the efferent vessels that accompany the vas deferens.

The testis of the dog, cat, bull, pig, rabbit, &c., have a central corpus Highmori; that of the mole, hedgehog, and bat a peripheral one; while that of the rat and mouse have none (Messing).

322. The **framework**.— From the anterior margin of the corpus Highmori spring numerous septa of connective tissue, which, passing in a radiating direction towards the albuginea, with which they form a continuity, subdivide the testis into a large number of long, conical compartments, or lobules, the basis of which is situated at the tunica albuginea, the apex at the corpus Highmori. Kölliker mentions that non-striped muscular tissue occurs in these septa.

From these septa thin connective tissue lamellæ pass into the compartments, and they form the supporting tissue for the blood-vessels and also represent the interstitial connective tissue between the seminal tubules.

This intertubular or interstitial tissue is distinctly lamellated, the lamellæ being of different thicknesses, and consisting of thin bundles of fibrous connective tissue—arranged more or less as fenestrated membranes—and endotheloid connective plates on their surface. Between the lamellæ are left spaces, and these form, through the fenestræ or holes of the lamellæ, an intercommunicating system of lymph spaces—being, in fact, the rootlets of the lymphatics (Ludwig and Tomsa).

Within the lamellæ are found peculiar cells, which are much larger than lymph-cells, and which, in some instances (*e.g.*, guinea-pig), include pigment granules. They contain a spherical nucleus. In man, in dog, cat, sheep, especially boar, these cells form large, continuous groups—plates and cylinders—and the cells are polyhedral, and exactly similar to epithelial cells. They are separated from one another within the group by a thin interstitial cement-substance. Their resemblance with epithelium is complete. They are remnants of the epithelial masses of the Wolffian body of the fœtus.

323. The **seminal tubules** (Fig. 137). — Within each compartment, above mentioned, lie

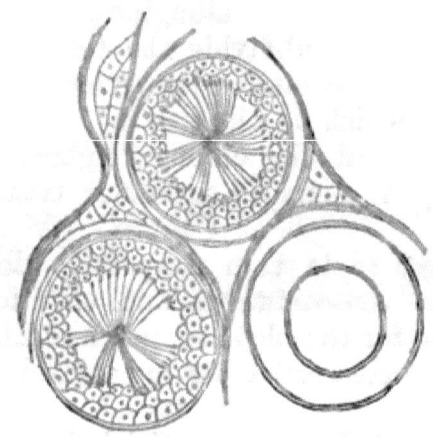

Fig. 137.—From a Section through the Testis of Dog.

Showing three seminal tubules in cross section. In two of these the lining epithelium—seminal cells—is shown, and bundles of spermatozoa projecting into the lumen of the tubules. Between the tubules is connective tissue containing groups of polyhedral epithelial-like cells.

numerous *seminal tubules*, twisted and convoluted in many ways, and extending from the periphery to near the corpus Highmori. The tubules, as a rule, are rarely branched; but in the young state, and especially towards the periphery, branching is not uncommon.

Each seminal tubule consists of a membrana propria, a lining epithelium, and a lumen. The membrana propria is a hyaline membrane, with oval nuclei at regular intervals. In man it is thick and lamellated, several such nucleated membranes being superimposed over one another. The lumen is in all tubes distinct and relatively large. The lining epithelium, or the *seminal cells*, differ in the adult in different tubules, and even in different parts of the same tubule, being dependent on the state of secretion.

324. Before puberty all tubules are uniform in this respect, being lined with two or three layers of polyhedral epithelial cells, each with a spherical nucleus. After puberty, however, the following different types can be distinguished:—

(a) Tubules or parts of tubules similar to those of the young state—viz., several layers of polyhedral epithelial cells lining the membrana propria. These are considered as (*a*) the outer and (*b*) the inner seminal cells. The former are next to the membrana propria; they are polyhedral in shape, transparent, and the nucleus of some of them is in the process of karyokinesis or indirect division (see par. 8); others include an oval transparent nucleus. The inner seminal cells generally form two or three layers, and are more loosely connected with one another than the outer seminal cells, and therefore possess a more rounded appearance. Between these a nucleated reticulum of fine fibres is sometimes noticed, the germ reticulum of von Ebner. But this is merely a supporting tissue, and has nothing to do with the germination of the cells or the spermatozoa (Merkel). The inner seminal cells show very abundantly the process of indirect division or karyokinesis, almost all being seen in one or other phase of it.

325. In consequence of this, numerous small spherical daughter-cells are formed; these lie nearest the lumen, and are very loosely connected with one another. It is these which are transformed into spermatozoa, and hence are appropriately called *spermatoblasts* (Fig. 137).

Amongst the seminal cells, especially of cat and

dog, are found occasionally, but not very commonly, large multinuclear cells, the nuclei of which are also in one or the other stage of karyokinesis.

(b) The innermost cells, *i.e.*, the spermatoblasts, become pear-shaped, the nucleus being situated at the

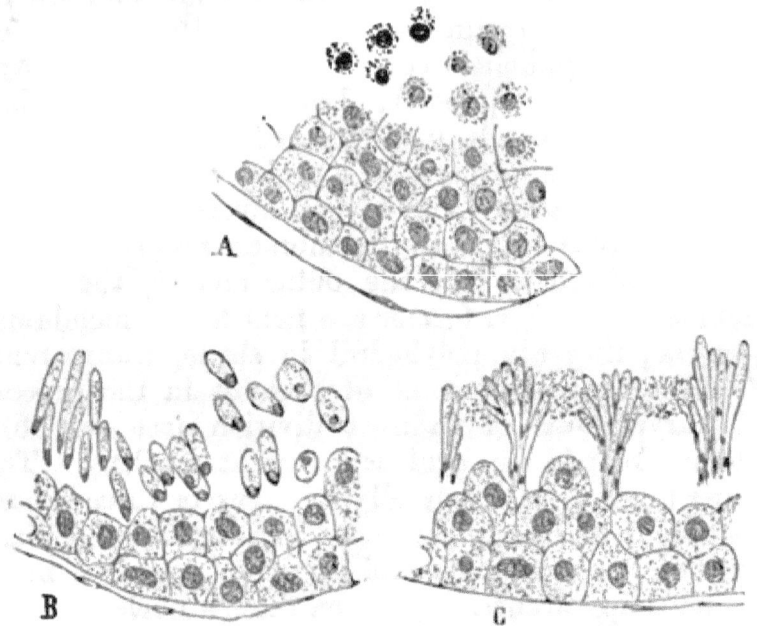

Fig. 138.—From a Section through the Testis of Dog, showing portions of three Seminal Tubules.

A, Seminal epithelial cells and numerous small cells loosely arranged; B, the small cells or spermatoblasts converted into spermatozoa; C, groups of these in a further stage of development. (Atlas.)

thinner extremity, becoming at the same time flattened and homogeneous (Fig. 138). The elongation of the spermatoblasts gradually proceeds, and in consequence of this we find numerous elongated, club-shaped spermatoblasts, each with a flattened nucleus at the thin end. These are the young spermatozoa, the nucleated extremity being the head.

(c) At the same time these young spermatozoa become grouped together by an interstitial granular substance, in peculiar fan-shaped groups; in these groups the head,—*i.e.*, the thin end containing the flattened homogeneous nucleus,—is directed towards the inner seminal cells, while the opposite extremity is directed into the lumen of the tube. Meanwhile the inner seminal cells continue to divide, and thus the groups of **young spermatozoa** get more and more buried, as it were, between them.

326. The original cell-body of the spermatoblasts goes on elongating until its protoplasm is almost, but not quite, used to form the rod-shaped *middle piece* (Schweigger Seidel) of the spermatozoa; from the distal end of this, a thin long hair-like filament, called the *tail*, grows out. Where this joins the end of the middle piece, there is, even for some time afterwards, a last remnant of the granular cell-body of the original spermatoblast to be noticed.

When the granular interstitial substance holding together the spermatozoa of a group has become disintegrated, the spermatozoa are isolated. While this development of the spermatozoa goes on, the inner seminal cells continue to produce spermatoblasts, which in their turn are converted into spermatozoa.

327. **Spermatozoa** (Fig. 139).—Fully formed spermatozoa of man and mammals consist of a homogeneous flattened and slightly convex-concave *head* (the nucleus of the original spermatoblast), a rod-shaped *middle piece* (derived directly from the cell body of the spermatoblast), and a long hair-like *tail*. While living, the spermatozoa show very rapid oscillatory and propelling movement, the tail acting as a flagellum or cilium; its movements are spiral.

In the newt there is a fine spiral thread attached to the end of the long, curved, spike-like head, and by a hyaline membrane it is fixed to the middle piece; it extends beyond this as the long thin *tail*. Also in the mammalian and human spermatozoa, a similar spiral thread, closely attached to the middle piece, and terminating as the tail, has been observed (H. Gibbes).

328. The seminal tubules of each compartment or lobule empty themselves into a short, more or less straight, tubule—the *vas rectum*. This is narrower

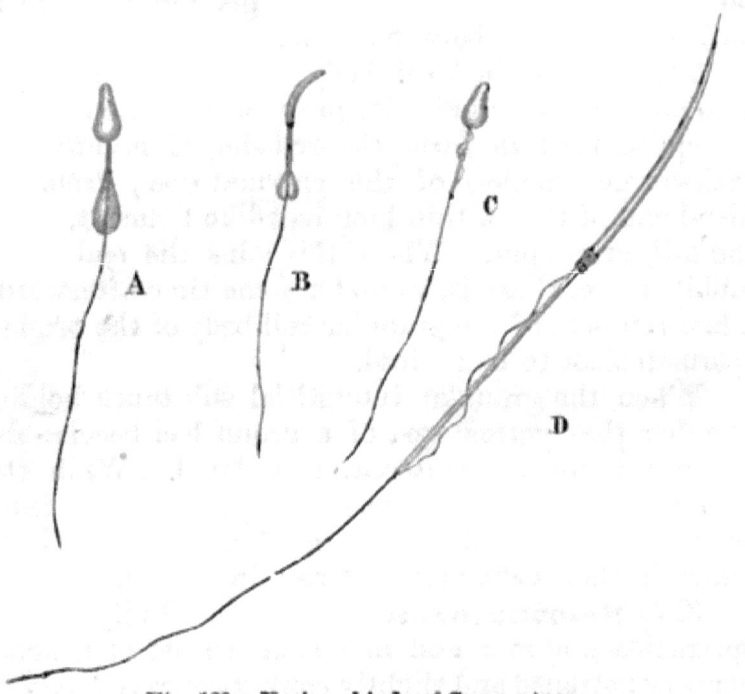

Fig. 139.—Various kinds of Spermatozoa.

A, Spermatozoon of guinea-pig not yet completely ripe; B, the same seen sideways, the head of the spermatozoon is flattened from side to side; C, a spermatozoon of the horse; D, a spermatozoon of the newt.

than the seminal tubule, and is lined with a single layer of polyhedral or short columnar epithelial cells. The vasa recta form, in the corpus Highmori, a dense network of tubular channels, which are irregular in diameter, being at one place narrow clefts, at another wide tubes, but never so wide as the seminal tubules; this network of channels is the *rete testis*.

329. (2) The **epididymis.**—From the rete testis we pass into the *vasa efferentia*, each being a tube

wider than those of the rete testis, and each leading into a *conical* network of coiled tubes. These are the *coni vasculosi*. The sum total of all the coni vasculosi forms the globus major or head of the epididymis.

330. The vasa efferentia and the tubes of the coni vasculosi are about the size of the seminal tubules, but, unlike them, are lined with a layer of beautiful columnar epithelial cells, with a bundle of cilia (Fig. 140). Outside these is generally a layer, more or less continuous, of small polyhedral cells. The substance of the columnar cells is distinctly longitudinally fibrillated. The membrana propria is thickened by the presence of a circular layer of non-striped muscular fibres. The rest, *i.e.*, the globus minor, or tail of the epididymis, is made up of a continuation of the tubes of the globus major, the tubes diminishing gradually in number by fusion, and at the same time thereby becoming larger. The columnar epithelial cells, facing the lumen of the tubes of the globus minor, are possessed of cilia of unusual length.

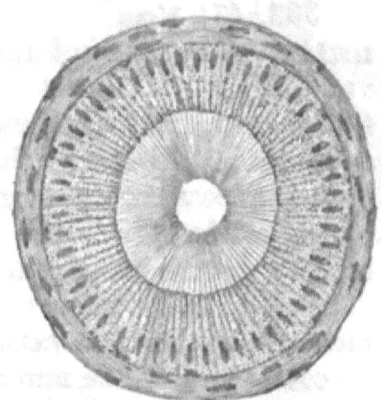

Fig. 140.—A tubule of the epididymis in cross section.

The wall of the tubule is made up of a thick layer of concentrically arranged non-striped muscular tissue, a layer of columnar epithelial cells with extraordinarily long cilia projecting into the lumen of the tube.

The tubes of the epididymis are separated from one another by a larger amount of connective tissue than those of the testis.

The tubes of the organ of Giraldé, situated in the beginning of the funiculus spermaticus, are lined with columnar ciliated epithelium. So is also the pedunculated hydatid of Morgagni attached to the globus major.

331. The seminal tubules and the tubes of the epididymis are entwined by a rich network of capillary blood-vessels. Between the tubes of the testis and epididymis are lymph spaces, forming an intercommunicating system, and emptying themselves into the superficial networks of lymphatics, *i.e.*, those of the albuginea; the arrangement of these networks is somewhat different in the testis and epididymis.

332. (3) **Vas deferens and vesiculæ seminales.**—The tubes of the globus minor open into the vas deferens. This is of course much larger than the former, and is lined with stratified columnar epithelium. Underneath this is a dense connective tissue mucosa, containing a rich network of capillary blood-vessels. Beneath this mucosa is a thin sub-mucous tissue, which in the Ampulla is better developed than in other parts, and therefore allows the mucous membrane to become folded. Outside the sub-mucous tissue is the muscular coat, which consists of non-striped muscular tissue, arranged as an inner circular and an outer longitudinal stratum. At the commencement of the vas deferens there is in addition an inner longitudinal layer. There is finally a fibrous tissue adventitia. This contains longitudinal bundles of non-striped muscular tissue, known as the *cremaster internus* (Henle). A rich plexus of veins—plexus pampiniformis—and a rich plexus of lymphatic trunks, are situated in the connective tissue of the spermatic cord. The plexus spermaticus consists of larger and smaller nerve-trunks, with which are connected small groups of ganglion cells and also large ganglionic swellings.

333. In the *vesiculæ seminales* we meet with exactly the same layers as constitute the wall of the vas deferens, but they are thinner. This refers especially to the mucosa and the muscular coat. The former is placed in numerous folds. The latter con-

sists of an inner and outer longitudinal and a middle circular stratum. The ganglia in connection with the nerve-trunks of the adventitia are very numerous.

334. In the *ductus ejaculatorii* we find a lining of columnar epithelial cells; outside of this is a delicate mucosa and a muscular coat, the latter consisting of an inner thicker longitudinal and an outer thinner circular stratum of non-striped muscular tissue.

When passing into the vesicula prostatica the columnar epithelium is gradually replaced by stratified pavement epithelium.

335. (4) The **prostate gland.** — Like other glands, the prostate consists of a framework and the gland tissue proper or the parenchyma.

The *framework*, unlike that of other glands, is essentially muscular, being composed of bundles of non-striped muscular tissue, with a relatively small admixture of fibrous connective tissue. The latter is chiefly limited to the outer capsule and the thin septa passing inwards, whereas the non-striped muscular tissue surrounds and separates the individual gland alveoli.

336. The **parenchyma** consists of the chief ducts, which open at the base of and near the colliculus seminalis, and of the secondary ducts, minor branches of the former, which ultimately lead into the alveoli. These are longer or shorter, wavy or convoluted branched tubes with numerous saccular or club-shaped branches. The alveoli and ducts are limited by a membrana propria, have a distinct lumen, and are lined with columnar epithelium. In the alveoli there is only a single layer of beautiful columnar epithelial cells, the substance of which is distinctly and longitudinally striated. In the ducts there is an inner layer of short columnar cells, and an outer one of small cubical, polyhedral or spindle-shaped cells.

At the mouth of the ducts the stratified pavement

epithelium of the pars prostatica of the urethra passes a short distance into the duct.

The alveoli are surrounded by **dense networks of capillary blood-vessels.**

In the peripheral portion of the gland numerous ganglia are interposed in the rich plexus of **nerves.** Also Pacinian corpuscles are to be met with.

337. (5) The **urethra.**—The mucous membrane of the male urethra is lined with simple **columnar epithelium,** except at the commencement—the pars prostatica—and at the end—the fossa navicularis—where it is stratified pavement epithelium.

The mucous membrane is fibrous tissue with very numerous elastic fibres. Outside of it is a muscular coat composed of non-striped muscular tissue, and arranged as an inner circular and an outer longitudinal stratum, except in the pars prostatica and pars membranacea, where it is chiefly longitudinal. In the latter portion the muscular bundles pass also into the mucous membrane, where they follow a longitudinal course between large veins **arranged in a longitudinal plexus.** These veins empty themselves into **small veins** outside. This plexus of large veins with the muscular tissue between **represents** a rudiment of a cavernous tissue (Henle).

The mucous membrane forms peculiar folds surrounding the lacunæ Morgagni. There are small mucous glands, lined with columnar epithelium, embedded in the mucous membrane; they open into the cavity of the urethra and are known as Littré's glands.

338. (6) The **glands of Cowper.**—Each **gland** of Cowper is a large compound tubular gland, which, **as regards** structure **of ducts and alveoli,** resembles a mucous **gland. The wall of the chief** ducts possesses a large amount of longitudinally-arranged non-striped muscular tissue. The epithelium lining the ducts is composed of columnar cells. The alveoli possess a large lumen and are lined with columnar mucous

cells, the outer portion of the cell being distinctly striated (Langerhans). In the cell the reticulum is also distinct. In this respect the alveoli completely resemble those of the sub-maxillary of the dog, but there are no real crescents in the alveoli of Cowper's gland.

339. (7) The **corpus spongiosum.**—The corpus spongiosum of the urethra is a continuation of the rudimentary corpus cavernosum above-mentioned in connection with the pars membranacea of the urethra. It is essentially a plexus of large veins arranged chiefly longitudinally and leading into small efferent veins. Between the large veins are bundles of non-striped muscular tissue. The capillary blood-vessels of the **mucous membrane of the urethra open into the veins of the** plexus. The outer portion of the corpus spongiosum, including **the bulbus urethræ,** shows, however, numerous venous sinuses, real cavernæ, into which open capillary blood-vessels.

340. The **glans penis** is of exactly the same structure as the corpus spongiosum. The outer surface is covered with a delicate fibrous tissue membrane, which on its free surface bears minute papillæ, extending into the stratified pavement epithelium. At the corona glandis exist small sebaceous follicles, the glands **of** Tyson; they are continued from the inner lamella **of** the **prepuce, where they abound.** The papillæ of the glans contain loops of **capillary blood-vessels.** Plexuses of non-medullated **nerve-fibres are** found underneath the epithelium of **the** surface of the glans. With these are connected the end bulbs described in a former chapter as the genital nerve-end corpuscles.

341. (8) The **corpora cavernosa penis.**—Each corpus cavernosum is enveloped in a fibrous capsule, the albuginea, made up of lamellæ of fibrous connective tissue. Numerous Pacinian corpuscles are

met with around it. The **matrix** of the corpus cavernosum consists of trabeculæ of fibrous tissue, between which pass bundles of non-striped muscular tissue all in different directions. Innumerable cavernæ or sinuses, intercommunicating with one another, are present in this matrix, capable of such considerable repletion, that in the maximum degree of this state the sinuses are almost in contact, and the trabeculæ compressed into very delicate septa. The sinuses are lined with a single layer of flattened endothelial plates, and their wall in many places is strengthened by the bundles of non-striped muscular tissue. The sinuses during erection become filled with blood, being directly continuous with capillary blood-vessels. These are derived from the arterial branches which take their course in the above trabeculæ of the matrix. The blood passes from the sinuses into small efferent veins. But the blood passes also directly from the capillaries into the efferent veins, and this is the course the blood takes under passive conditions, while during erection it passes chiefly into the above sinuses.

342. In the peripheral part of the corpus cavernosum there exists a direct communication between the sinuses and minute arteries (Langer), but in the rest the arteries do not directly communicate with the sinuses except through the capillary blood-vessels. In the passive state of the corpus cavernosum, the muscular trabeculæ forming part of the matrix are contracted, and the minute arterial branches embedded in them are therefore much coiled up; these are the arteriæ helicinæ.

CHAPTER XXXII.

THE FEMALE GENITAL ORGANS.

343. (1) THE **ovary** (Fig. 141.)—In the ovary, as in other glands, the framework is to be distinguished from the parenchyma. In the part of the ovary next to the hilum there are numerous blood-vessels, in a loose fibrous connective tissue, with numerous longitudinal bundles of non-striped muscular tissue directly continuous with the same tissues of the ligamentum latum. This portion of the ovary is the zona vasculosa (Waldeyer). All parts of the zona vasculosa—*i.e.*, the bundles of fibrous connective tissue, the blood-vessels, and the bundles of non-striped muscular tissue—are traceable into the parenchyma. The stroma of this latter, however, is made up of bundles of shorter

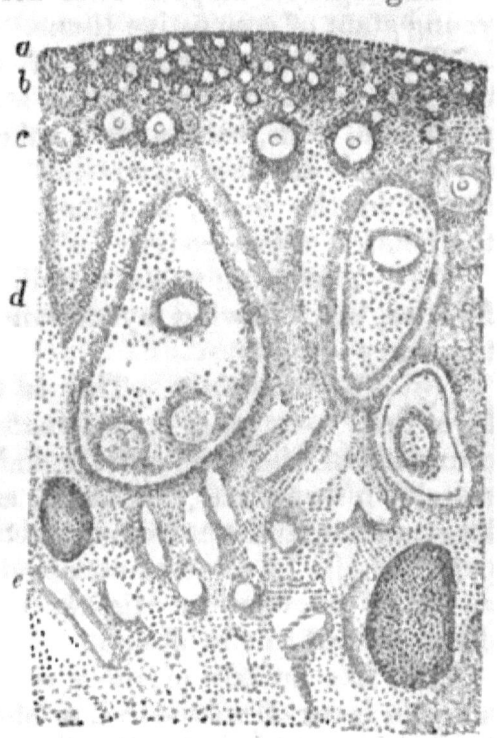

Fig. 141.—Vertical Section through the Ovary of half-grown Cat.

a, The albuginea; the germinal epithelium is not distinguishable owing to the low power under which the section is supposed to be viewed; *b*, the layer of smallest Graafian follicles and ova; *c*, the medium-sized follicles; *d*, the layer of large follicles; *e*, the zona vasculosa. (Atlas.)

or longer transparent spindle-shaped cells, each with an oval nucleus. These bundles of spindle-shaped cells form, by crossing and interlacing, a tolerably dense tissue, in which lie embedded in special arrangements the Graafian follicles. Around the larger examples of the latter the spindle-shaped cells form more or less concentric layers. In the human ovary bundles of fibrous tissue are also met with.

The spindle-shaped cells are most probably a young state of connective tissue.

Between these bundles of spindle-shaped cells occur cylindrical or irregular streaks or groups of polyhedral cells, each with a spherical nucleus; they correspond to the interstitial epithelial cells mentioned in the testis, and they are also derived from the fœtal Wolffian body.

344. According to the distribution of the Graafian follicles, the following layers can be distinguished in the ovary :—

(a) *The albuginea.* This is the most peripheral layer not containing any Graafian follicles. It is composed of the bundles of spindle-shaped cells, intimately interwoven. In man, an outer and inner longitudinal, and a middle circular, layer can be made out (Henle). In some mammals an outer longitudinal, an inner circular, or slightly oblique layer can be distinguished in the albuginea.

The free surface of the albuginea is covered with a single layer of polyhedral, or short columnar granular-looking epithelial cells, the *germinal epithelium* (Waldeyer). This epithelium, in its shape and aspect, forms a marked contrast to the transparent, flattened, endothelial plates covering the ligamentum latum.

345. (b) The *cortical layer* (Schrön). This is a layer containing the smallest Graafian follicles, either aggregated as a more or less continuous layer (cat and rabbit), or in small groups (human), separated by the

stroma. These follicles are spherical or slightly oval, of about $\frac{1}{1000}$ inch in diameter, and each of them is limited by a delicate *membrana propria*. Inside of this is a layer of flattened, transparent, epithelial cells, each with an oval, flattened nucleus; this is the *membrana granulosa*. The space within the follicle is occupied by, and filled up with, a spherical cell—the ovum cell, or *ovum*. This is composed of a granular-looking protoplasm, and in this is a big spherical, or slightly oval, nucleus—the *germinal vesicle*. The substance of this is either a fine reticulum, limited by a delicate membrane, with one or more nucleoli or *germinal spots*, or it is in one of the phases of indirect division or karyokinesis, thus indicating division of the ovum.

346. (c) From this cortical layer to the zona vasculosa we find embedded in the stroma isolated Graafian follicles, of various sizes, increasing from the former to the latter. The biggest follicles measure in diameter about $\frac{1}{20}$ inch. Those of the middle layers are of medium size (Fig. 142). In them we find inside the membrana propria the membrana granulosa, made up of a single layer of transparent, columnar, epithelial cells. The ovum, larger than in the small cortical follicles, fills out the cavity of the follicle, and is limited by a thin hyaline cuticle—the *zona pellucida*. This appears as an excretion of the cells of the membrana granulosa. The protoplasm of the ovum is fibrillated. The part surrounding the germinal vesicle is more transparent, and stains differently in osmic acid than the peripheral part. The big nucleus, or germinal vesicle, is limited by a distinct membrane, and inside

Fig. 142.—A Small Graafian follicle, from the Ovary of Cat.

The follicle is lined with a layer of columnar epithelial cells—the membrana granulosa. The ovum fills out the cavity of the follicle; it is surrounded by a thin zona pellucida, and it includes a germinal vesicle or nucleus with the intranuclear reticulum. (Atlas.)

this membrane is a reticulum with generally one big nucleolus or *germinal spot*.

Between these medium-sized follicles and the small follicles of the cortical layer we find all intermediate degrees as regards size of the follicle and the ovum, and especially as regards the shape of the cells of the membrana granulosa, the intermediate sizes of follicles being lined by a granulosa made up of a layer of polyhedral epithelial cells.

Fig. 143.—A large Graafian Follicle of the Ovary of Cat.

The follicle is limited by a capsule, the theca folliculi; the membrana granulosa is composed of several layers of epithelial cells. The ovum with its distinct hyaline zona pellucida is embedded in the epithelial cells of the discus proligerus. The cavity of the follicle is filled with fluid, the liquor folliculi.

347. The deeper Graafian follicles, *i.e.*, those that are to be regarded as big follicles, contain an ovum—occasionally, two or even three ova—which is similar to that of the previous follicles, except that it is larger, and its zona pellucida thicker. The ovum does not fill out the whole cavity of the follicle, since at one side, between it and the membrana granulosa, there is an albuminous fluid, the rudiment of the liquor folliculi.

348. The biggest or most advanced follicles are of great size, easily visible by the naked eye, and contain a large quantity of this liquor folliculi (Fig. 143). In fact, the ovum occupies only a small part of the

cavity of the follicle. The ovum is big, surrounded by a thick zona pellucida, is situated at one side, surrounded by the *discus proligerus*. This consists of layers of polyhedral cells, except the cells immediately around the zona pellucida, which are columnar. The ovum with its discus proligerus is connected with the membrana granulosa. This latter consists of *stratified pavement epithelium* forming the entire lining of the follicle. The outermost layer of cells is columnar. The membrana propria of these big follicles is strengthened by concentric layers of the stroma cells, and this represents the *tunica fibrosa* (Henle) or outer coat of the follicle—theca folliculi externa. Numerous blood capillaries connected into a network surround the big follicles.

In those follicles that contain a greater or smaller amount of the liquor folliculi, we notice in the fluid a variable number of detached granulosa cells in various stages of vacuolation, maceration, and disintegration.

349. In connection with the medium-sized and large Graafian follicles are seen occasionally smaller or larger solid cylindrical or irregularly-shaped outgrowths of the membrana granulosa and membrana propria; they indicate a new formation of Graafian follicles, some containing a new ovum. When these side branches become by active growth converted into larger follicles, they may remain in continuity with the parent follicle, or may be constricted off altogether. In the first case, we have one large follicle with two or three ova, according as a parent follicle has given origin to one or two new outgrowths.

Amongst the epithelial cells constituting the stratified membrana granulosa of the ripe follicles we notice a nucleated reticulum.

Many follicles reach ripeness, as far as size and constituent elements are concerned, long before puberty, and they are subject to degeneration; but this process of degeneration involves also follicles of smaller sizes.

350. Before menstruation, generally one, occasionally two or more, of the ripe follicles become very hyperæmic. They grow, in consequence, very rapidly in size; their liquor folliculi increases to such a degree that they reach the surface of the ovary; finally—*i.e.*, during **menstruation**—they burst at a superficial point; the ovum, with its discus proligerus, is ejected, and brought into the abdominal ostium of the oviduct. The cavity of the follicle collapses, and a certain amount of blood, derived from the broken capillaries of the wall of the follicle, is effused into it. The follicle is converted into a *corpus luteum*, by an active multiplication of the cells of the granulosa. New capillaries with connective tissue cells derived from the theca folliculi externa gradually grow into the interior, *i.e.*, between the cells of the granulosa. This growth gradually fills the follicle, except the centre; this contains blood-pigment in the shape of granules, chiefly contained in large cells, and a few new blood-vessels, the blood-pigment being the remains of the original blood effused into the follicle. But, ultimately, the pigment all disappears, and a sort of gelatinous tissue occupies the centre, while the periphery—*i.e.*, the greater part of the follicle—is made up of the hypertrophied granulosa, with young capillary vessels between its cells. The granulosa cells undergo fatty degeneration, becoming filled with several small fat globules, which gradually become confluent into a big globule. In this state the corpus luteum is complete, and has reached the height of its progressive growth. The tissue is then gradually absorbed, and cicatrical tissue is left. When this shrinks it produces a shrinking of the corpus luteum. This represents the last stage in the life of a Graafian follicle. The corpus luteum of Graafian

follicles, of which the ovum has been impregnated, grows to a much larger size than under other conditions, the granulosa becoming by overgrowth much folded.

351. **Development of the ovary and Graafian follicles.**—The germinal epithelium of the surface of the fœtal ovary at an early stage undergoes rapid multiplication, in consequence of which the epithelium becomes greatly thickened. The vascular stroma of the ovary at the same time increases, and permeates the thickened germinal epithelium. The two tissues in fact undergo mutual ingrowth, as is the case in the development of all glands—viz., the epithelial or glandular part suffers mutual ingrowth with the vascular connective tissue stroma.

In the case of the ovary, larger and smaller islands or nests (Balfour) of epithelial cells are thus gradually differentiated off from the superficial epithelium. These nests are largest in the depth, and smallest near the surface. They remain in connection with one another and with the surface for a considerable period. Even some time after birth some of the superficial nests are still connected with the surface epithelium, and with one another (Fig. 143A). These correspond to the ovarial tubes (Pflüger). While in the rabbit these nests are solid collections, in the dog they soon assume the character of tubular structures (Pflüger, Schäfer). The cells constituting the nests undergo multiplication (by karyokinesis), in consequence of which the nests increase in size, and even new nests may be constricted off from old ones (*see also above*).

352. At the earliest stages we notice in the germinal epithelium some of the cells becoming enlarged in their cell-body, and especially their nucleus; these represent the *primitive ova*. When the germinal epithelium undergoes the thickening above

mentioned, and when this thickened epithelium separates into the nests and ovarial tubes, there is a continued formation of primitive ova—*i.e.*, cells of the nests undergo the enlargement of cell-body and

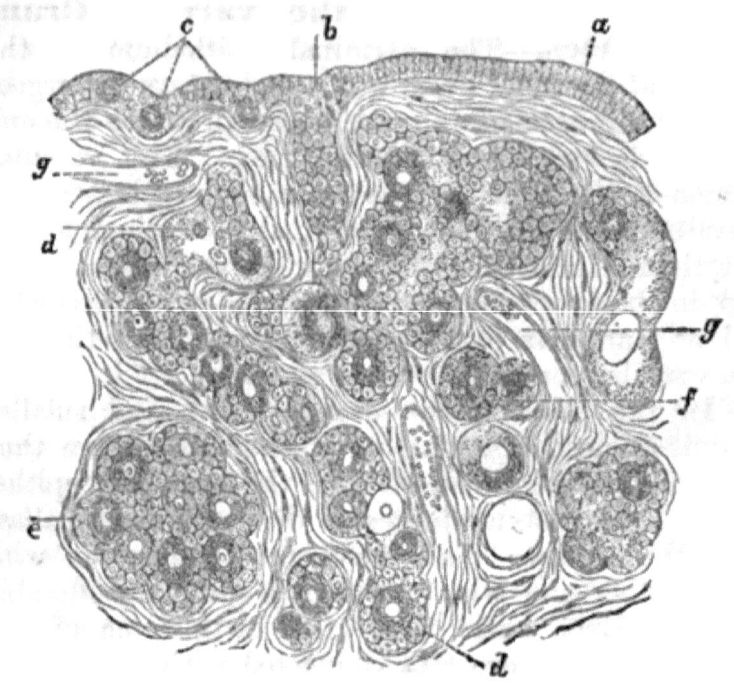

Fig. 143A.—From a Vertical Section through the Ovary of a new-born Child.

a, Germinal epithelium; *b*, ovarian tube; *c*, primitive ova; *d*, longer tubes becoming constricted off into several Graafian follicles; *e*, large nests; *f*, isolated finished Graafian follicles; *g*, blood-vessels. (Waldeyer, in Stricker's Manual.)

nucleus, by which they are converted into primitive ova. Like the other epithelial cells, the primitive ova of the nests and ovarial tubes undergo division into two or even more primitive ova after the mode of karyokinesis (Balfour). Thus each nest contains a series of ova.

353. The ordinary small epithelial cells of the nests and ovarial tubes serve to form the membrana granulosa of the Graafian follicles. According to

the number of ova in a nest or in an ovarial tube a subdivision takes place in so many Graafian follicles, each consisting of one ovum with a more or less complete investment of small epithelial cells—*i.e.*, a membrana granulosa. This subdivision is brought about by the ingrowth of the stroma into the nests.

The superficial nests being the smallest, as above stated, form the cortical layer of the small Graafian follicles; the deeper ones give origin to larger follicles. Thus we see that the ovum and the cells of the membrana granulosa are derived from the primary germinal epithelium; all other parts—membrana propria, theca externa, stroma, and vessels—are derived from the fœtal stroma.

There is a good deal of evidence to show that ova and Graafian follicles are, as a rule, reproduced after birth (Pflüger, Kölliker), although other observers (Bischoff, Waldeyer) hold the opposite view.

354. (2) The **oviduct.**—The oviduct consists of a lining epithelium, a mucous membrane, a muscular coat, and an outer fibrous coat—the serous covering, or peritoneum. The epithelium is columnar and ciliated. The mucous membrane is much folded; it is a connective tissue membrane with networks of capillary blood-vessels. In man and mammals there are no proper glands present, although there are seen appearances in sections which seem to indicate the existence of short gland tubes; but these appearances are explained by the folds of the mucous membrane. The muscular coat is composed of non-striped muscular tissue of a pre-eminently circular arrangement; in the outer part there are a few oblique and longitudinal bundles. The serous covering contains numerous elastic fibrils in a connective tissue matrix.

355. (3) The **uterus.**—The epithelium lining the cavity of the uterus is a single layer of columnar

cells, each with a bundle of cilia on their free surface. These are very easily detached, and therefore difficult to find in a hardened and preserved specimen. But in the fresh and well-preserved human uterus (Friedländer), as well as in that of mammals, the cells are distinctly ciliated. The whole canal of the cervix is also lined with ciliated epithelium, but in children, according to Lott, only beginning from the middle. The surface of the portio vaginalis uteri is, like that of the vagina, covered with stratified pavement epithelium.

356. The mucous membrane of the cervix is different from that of the fundus. In the former it is a fibrous tissue possessed of permanent folds—the palmæ plicatæ. Few thin bundles of non-striped muscular tissue penetrate from the outer muscular coat into the mucous membrane. Between the palmæ plicatæ are the openings of minute gland-tubes, more or less cylindrical in shape. They possess a membrana propria and a distinct lumen lined with a single layer of columnar epithelial cells, which, according to some, are ciliated in the new-born child, but, according to Friedländer, non-ciliated. Goblet-cells are met with amongst the lining epithelium. Several observers (Kölliker, Hennig, Tyler Smith, and others) maintain the existence of minute, thin, and long vascular papillæ projecting above the general surface of the mucous membrane in the lower part of the cervix; these apparent papillæ are, however, only due to sections through the folds of the mucous membrane. The mucous membrane of the fundus is a spongy plexus of fine bundles of fibrous tissue, covered or lined respectively with numerous small endothelial plates, each with an oval flattened nucleus. The spaces of this spongy substance are lymph-spaces, and contain the glands and the blood-vessels (Leopold).

357. The glands — **glandulæ uterinæ** — are

short tubular glands. They occur in the new-born child chiefly at the sides; during puberty their number and their size increase considerably, new glands being formed by the ingrowth of the surface epithelium into the mucous membrane (Kundrat and Engelmann). During menstruation, and especially during pregnancy, they greatly increase in length. They are more or less wavy and branched at the bottom. A delicate membrana propria forms the boundary of the tube; a distinct lumen is seen in the middle, and this is lined with a single layer of ciliated columnar epithelium (Allen Thomson, Nyländer, Friedländer, and others).

358. During menstruation the thickness of the mucous membrane increases, the epithelium of the surface and of the greater part of the glands being destroyed by fatty degeneration, and finally altogether detached. Afterwards its restitution takes place from the remnant in the depth of the glands. But according to J. Williams and also Wyder, the greater part of the mucous membrane, in addition to the epithelium, is destroyed during menstruation.

The muscular coat forms the thickest part of the wall of the uterus; it is composed entirely of the non-striped variety.

In the cornua uteri of mammals the muscular coat is generally composed of an inner thicker circular and an **outer thinner** longitudinal stratum, a few oblique bundles **passing** from the latter into the former. In the human uterus the muscular coat is composed of an outer thin longitudinal, a middle thick layer of circular bundles, and an inner thick one of oblique and circular bundles. Within these layers the bundles form plexuses.

359. The **arterioles** in the cervix and their capillaries are distinguished by the great thickness of their wall. The mucous membrane contains the *capillary networks*. These discharge their blood into veins

situated in the muscular coat. There the *veins* are very numerous, and arranged in *dense plexuses*, those of the outer and inner stratum being smaller than those of the middle stratum, where they correspond to huge irregular sinuses, the bundles of muscular tissue of the muscular coat giving special support to these sinuses. Hence the plexus of venous sinuses of the middle stratum represents a sort of *cavernous tissue*.

360. The **lymphatics** are very numerous; in the intermuscular connective tissue of the muscular coat are lymph sinuses and lymph clefts forming an intercommunicating system; they take up the lymph sinuses of the mucous membrane above mentioned, and on the other hand lead into a plexus of lymphatic vessels with valves, situated in the subserous connective tissue.

The **nerves** entering the mucous membrane are connected with ganglia. According to Lindgren, there is in the mucous membrane a plexus of non-medullated nerve-fibres, which, near the epithelium, break up into their constituent primitive fibrillæ.

361. (4) The **vagina.**—The epithelium lining the mucous membrane is a thick, stratified, pavement epithelium. The superficial part of the mucous membrane—*i.e.*, the mucosa—is a dense, fibrous, connective tissue with numerous networks of elastic fibres; it projects into the epithelium in the shape of numerous long, single or divided papillæ, each with a simple or complex loop of capillary blood-vessels. The mucosa with the covering epithelium projects above the general surface in the shape of longer or shorter, conical or irregular, pointed or blunt, permanent folds—the rugæ. These contain a plexus of large veins, between which are bundles of non-striped muscular tissue; hence they resemble a sort of cavernous tissue.

Outside of the mucosa is the loose submucosa containing a second venous plexus; its meshes are elongated and parallel to the long axis of the vagina. Outside of the submucous tissue is the muscular coat, consisting of an inner circular and an outer longitudinal stratum of non-striped muscular tissue. Oblique bundles pass from one stratum into the other. From the circular stratum bundles may be traced into the submucosa and mucosa. A layer of fibrous tissue forms the outer boundary of the wall of the vagina, and in it is the most conspicuous plexus of veins, the plexus venosus vaginalis. This plexus also contains bundles of non-striped muscular tissue, and therefore resembles a cavernous tissue (Gussenbaur.) It is not quite definitely ascertained whether or not there are secreting glands in the mucous membrane of the vagina. Von Preuschen and also Hennig described tubular glands in the upper part of the fornix and in the introitus.

The lymphatics form plexuses in the mucosa, submucosa, and the muscular coat. The first are small vessels, the second are larger than the third and possess valves. The efferent vessels form a rich plexus of large trunks with saccular dilatations in the outer fibrous coat.

There are in the mucous membrane solitary lymph follicles and diffuse adenoid tissue (Loevenstein).

Numerous ganglia are contained in the nerve plexus belonging to the muscular coat.

End bulbs in connection with the nerve-fibres of the mucosa have been mentioned in Chapter XV.

362. (5) The **urethra.**—The structure of the female urethra is similar to that of the male, except that the lining epithelium is a sort of stratified transitional epithelium, the superficial cells being short, columnar, or club-shaped; underneath this layer are several layers of polyhedral, or cubical cells. Near the orificium externum the epithelium is stratified pavement epithelium.

The muscular coat is composed of an inner longitudinal, and an outer circular, layer of non-striped muscular tissue.

363. (6) The **nymphæ, clitoris,** and **vestibulum.**—These are lined with thick stratified epithelium, underneath is a fibrous connective tissue mucous membrane, extending into the epithelium in the shape of cylindrical papillæ with capillary loops and nerve-endings (end bulbs). The nymphæ contain large sebaceous follicles, but no hairs.

The nymphæ contain a plexus of large veins with bundles of non-striped muscular tissue; hence it resembles a cavernous tissue (Gassenbaur). The corpora cavernosa of the clitoris, the glans clitoridis, and the bulbi vestibuli, correspond to the analogous parts in the penis of the male. The glands of Bartholin correspond in structure to the glands of Cowper in the male.

CHAPTER XXXIII.

THE MAMMARY GLAND.

364. THIS, like other glands, consists of a framework and parenchyma. The former is lamellar fibrous connective tissue subdividing the latter into lobes and lobules and containing a certain amount of elastic fibres. In some animals (rabbit, guinea-pig) there are also small bundles of non-striped muscular tissue. From the interlobular septa fine bundles of fibrous tissue with branched connective tissue corpuscles pass between the alveoli of the gland substance. The amount of this interalveolar tissue varies in different places, but in the active gland is always relatively scanty.

Migratory or lymph corpuscles are to be met with in the interalveolar connective tissue of both active and resting glands. In the latter they are more numerous than in the former. According to Creighton, they are derived, in the resting gland, from the epithelium of the gland alveoli. Granular large yellow (pigmented) nucleated cells occur in the connective tissue, and also in the alveoli of the resting gland, and Creighton considers them both identical, and derived from the alveolar epithelium. And according to this author, the production of these cells, would constitute the principal function of the resting gland.

The large *ducts* as they pass from the gland to the nipple, acquire a thick sheath, containing bundles of non-striped muscular tissue. These latter are derived from the bundles of non-striped muscular tissue present in the skin of the nipple of the breast.

The small ducts in the lobules of the gland tissue possess a membrana propria, and a lining—a single layer of longer or shorter columnar epithelial cells.

The terminal branches of the ducts, *i.e.*, just before these latter pass into the alveoli, are lined with a single layer of flattened pavement epithelium cells; they are analogous to the intermediary portion of the ducts of the salivary glands (see Chap. XXII.).

365. Each of these terminal branches divides and takes up several *alveoli* (Fig. 144). These are wavy tubes, saccular or flask-shaped. The alveoli are larger in diameter than the intralobular ducts. Each alveolus in the active gland has a relatively large cavity, varying in different alveoli; it is lined with a single layer of polyhedral, granular-looking, or short columnar epithelial cells, each with a spherical nucleus; a membrana propria forms the outer limit. This membrana propria, like that of the salivary, lachrymal and other glands, is a basket-work of branched cells.

In the active gland each epithelial cell is capable of forming in its interior one or more smaller or larger oil globules. These may, and generally do, become confluent, and, pressing the nucleus towards one side of the cell, give to the latter the resemblance of a fat-cell. The oil globules are finally ejected by the cell-protoplasm into the lumen of the alveolus, and repre-

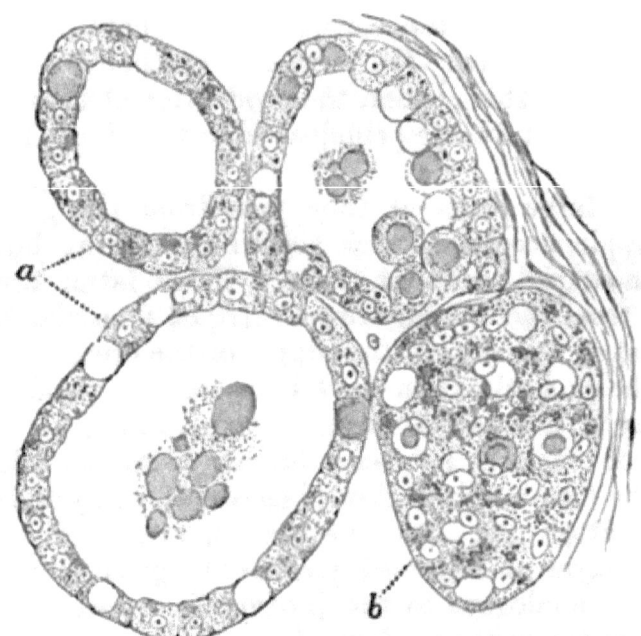

Fig. 144.—From a Section through the Mammary Gland of Cat in a late stage of Pregnancy.

a, The epithelial cells lining the gland-alveoli, seen in profile; *b*, the same, seen from the surface. Many epithelial cells contain an oil globule. In the cavity of some of the alveoli are milk globules and granular matter. (Atlas.)

sent now the *milk globules*. The cell resumes its former solid character, and commences again to form oil globules in its protoplasm. The epithelial cells, as long as the secretion of milk lasts, go on again and again forming oil globules in the above manner without being themselves destroyed (Langer). These milk globules, when in the lumen of the alveoli, are enveloped in a delicate cuticle—the albumin membrane of Ascherson. This membrane they receive from the cell protoplasm.

According to the state of secretion, most epithelial cells lining an alveolus may be in the condition of forming oil globules, or only some of them; and according to the rate in which milk globules are formed and carried away, the alveoli differ in the number of milk globules they contain.

According to Schmid, the epithelial cells, after having secreted milk globules for some time, finally break up, and are replaced by new epithelial cells derived by the division of the other still active epithelial cells.

366. The resting gland, *i.e.*, the gland of a non-pregnant or non-suckling individual, contains, comparatively speaking, few alveoli, but a great deal of fibrous connective tissue; the alveoli are all solid cylinders, containing within the limiting membrana propria masses of polyhedral granular-looking epithelial cells. During pregnancy these solid alveoli undergo rapid multiplication, elongation, and thickening, owing to the rapid division of the epithelial cells.

Finally, when milk secretion commences, the cells occupying the central part of the alveolus undergo the fatty degeneration just like the peripheral cells, but they, *i.e.*, the central cells, are eliminated, while the peripheral ones remain. These central cells are the *colostrum corpuscles*, and consequently they are found in the milk of the first few days only.

367. Ordinary milk contains no colostrum corpuscles, but only milk globules of many various sizes, from the size of a granule to that of a globule several times as big as an epithelial cell of an alveolus of the milk gland. These large drops are produced by fusion of small globules after having passed out of the alveoli. Each milk globule is an oil globule surrounded, as

stated above, by a thin albuminous envelope—Ascherson's membrane. The small bits of granular substance met with here and there, are probably the remains of broken-down protoplasm of epithelial cells.

368. Each gland alveolus is surrounded by a dense *network of capillary blood-vessels*. The alveoli are surrounded by *lymph spaces* like those in the salivary glands (Coyne) and these spaces lead into *networks of lymphatic vessels* of the interlobular connective tissue.

CHAPTER XXXIV.

THE SKIN.

369. THE skin consists of the following layers (Fig. 145):—(1) the epidermis; (2) the corium, or cutis vera, with the papillæ; (3) the subcutaneous tissue, with the adipose layer or the adipose tissue.

370. (1) The **epidermis** (Fig. 14), in all its constituent elements, has been minutely described in Chapter III. Its thickness varies in different parts, and is chiefly dependent on the variable thickness of the stratum corneum. This is of great thickness in the palm of the hand and the sole of the foot. The stratum Malpighii fits into the depressions between the papillæ of the corium as the interpapillary processes. The presence of prickle cells, of pigment granules, and of branched interstitial nucleated cells, &c., has been mentioned in Chapter III.

There occur in the stratum Malpighii migratory cells of granular aspect; they appear to migrate from the papillary layer of the corium into the stratum Malpighii (Biesiadecki).

371. (2) The **corium** is a dense feltwork of bundles of fibrous connective tissue, with a large

admixture of networks of elastic fibres. From the surface of the corium project small conical or cylindrical *papillæ*. These are best developed in those parts where the skin is thick, *e.g.*, volar side of hand and foot, scalp, lips of mouth, &c. Between the surface of the corium and the epidermis there is a basement membrane. Migratory cells, with and

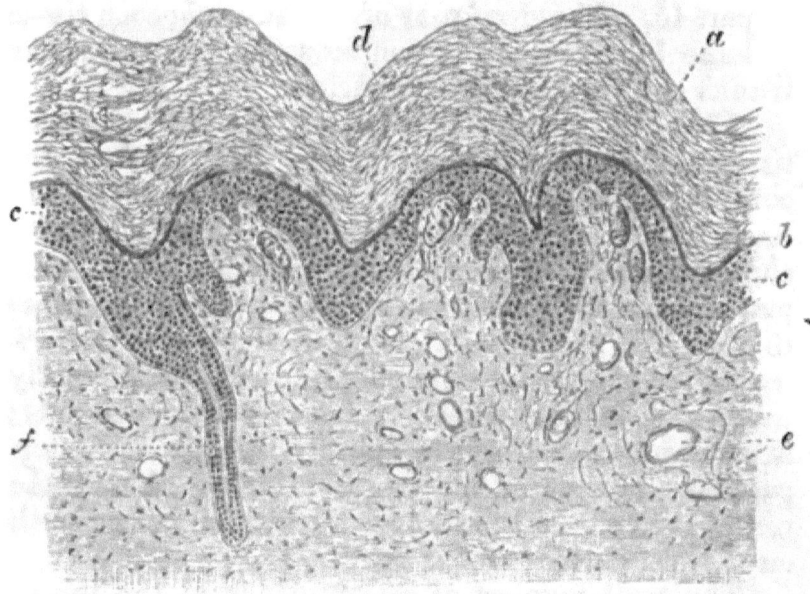

Fig. 145.—Vertical Section through the Skin of Human Finger.

a, Stratum corneum ; *b*, stratum lucidum ; *c*, stratum Malpighii ; *d*, Meissner's, or tactile corpuscles ; *e*, blood-vessels cut across ; *f*, a sudoriferous canal or duct.

without pigment granules in their interior, are met with, especially in the superficial part of the corium ; they, as well as the fixed or branched connective tissue corpuscles (see par. 40), and other structures, as vessels and nerves, lie in the interfascicular spaces.

372. (3) The superficial part of the **subcutaneous tissue** insensibly merges into the deep part of the corium ; it consists of bundles of fibrous connective tissue aggregated into trabeculæ crossing one another

and interlacing in a complex manner. Numerous elastic fibres are attached to these trabeculæ. It contains groups of fat cells, in many places arranged as more or less continuous lobules of fat tissue, forming the *stratum adiposum*. These lobules are separated by septa of fibrous connective tissue; their structure and development, and the distribution of the blood-vessels amongst the fat cells, have been described in par. 45. The deep part of the subcutaneous tissue is loose in texture, and contains the large vascular trunks and the big nerve branches.

373. The superficial part of the subcutaneous tissue, or, as some have it, the deep part of the corium, contains the sudoriparous or *sweat glands*. Each gland is a single tube coiled up into a dense clump of about $\frac{1}{80}$ of an inch in diameter—in some places, as in the axilla, reaching as much as six times this size. From each gland a duct—the *sudoriferous canal*—passes through the corium in a slightly wavy and vertical direction towards the epidermis; it penetrates more or less spirally through the interpapillary process of the stratum Malpighii and the rest of the epidermis, and appears with an open mouth on the free surface of the skin.

The total number of sweat glands in the human skin has been computed by Krause to be over two millions; but it varies greatly in different parts of the body, the largest number occurring in the palm of the hand, the next in the sole of the foot, the next on the dorsum of the hand and foot, and the smallest in the skin of the dorsum of the trunk.

374. The sudoriferous canal and the coiled tube possess a distinct lumen; this is lined with a delicate cuticle, especially marked in the sudoriferous canal and in the commencement of the coiled tube. In the epidermis the lumen bordered by this cuticle is all that is present of the sudoriferous canal. It receives

a continuation from the middle layer of the stratum Malpighii and from the basement membrane; the former is the lining epithelium, the latter the limiting membrana propria of the sudoriferous canal. The epithelium consists of two or three layers of small polyhedral cells, each with a spherical or oval nucleus.

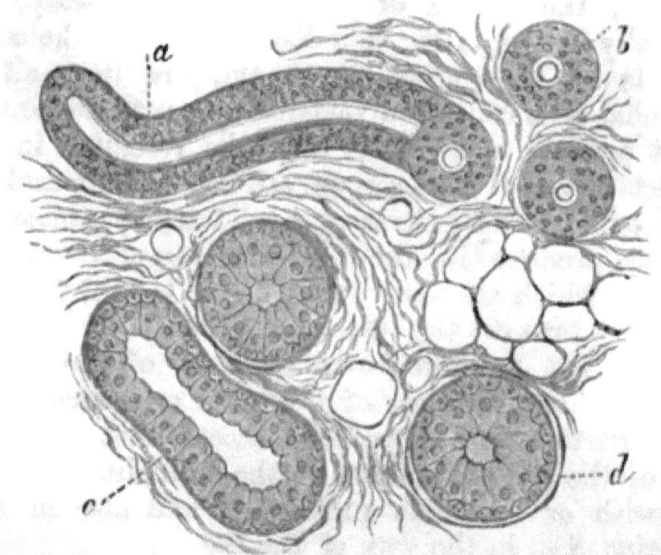

Fig. 146.—From a Section through Human Skin, showing the Sweat gland tubes cut in various directions.

a, First part of the coiled tube seen in longitudinal section ; *b*, the same seen in cross section ; *c*, the distal part seen in longitudinal section ; *d*, the same seen in cross section. (Atlas.)

375. The structure of the sudoriferous canal is then—a limiting membrana propria, an epithelium composed of two or three layers of polyhedral cells, an internal delicate membrane, and, finally, the central cavity, or lumen.

The first part—about one-third or one-fourth—of the coiled tube (Fig. 146) is of the same structure, and is directly continuous with the sudoriferous canal, with which it is identical, not only in structure, but in size. The remainder of the coiled tube—*i.e.*, the distal part—is larger in diameter, and differs in these essential respects, that its epithelium is a single layer of transparent columnar cells, and that there is between this and the limiting membrana propria a *layer of non-striped muscle cells* (Kölliker) arranged parallel with the long axis of the tube. In some places, as in the palm of the hand and foot, in the scrotum, the nipple of the breast, the scalp, but especially in the axilla, this distal portion of the coiled tube is of very great length and breadth, and its epithelial cells contain a variable amount of granules.

It appears to me that the cells resemble in this respect those of the serous salivary glands and the chief cells of the gastric glands (Langley), inasmuch as they produce in their interior larger or smaller granules which are used up during secretion, from the periphery towards the lumen.

376. The **ceruminous glands** of the meatus auditorius externus are of the same structure as the distal portion just described, except that the inner part of the cell protoplasm of the epithelium contains yellowish or brownish pigment, found also in their secretion, *i.e.*, in the wax of the ear.

Around the anus there is an elliptical zone, in the skin of which are found large coiled gland tubes—the *circumanal glands* of A. Gay—which are identical in structure with the distal portion of the sweat gland tubes.

377. The sweat glands develop as a solid cylindrical outgrowth of the stratum Malpighii of the epidermis, which gradually elongates till it reaches the superficial part of the subcutaneous tissue, where it commences to coil. The lumen of the tube is of later appearance. The membrana propria is derived from the tissue of the cutis, but the epithelium and muscular layer are both derived from the original outgrowth of the epidermis.

378. The **hair-follicles** (Fig. 147).—The skin

almost everywhere contains cylindrical *follicles*, planted more or less near to one another, and in groups. In each of them is fixed the *root of a hair;* that part

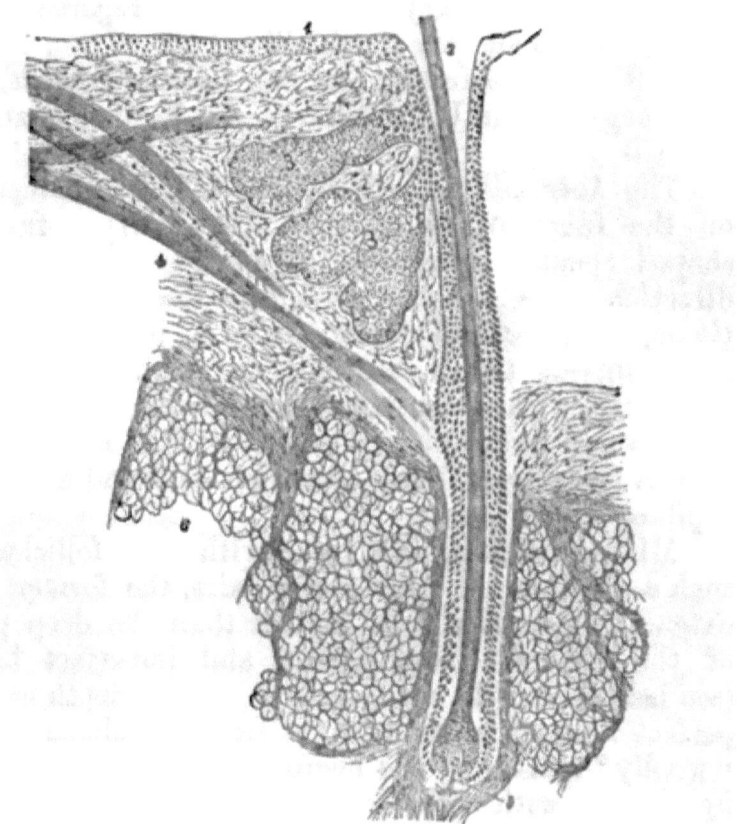

Fig. 147.—Longitudinal Section through a Human Hair.
1, Epidermis; 2, mouth of hair follicle; 3, sebaceous follicle; 4, musculus arrector pili; 5, papilla of hair; 6, adipose tissue. (Atlas.)

of the hair which projects beyond the general surface of the skin is the *shaft*.

A very few places contain no hair-follicles, such, for instance, as the volar side of the hand and foot, and the skin of the penis.

In size, the hairs and hair-follicles differ in different parts. Those of the scalp, the cilia of the eyelids, the hairs of the axilla and pubic region, those of the male whiskers and moustache, are coarse and thick, while the hairs of other places—*e.g.*, the skin-surface of the eyelids, the middle of the arm and forearm, &c.—are very minute : but, as regards structure, they are all very much alike.

379. A complete hair and hair-follicle—that is, the papillary hair of Unna—shows the following structure :—

The *hair-follicle*. Each hair-follicle commences on the free surface of the skin with a funnel-shaped opening or *mouth;* it passes in an *oblique* direction through the corium into the subcutaneous tissue, in whose middle strata—*i.e.*, in the stratum adiposum—it terminates with a slightly enlarged extremity, with which it is invaginated over a relatively small fungus-shaped *papilla*. This latter is of fibrous tissue, containing numerous cells and a loop of capillary blood-vessels.

Minute hairs do not reach with their follicles to such a depth as the large coarse hairs, the former not extending generally much farther than the deep part of the corium. Degenerating and imperfect hairs (see below) also do not reach to such a depth as the perfect large hair-follicles. In individuals with " woolly " hair—*e.g.*, the negro race (C. Stewart), and in animals with " woolly " hair, such as the fleece of sheep—the deep extremity of the hair-follicle is curved, sometimes even slightly upwards.

380. The structure of a hair-follicle is as follows (Fig. 148) : There is an outer coat composed of fibrous tissue ; this is the fibrous coat of the *hair-sac*. It is merely a condensation of the surrounding fibrous tissue, and is continuous with the papilla at the extremity of the hair-follicle. About the end of the hair-follicle, or sometimes as much as in the lower fourth, there is inside of this

fibrous layer of the hair-sac a single continuous layer of transversely or *circularly-arranged spindle-shaped cells*, each with an oval flattened or staff-shaped nucleus, completely resembling, and generally considered to be, non-striped muscle cells. Inside of this layer of the hair-sac is a glassy-looking, hyaline, basement membrane, which is not very distinct in minute hairs, but is sufficiently conspicuous in large adult hair-follicles. This *glassy membrane*, as it is called, is a direct continuation of the basement membrane of the surface of the corium, and it can be traced as a delicate membrane also over the surface of the hair-papilla.

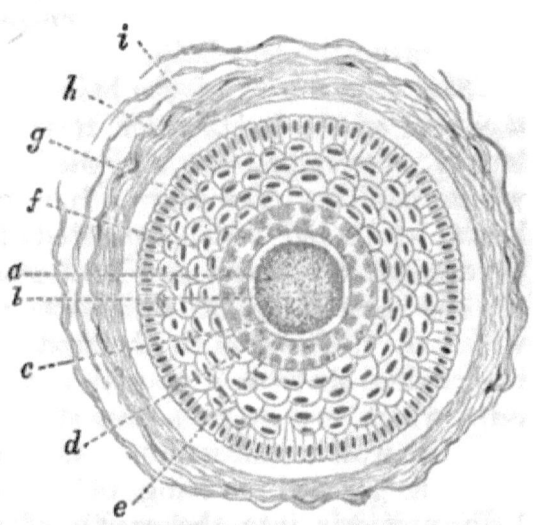

Fig. 148.—Cross Section through a Human Hair and Hair Follicle.

a, Marrow of hair; *b*, cortex of hair; *c*, cuticle of hair; *d*, Huxley's layer of inner root-sheath; *e*, Henle's layer of inner root-sheath; *f*, outer root-sheath; *g*, glassy membrane; *h*, fibrous coat of hair sac; *i*, lymph spaces in the same.

381. Next to the glassy membrane is the *outer root-sheath*, the most conspicuous part of the hair-follicle. It consists of a thick stratified epithelium of exactly the same nature as the stratum Malpighii of the epidermis, with which it is directly continuous, and from which it is developed. In the outer root-sheath the layer of cells next to the glassy membrane is columnar, just like the deepest layer of cells in the stratum Malpighii; then follow inwards several layers of polyhedral cells; and, finally, flattened nucleated scales form the innermost boundary of the outer root-

sheath. The stratum granulosum of the stratum Malpighii is not continued beyond the mouth of the hair-follicle, but there it is generally very marked. The outer root-sheath becomes greatly attenuated at the papilla—in fact, is there continuous with the cells constituting the hair-bulb.

382. The centre of the hair-follicle is occupied by the *root of the hair*, which terminates with an enlarged extremity—the *hair-bulb;* this grasps the whole papilla. The hair-bulb is composed of polyhedral epithelial cells, separated from one another by cement substance, and continuous with the cells of the extremity of the outer root-sheath, from which they originate in the first instance; just over the papilla there is a special row of short columnar cells, which are in an active state of multiplication, and by which continuously new cells are formed. Thus a gradual shifting of the cells of the hair-bulb upwards into the cavity of the hair-follicle— *i.e.*, the hair—takes place; but at the same time these progressing cells become elongated, spindle-shaped, and constitute the cells of the *hair substance*, except in the very centre, where they remain polyhedral, so as to represent the cells of the *marrow of the hair*, and in the periphery, where they remain more or less polyhedral, so as to form the *inner root-sheath*.

383. **The root of the hair,** except at the hair-bulb, shows the following parts: The *substance of the hair, the cuticle*, and the *inner root-sheath*. The substance of the hair is composed of the *hair fibres*, *i.e.*, long thin fibres, or narrow long scales, each composed of hyaline horny substance, and possessed of a thin staff-shaped remnant of a nucleus. These are held together by a certain amount of interstitial cement substance. Towards the bulb they gradually change into the spindle-shaped cells above men-

tioned. They can be isolated by strong acids and alkalies. In pigmented hairs there occur numerous *pigment granules between* the hair fibres, but also *diffused pigment in* their substance. The same is noticed with reference to the hair-bulb—viz., pigment granules being present in the intercellular cement, and pigment also in the cell substance. In the centre of many hairs is a cylindrical space, containing generally one row of polyhedral cells, which are, to a great extent, filled with air, and, in pigmented hair, also with pigment granules.

384. On the surface of the hair substance is a thin cuticle, a single layer of horny non-nucleated hyaline scales arranged more or less transversely; they are imbricated, and, according to the degree of imbrication, the cuticle shows more or less marked projections, which give to the circumference of the hair the appearance of minute teeth, like those of a saw.

385. The inner root-sheath in well-formed, thick hairs, is very distinct, and consists of a delicate *cuticle* next to the cuticle of the hair; then an *inner*, or *Huxley's, layer*, which is a single, or sometimes double, layer of horny cubical cells, each with a remnant of a nucleus; and, finally, an *outer*, or *Henle's, layer*—a single layer of non-nucleated horny cubical cells.

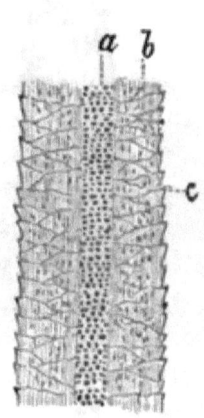

Fig. 149.—Longitudinal View of the Shaft of a Pigmented Human Hair.

a, Marrow of hair; *b*, fibres of hair substance; *c*, cuticle.

The **shaft of the hair** (Fig. 149), or the part projecting over the free surface of the skin, is of exactly the same structure as the root, except that it possesses no inner root-sheath.

386. As mentioned above, at the hair-bulb the polyhedral cells constituting this latter gradually pass into the different parts of the hair—*i.e.*, marrow-sub-

stance, cuticle, and inner root-sheath—and the continual new production of cells over the papilla causes a gradual progression and conversion of the cells, and a corresponding growth in length of the hair shaft.

Pigmented hairs, as mentioned above, contain pigment granules between—*i.e.*, in the interstitial substance cementing together — the hair fibres, and diffuse pigment in their substance. According to the amount of these pigments, but especially of the interstitial pigment granules (Pincus), the colour of the hair is of a greater or lesser dark tint. In red hairs there is chiefly diffuse pigment. In white or fair hairs neither the one nor the other pigment is present; in grey there is air at least in the superficial layers of the hair substance, besides absence of pigment.

Sleek hairs are circular, curly oval, in cross-section.

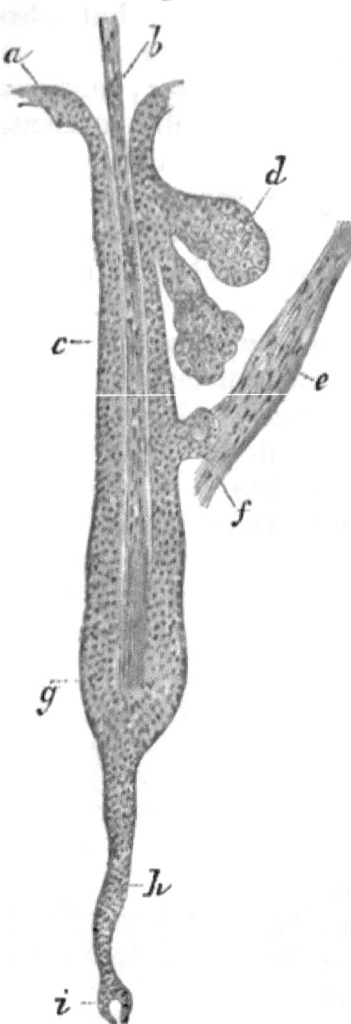

Fig. 150.—From a Section through Human Scalp, showing a degenerating Hair.

a, The epidermis; *b*, the hair; *c*, the outer root-sheath of the hair follicle; *d*, the sebaceous follicle; *e*, the arrector pili; *f*, a cyst grown out of the outer root-sheath; *g*, the hair-knob; *h*, the new outgrowth of the outer root-sheath; *i*, the new papilla. (Atlas.)

387. **New formation of hair** (Fig. 150). — Every hair, be it fine and short or thick and long, under normal conditions, has only a limited exis-

tence, for its hair-follicle, including the papilla, sooner or later undergoes degeneration, and subsequent to this a new papilla and a new hair are formed in its place. What happens is this—the lower part of the hair-follicle, including the papilla and hair-bulb, degenerates and is gradually absorbed. Then there is left only the upper part of the follicle, and in the centre of this is the remainder—*i.e.*, non-degenerated portion—of the hair root. The fibres of this are at the extremity fringed out and lost amongst the cells of the outer root-sheath of the follicle. This represents the *hair-knob* (Henle). Now, from the outer root-sheath a cylindrical outgrowth of epithelial cells into the depth takes place; against the extremity of this a new papilla is made. In connection with this new papilla, and in the centre of that cylindrical outgrowth, a new hair and hair-bulb are formed, and as these gradually grow outwards towards the surface they lift, or rather push, the old hair—*i.e.*, the hair-knob—out of the follicle. The outer part of the follicle of the old hair persists.

Thus we find in all parts of the skin where hairs occur, complete or papillary hairs side by side with degenerating hairs, or hair-knobs.

388. **Development of hair.**—In the human fœtus the hair-follicles make their first appearance about the end of the third month, as solid cylindrical outgrowths from the stratum Malpighii. This is the rudiment of the outer root-sheath. After having penetrated a short distance into the corium, this latter becomes condensed around it as the rudiment of the hair-sac, and at the distal extremity forms the papilla growing against the outer root-sheath and invaginating it. In connection with the papilla a rapid multiplication of the epithelial cells of this extremity of the outer root-sheath takes place, and this forms the hair-bulb, by the multiplication of whose cells the hair

and the inner root-sheath are formed. As growth and multiplication proceed at the hair-bulb, so the new hair, with its distal-pointed end, gradually reaches the outer surface. It does not at once penetrate the epidermis, but remains growing and burrowing its way for some time in the stratum corneum of the epidermis in a more or less horizontal direction.

389. In many mammals occur, amongst ordinary hairs, special large hairs, with huge hair-follicles planted deeply into the subcutaneous tissue; such are the big hairs in the skin about the lips of the mouth in the dog, cat, rabbit, guinea-pig, mouse and rat, &c. These are the *tactile hairs*. Their hair-follicle possesses a thick hair-sac, in which are contained large sinuses intercommunicating with one another and with the blood system; these sinuses are separated by trabeculæ of non-striped muscular tissue, and represent, therefore, a cavernous tissue. The papilla is very huge, and so is the outer root-sheath and the hair-root in all its parts. There are vast numbers of nerve-fibres, distributed and terminating amongst the cells of the outer root-sheath (Arnstein).

390. With each hair-follicle is connected one or two *sebaceous follicles*. These consist of several flask-shaped or oblong alveoli, joining into a common short duct opening into the hair-follicle near the surface— *i.e.*, that part called the neck of the hair-follicle.

The alveoli have a limiting membrana propria; next to this is a layer of small polyhedral, granular-looking, epithelial cells, each with a spherical or oval nucleus; next to this, and filling the entire space of the alveolus, are large polyhedral cells, each with a spherical nucleus; the cell-substance is filled with minute oil-globules, between which is left a sort of honeycombed reticulated stroma. The cells nearer to the centre of the alveolus are the largest. Towards the duct they become shrivelled up into an amorphous mass. The duct itself is a continuation of the outer root-sheath.

As multiplication goes on in the marginal layer of epithelial cells—*i.e.*, those next the membrana propria—the products of this multiplication are gradually shifted forward towards the duct, and through this into the neck and mouth of the hair-follicle, where they constitute the elements of *sebum*.

There is a very characteristic misproportion between the size of the hair-follicle and that of the sebaceous gland in the embryo and new-born, the sebaceous gland being there so large that it forms the most conspicuous part, the minute hairs (lanugo) being situated, as it were, in the duct of the sebaceous follicle.

391. In connection with each hair-follicle, especially where they are of good size—as in the scalp—there is a bundle, or rather group of bundles, of non-striped muscular tissue; this is the *arrector pili*. It is inserted in the hair-sac near the bulbous portion of the hair-follicle, and passes in an oblique direction towards the surface of the corium, grasping, as it were, on its way the sebaceous follicle, and terminating near the papillary layer of the surface of the corium. The arrector pili forms with the hair-follicle an acute angle—this latter being planted into the skin in an oblique direction, as mentioned above—and consequently, when the arrector contracts, it has the effect of raising the hair follicle and hair (cutis anserina—" goose's skin"), and of making the hair assume a more upright position (causes it, as we say, to "stand on end"). At the same time, it compresses the sebaceous follicle, and thus facilitates the discharge of the sebum.

392. The corium of the scrotum, of the nipple of the breast, of the labia pudendi majora, and of the penis, contains numbers of bundles of non-striped

muscular tissue (Kölliker), independent of the hairs; these run in an oblique and horizontal direction, and form plexuses.

393. The **nails** (Fig. 151).—We distinguish the *body* of the nail from the *free margin* and from the

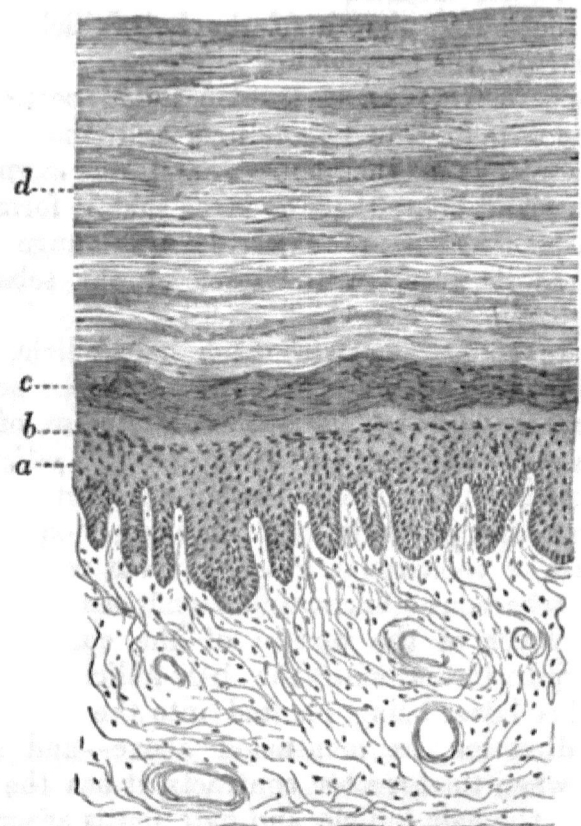

Fig. 151.—Vertical Section through the Human Nail and Nail-bed.
a, Stratum Malpighii of nail-bed; *b*, stratum granulosum of nail-bed; *c*, the deep layers of the nail substance; *d*, the superficial layers of same.

root; the body is the nail proper, and is fixed on to the *nail-bed*, while the nail-root is fixed on the nail-matrix—*i.e.*, the posterior part of the nail-bed. The nail is inserted, with the greater part of its lateral and the posterior margin, in the *nail-groove*, a fold by which the nail-matrix passes into the surrounding skin.

394. The *substance* of the nail is made up of a large number of strata of homogeneous horny scales— the *nail-cells*—each with a staff-shaped remnant of a nucleus.

The corium of the nail-bed is highly vascular; it is firmly fixed by stiff bands of fibrous tissue on the subjacent periosteum; it is covered with a stratum Malpighii of the usual description, except that the stratum granulosum is absent in the nail-matrix, but is present in a rudimentary state in the rest of the nail-bed. The nail itself represents the stratum lucidum, of course of exaggerated thickness, situated over the stratum Malpighii of the nail-bed. There is no stratum corneum over the nail.

The stratum Malpighii and corium of the nail-bed are placed into permanent minute folds, and the nail possesses on its lower surface corresponding linear indentations.

395. In the fœtal nail-bed the stratum Malpighii is covered with the usual stratum lucidum and stratum corneum, but the former is the larger; by a rapid multiplication of the cells of the stratum Malpighii, and a conversion of its superficial cells into the scales of the stratum lucidum, the fœtal nail is produced. At this early stage the nail is covered by stratum corneum. By the end of the fifth month the margin breaks through this stratum corneum, and by the seventh month the greater part has become clear of it.

396. The **blood-vessels of the skin.**—The blood-vessels are arranged in different systems for the different parts of the skin (Tomsa):—

(*a*) There is, first, the vascular system of the adipose tissue, differing in no way from the distribution of blood-vessels in fat tissue of other places.

(b) Then there is the vascular system of the hair-follicles. The papilla has a capillary loop, or rather a minute arteriole, a capillary loop, and a descending vein, and the fibrous tissue of the hair-sac possesses capillaries arranged as a network with elongated meshes, with its afferent arteriole and efferent vein.

(c) The sebaceous follicle has its afferent arteriole and efferent vein, and capillary networks surrounding the alveoli of the gland. The arrector pili and other bundles of non-striped muscular tissue possess capillary networks with elongated meshes.

(d) The sweat-glands have an afferent arteriole, from which proceeds a very rich network of capillaries, twining and twisting round the gland-tube. The duct possesses its separate afferent arteriole and capillaries, forming elongated meshes.

(e) The last arterial branches are those that reach the surface of the corium, and there break up into a dense capillary network with loops for the papillæ. In connection with these capillaries is a rich plexus of veins in the superficial layer of the corium.

(f) In the nail-bed are dense networks of capillaries, with loops for the above-named folds.

397. The **lymphatics.**—There are networks of lymphatic vessels in all strata of the skin; they are, more or less, of horizontal expansion, with oblique branches passing between them. Their wall is a single layer of endothelial cells, and some of them possess valves. Those of the surface of the corium take up lymphatics of the papillæ. The subcutaneous lymphatics are the biggest. The fat tissue, the sweat-glands, and the hair-follicles possess their own lymphatic clefts and sinuses. The interfascicular spaces of the corium and subcutaneous tissue are directly continuous with the lymphatic vessels in these parts.

398. The **nerves.**—The nerve-branches break up

into a dense plexus of fine nerve-fibres in the superficial layer of the corium. This plexus extends horizontally, and gives off numerous elementary fibrils to the stratum Malpighii, in which they ascend vertically and in a more or less wavy fashion towards the stratum lucidum (Langerhans, Podkopaeff, Eberth, Eimer, Ranvier, and others). According to some, they terminate with a minute swelling; according to others, they form networks; but they always remain *between* the epithelial cells.

The subcutaneous nerve-branches of some places —palm of hand and foot and skin of penis—give off single medullated nerve-fibres, terminating in a Pacinian corpuscle, mentioned in a former chapter. In the volar side of the fingers and toes there occur in some of the papillæ of the corium the tactile or Meissner's corpuscles, each connected with one or two medullated nerve-fibres, as described in a previous chapter. The outer root-sheath of the hair-follicles contains the terminations of fine nerve-fibres, in the shape of primitive fibrillæ (Jobert, Bonnet, and Arnstein). According to Jobert, the nerve-fibres entwine the hair-follicle in circular turns. The tactile hairs possess a greater supply of nerves than the ordinary hair-follicles.

CHAPTER XXXV.

THE CONJUNCTIVA AND ITS GLANDS.

399. (1) THE **eyelids.**—The outer layer of the eyelids is *skin* of ordinary description; the inner is a delicate, highly vascular membrane—the *conjunctiva palpebræ*. This includes a firm plate—the *tarsal-plate*

—which is not cartilage, but very dense, white, fibrous tissue. In it lie embedded the *Meibomian glands*. These extend in each eyelid in a vertical direction from the distal margin of the tarsal-plate to the free margin of the eyelid; in the posterior angle of this margin lies the opening or mouth of each of the Meibomian glands.

The *duct* of a Meibomian gland is lined with a continuation of the stratified pavement epithelium, lining the free margin of the lid; it passes in the tarsal-plate toward its distal margin, and takes up on all sides short minute ducts, each of which becomes enlarged into a spherical, saccular, or flask-shaped *alveolus*. This is *identical in structure and secretion with the alveoli of the sebaceous follicles of the skin*.

400. The conjunctival layer is separated from the subcutaneous tissue of the skin-layer of the eyelid by the *bundles of the sphincter orbicularis*—striped muscular tissue. Some bundles of this extend near the free margin of the lid, and represent what is known as the musculus ciliaris Riolani. This sends bundles around the mouth of the Meibomian ducts.

401. At the anterior angle of the free margin of the lid are the eyelashes or *cilia*, remarkable for their thickness and rapid reproduction. Near the cilia, but towards the Meibomian ducts, open the ducts of peculiar large glands—the *glands of Mohl*. Each of these is a wavy or spiral tube, passing in a vertical direction from the margin of the lid towards its distal part; it completely coincides in structure to the large portion of a sweat gland—*i.e.*, that part containing a columnar epithelial lining, and between this and the membrana propria a longitudinal layer of non-striped muscular cells.

The free margin is covered, as mentioned above, with stratified pavement epithelium, into which the mucous membrane extends in the shape of minute papillæ. In the conjunctiva palpebræ the epithelium is thin, but stratified pavement epithelium; there are no papillæ, but the sub-epithelial mucosa—that is, the layer situated between the epithelium of the surface and the tarsal-plate—contains a dense network of capillary blood-vessels.

402. Passing from the eyelids on to the eyeball, we have the continuation of the conjunctiva palpebræ —*i.e.*, the fornix conjunctivæ—and, further, the conjunctiva fixed to the sclerotic, and terminating at the margin of the cornea—the conjunctiva bulbi. The epithelium covering the conjunctiva fornicis and conjunctiva bulbi is stratified epithelium, the superficial cells being short columnar; next to the fornix the superficial cells are beautiful columnar, and the mucosa underneath the epithelium is placed in regular folds (Stieda, Waldeyer). Towards the cornea the epithelium of the conjunctiva assumes the character of stratified pavement epithelium, and minute papillæ extend into it from the mucosa.

403. The mucous membrane is fibrous tissue, containing the networks of capillary blood-vessels.

Into the fornix lead minute mucous glands, embedded in the conjunctiva fornicis; they are the *glands of Krause*. Similar glands exist in the distal portion of the tarsal-plate.

404. The **blood-vessels** of the conjunctiva terminate as the capillary network of the superficial layer of the mucosa, and as capillary networks for the Meibomian glands, Krause's gland, &c. Around the corneal margin the conjunctival vessels are particularly dense, and loops of capillaries extend from it into the very margin of the cornea.

405. The **lymphatics** form a superficial and deep network. Both are connected by short branches. The deep vessels are possessed of valves. The super-

ficial plexus is densest at the limbus corneæ, and they are in direct connection with the interfascicular lymph clefts, both of the sclerotic and cornea. In the margin of the lid the superficial lymphatics of the skin anastomose with those of the conjunctiva.

Lymph follicles occur in groups in the conjunctiva of many mammals about the inner angle of the eye. In the lower eyelid of cattle they are very conspicuous, and known as the glands of Bruch. They are also well-marked in the third lid of many mammals.

According to Stieda and Morano, isolated lymph follicles occur also in the human conjunctiva.

406. The **nerves** are very numerous in the conjunctiva; they form plexuses of non-medullated fibres underneath the epithelium. From these plexuses fine fibrils pass into the epithelium of the surface, between whose cells they terminate as a network (Helfreich, Morano). End bulbs of Krause occur in great numbers in man and calf. They have been mentioned in a former chapter.

407. (2) **The lachrymal glands** are identical in structure with the serous or true salivary glands. The arrangement of the connective tissue stroma, the nature and structure of the ducts—especially of the intralobular ducts—and alveoli, the distribution of blood-vessels and lymphatics, are exactly the same as in the true salivary glands. Reichel has found that the epithelial cells lining the alveoli are well defined, conical or cylindrical, transparent and slightly granular during rest; but during secretion they grow smaller, more opaque and granular, their outlines not well defined, and the nucleus becomes more spherical and placed more centrally.

408. In most mammals there is in the inner

angle of the eye, and closely placed against the surface of the eyeball, a gland, called *Harder's gland*. According to Wendt, this is either a true serous gland, like the lachrymal—as in the ox, sheep, and pig—or it is identical in structure with a sebaceous gland, as in the mouse, rat, and guinea-pig; or it consists of two portions, one of which (white) is identical with a sebaceous, while the other (rose-coloured) is a true serous gland; such is the case in the rabbit and hare. According to Giacomini, a rudiment of Harder's gland exists also in the ape and man.

CHAPTER XXXVI.

THE CORNEA, SCLEROTIC, LIGAMENTUM PECTINATUM, AND CILIARY MUSCLE.

409. I. THE **cornea** (Fig. 152) of man and many mammals consists of the following layers, counting from front to back :—

(1) The *epithelium of the anterior surface* (see Fig. 15); this is a very transparent, stratified, pavement epithelium, such as has been described in par. 22. It is directly continuous with the epithelium of the conjunctiva, but it is more transparent; in dark pigmented eyes of mammals the epithelium of the conjunctiva is also pigmented. In these cases the pigment, as a rule, does not pass beyond the margin of the cornea.

410. (2) Next follows a homogeneous elastic membrane, *Bowman's membrane*, or elastica anterior. It is best shown in the human eye, but is present, even though only rudimentary, in the eye of mammals.

(3) Then follows the *ground substance*, or substantia propria, of the cornea. This is composed of lamellæ of bundles of fibrous connective tissue. Neighbouring lamellæ are connected with one another by oblique bundles.

The fibre bundles within each lamella run parallel to the surface of the cornea, but may cross one another under various angles.

In the anterior layer of the ground substance some of the bundles pass through several lamellæ in an oblique manner; they represent the fibræ arcuatæ.

The fibrils within the bundles, and the bundles, and the lamellæ of bundles are held together by an interstitial, albuminous, semi-fluid, cement substance, which, like other similar interstitial substances, belongs to the globulins, and is soluble in 10 per cent. saline solution (Schweigger Seidel). A few elastic

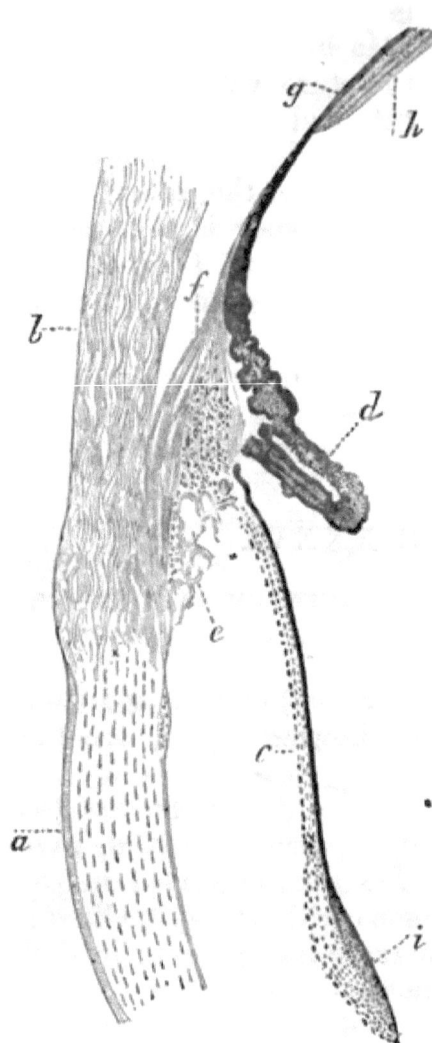

Fig. 152.—From a Vertical Section through the Membranes of the Eye of a Child.

a, Cornea; *b*, sclerotic; *c*, iris; *d*, processus ciliaris; *e*, ligamentum pectinatum; *f*, ciliary muscle, its meridional bundles; *g*, choroid membrane; *h*, the retina of the ora serrata; *i*, the sphincter pupillæ in cross section. (Atlas.)

fibrils are seen here and there. Between the lamellæ are left the lacunæ and canaliculi for the branched, flattened, nucleated *corneal corpuscles*, described in a previous chapter (Figs. 25, 26). They anastomose with one another within the same plane, and also, to a limited degree, with those of neighbouring planes.

411. (4) The **membrana Descemeti**, or elastica posterior, is a resistent elastic membrane, conspicuous by its thickness in all corneæ.

(5) The posterior surface of this membrane is covered with a mosaic of beautiful *polygonal endothelial cells*, each with an oval nucleus—the endothelium of Descemet's membrane. Under stimulation these cells contract. At first they appear slightly and numerously branched, but gradually their processes become longer and fewer, and ultimately they are reduced to minute clumps of nucleated protoplasm, each with a few long processes.

There are no blood-vessels in the normal cornea, except in fœtal life, when there is underneath the anterior epithelium a plexus of capillaries.

The lymphatics are represented as the intercommunicating lymph-canalicular system—*i.e.*, the lacunæ and canaliculi of the corneal corpuscles; and in connection with these are lymph channels lined with a continuous endothelium and containing the nerve bundles.

412. The **nerves** (Figs. 68, 69, 70) are distributed as the nerves of the anterior layers, and as those of the Descemet's membrane. The first form rich plexuses of fibrillated axis cylinders, with triangular nodal points (Cohnheim), in the anterior layers of the ground substance; from this plexus pass obliquely through Bowman's membrane short branches—the rami perforantes (Kölliker)—and these immediately

underneath the epithelium break up into their constituent primitive fibrils, the latter coming off the former brush-like (Cohnheim). These primitive fibrillæ ultimately ascend into the anterior epithelium (Hoyer, Cohnheim, and others), where they branch, and nearly reach the surface. They always run between the epithelial cells, and are connected into a network. According to some observers, they terminate with free ends, pointed or knobbed; but according to others, with whom I agree, these apparent free ends are not in reality free endings.

413. The nerves of Descemet's membrane form also a plexus of non-medullated fibres in the posterior layers of the ground substance; from them come off vast numbers of primitive fibrillæ, running a more or less straight and long course, crossing one another often under right angles; they give off very fine fibrils, which are closely associated with the corneal corpuscles, without, however, really becoming continuous with their protoplasm.

414. II. The **sclerotic** consists of lamellæ of tendinous tissue. The bundles of fibrous tissue are opaque as compared with those of the cornea, although they pass insensibly into them. There are lymph clefts between the lamellæ and trabeculæ, and in them lie the flattened connective tissue corpuscles, which, in the dark eyes of some mammals only, contain pigment granules. Numerous elastic fibrils are met with in the inner layers of the sclerotic.

415. Between the sclerotic and choroid **membrane** is a loose fibrous tissue, which acts also as the supporting tissue for the blood-vessels passing to and from the choroid. The part of this loose tissue next to the sclerotic, and forming part, as it were, of the sclerotic, contains, in dark eyes of mammals, numerous pigmented connective tissue corpuscles; it is then called *lamina fusca*. The rest—*i.e.*, next

to the choroid membrane—is the supra-choroidal tissue.

416. There are blood-vessels in the sclerotic, which belong to it: they are arterioles, capillaries, and veins; in addition to these are the vascular branches passing to and from the choroid.

417. III. The **ligamentum pectinatum iridis** (see Fig. 152) is a conical mass of spongy tissue, joining firmly the cornea and sclerotic to the iris and ciliary processes. It forms an intimate connection, on the one hand, with the junction of cornea and sclerotic, and on the other, with that of the iris and ciliary processes. This ligament is composed of trabeculæ and lamellæ of stiff elastic fibres, forming a continuity, on the one hand, with the lamina Descemeti of the cornea and the elastic fibres of the sclerotic, and on the other, with the **tissue of the ciliary border of the iris**. The trabeculæ anastomose, so as to form a honeycombed plexus, and the spaces in this plexus are lined with a layer of flattened endothelial cells, directly continued from the endothelium of Descemet's membrane, on the one hand, and with the layer of endothelial cells covering the anterior surface of the iris, on the other hand. In some mammals, the spaces in the ligamentum pectinatum at the iris end are very considerable, and are called the spaces of Fontana.

The interlamellar and interfascicular lymph-spaces of **the sclerotic form an intercommunicating** system.

The nerves form a dense plexus of non-medullated fibres in the tissue of the sclerotic (Helfreich).

At the point of junction of the cornea and sclerotic, but belonging to the latter, and in the immediate neighbourhood of the ligamentum pectinatum iridis, is a circular canal—the *canal of Schlemm;* this is lined with endothelium, and is considered by some

(Schwalbe) as a lymphatic canal; by others (Leber) as a venous vessel.

418. IV. The **ciliary muscle** (Fig. 152) or tensor choroideæ, is fixed to this ligamentum pectinatum; it is composed of bundles of non-striped muscular tissue. This muscle consists of two parts: (a) one of circular bundles nearest to the iris—this is the portio Mülleri; (b) the greater part is composed of radiating bundles, passing from the ligamentum pectinatum in a meridional direction for a considerable distance backwards into the tissue of the choroid membrane. It occupies the space between the ligamentum pectinatum, sclerotic, ciliary processes, and the adjoining portion of the choroid membrane. The bundles of the muscle are arranged more or less in lamellæ; within each lamella they form plexuses.

A rich plexus of non-medullated nerve-fibres, with groups of ganglion cells, belongs to the ciliary muscle.

CHAPTER XXXVII.

THE IRIS, CILIARY PROCESSES, AND CHOROIDEA.

419. I. THE **iris** consists of the following layers:—

(1) The endothelium of the anterior surface: transparent, flattened, or polyhedral cells, each with a spherical or slightly oval nucleus; in dark-coloured eyes of man and mammals brown pigment granules are contained in the cell-substance.

(2) A delicate hyaline basement membrane: it is continuous through the trabeculæ of the ligamentum pectinatum, with the membrana Descemeti of the cornea.

(3) The substantia propria: this is the ground-substance; it consists of fibrous connective tissue in bundles, accompanying the blood-vessels, which are very numerous in the tissue of the iris. Many connective tissue corpuscles are found in the substantia propria; they are more or less branched, and many of them contain, in all but albino and blue eyes, yellowish-brown pigment granules. The colour of the iris varies according to the number of these pigmented connective tissue cells, and to the amount of the pigment granules present in them.

(4) A hyaline delicate basement membrane limits the substantia propria at the posterior surface; this is an elastic membrane, and is continued over the ciliary processes and choroid as the *lamina vitrea*.

420. (5) The last layer is the epithelium of the posterior surface: this is a layer of polyhedral cells, filled with dark pigment granules, except in albinos, where there are no pigment granules. This endothelium is called the uvea, or tapetum nigrum. The interstitial cement substance between the cells is not pigmented, but transparent.

The name "uvea" is sometimes applied to the whole of the iris, ciliary processes, and choroid membrane.

In blue eyes the posterior epithelium is the only pigmented part of the iris, and so it is also in the iris of new-born children; hence, their eyes are blue. Such iris appears blue because its dull tissue is viewed on dark ground—*i.e.*, on the pigmented epithelium of the posterior surface.

421. Near the pupillary border the posterior section of the substantia propria contains a broad layer of circular bundles of non-striped muscular tissue: this is the *sphincter pupillæ*. In connection with this are bundles of non-striped muscular fibres, passing in a radiating direction towards the ciliary margin of the

iris: these are the bundles of the *dilatator pupillæ*, forming a sort of thin membrane near the posterior surface of the iris (Henle and others). At the ciliary margin the bundles take a circular direction, and form a plexus (Ivanoff).

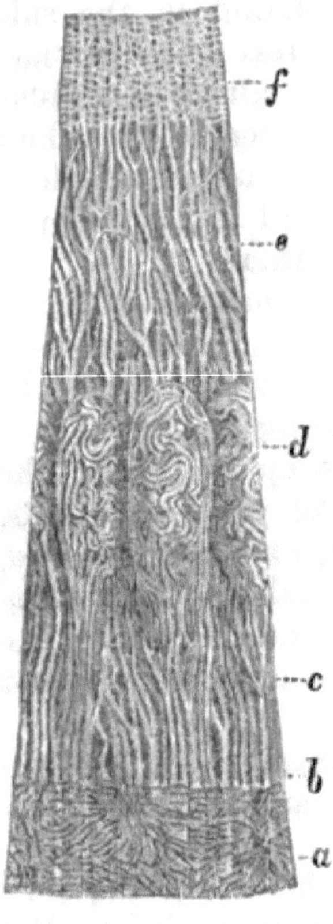

422. The **blood-vessels** (Fig. 153) of the iris are very numerous. The arteries are derived from the circulus arteriosus iridis major, situated at the ciliary margin of the iris, and from the arteries of the ciliary processes. These arteries run in a radiating direction towards the pupillary margin, where they terminate in a dense network of capillaries for the sphincter pupillæ. But there are also numerous capillary blood-vessels of a more or less longitudinal direction near the posterior surface of the iris. The veins accompany the arteries, and both are situated in the middle stratum of the substantia propria.

In the sheath of the blood-vessels are *lymph clefts* and *lymph sinuses;* there appear to be no other lymphatics.

423. The **nerve-fibres** are very numerous (Arnold, Formad), and in the outer or ciliary portion of the iris form a rich plexus, from which are derived, (*a*) networks of non-medullated fibres for the dilator pupillæ; (*b*) a network of fine

Fig. 153. — Blood-vessels (injected) of the Iris and Choroid Membrane of the Eye of a Child.

a, Capillaries of the choroid; *b*, ora serrata; *c*, blood-vessels derived from *d*, those of the ciliary processes, and from *e*, those of the iris; *f*, capillary network of the pupillary border. (Kölliker, after Arnold.)

non-medullated fibres for the anterior surface; and (c) a network of non-medullated fibres for the sphincter pupillæ.

The capillary blood-vessels are also accompanied by fine nerve-fibres (A. Meyer), and, according to Faber, there exist ganglion cells in these nerve networks.

424. II. The **ciliary processes** are similar in structure to the iris, except, of course, that they do not possess an anterior endothelium or an anterior basement membrane. The *substantia propria* is fibrous tissue with elastic fibres and numerous branched cells, pigmented in dark (but not in blue) eyes. The posterior basement membrane is very thick, and is called the *lamina vitrea;* in it may be detected bundles of fine fibrils. It possesses permanent folds arranged in a network (H. Müller). The inside of it is covered with a layer of pigmented polyhedral epithelium, the *tapetum nigrum:* the cells are polygonal when viewed from the surface. The individual cells are separated by thin lines of a transparent cement substance. This pigmented epithelium is covered with a layer of transparent columnar epitheloid cells, each with an oval nucleus. These are closely fixed on the tapetum nigrum, and represent a continuation of the retina over the ciliary processes: this is the *pars ciliaris retinæ* (Fig. 154).

425. The arterial branches for the ciliary processes and muscle are chiefly derived from the circulus arteriosus iridis major, and form a dense network of capillaries for the former; to each of these corresponds a conical group of capillaries (Fig. 153).

426. III. The **choroid** membrane consists— counting from outwards, *i.e.*, from the sclerotic, inwards, *i.e.*, towards the retina—of the following layers :—

(1) The membrana supra-choroidea. This is a continuation of the sclerotic, with which it is identical

in structure; the spaces between its lamellæ are lined with endothelium, and represent lymph spaces (Schwalbe).

(2) Next follows an elastic layer which contains networks of elastic fibres, the branches of the arteries

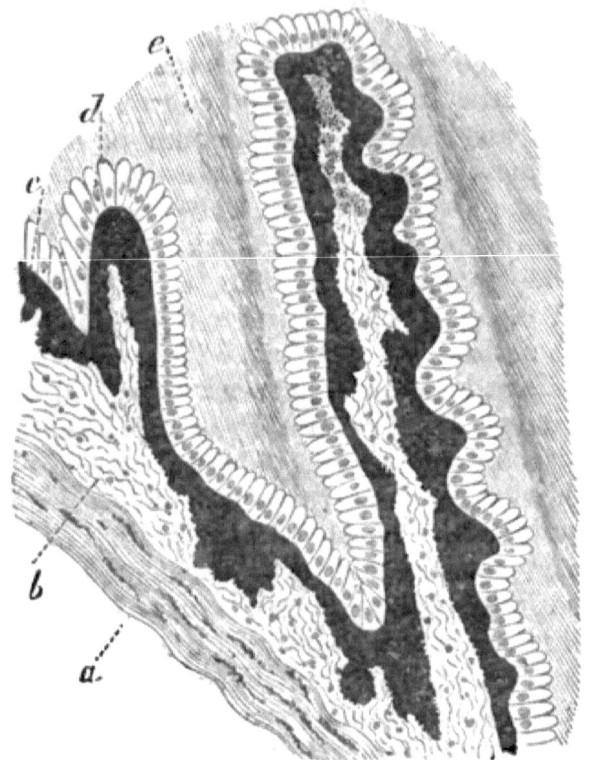

Fig. 154.—From a Vertical Section through the Ciliary Processes of the Ox's Eye.

a, Fibrous tissue with pigmented cells; b, loose fibrous tissue forming the proper membrane of the ciliary process; c, the pigmented epithelium covering the posterior surface of the ciliary process; d, the epitheloid cells, forming the pars ciliaris retinæ covering the back of the ciliary processes; e, Zonula Zinnii, with bundles of fibres. (Atlas.)

and veins, and, in its outer portion, pigmented cells (Fig. 155).

427. (3) Then follows the membrana choriocapillaris, a dense network of capillary blood-vessels embedded in a tissue containing numerous branched

and unbranched pigmented and unpigmented connective-tissue cells.

(4) The lamina vitrea ; and, finally,

(5) The tapetum nigrum, or the pigmented epithelium, which, however, is considered part of the retina. In the region of the ora serrata of the retina also—*i.e.*, next to the ciliary processes—this zone of the choroidea is lined with a layer of transparent, columnar, epitheloid cells, representing the pars ciliaris retinæ.

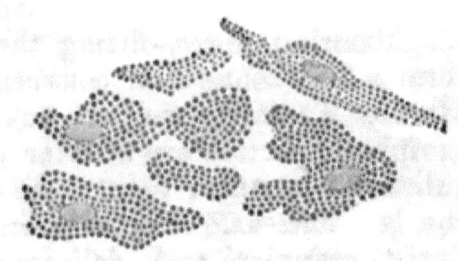

Fig. 155.—Pigmented Connective Tissue Cells of the Choroid Coat. (Atlas.)

428. The arteriæ ciliares breves and recurrentes, situated in the outer part of the choroidal tissue, form ultimately the dense networks of capillaries for the chorio-capillaries. The veins derived from this pass into the outer part of the choroid, where they anastomose so as to form the peculiar large veins, which are called the venæ vorticosæ.

CHAPTER XXXVIII.

THE LENS AND VITREOUS BODY.

429. (1) THE **lens** consists of a thick, firm, elastic capsule and of the lens substance. The former shows fine longitudinal striæ, and diminishes in thickness towards the posterior pole of the lens. The surface of the capsule facing the anterior surface of the lens-substance is lined with a single layer of polyhedral, granular-looking, epithelial cells, each with a spherical or oval nucleus. This epithelium stops as such at the

margin of the lens, where its cells, gradually elongating, pass into the lens-fibres. The nuclei of these lie in a curved plane belonging to the anterior half of the lens : this is the *nuclear zone*. The lens-substance consists of the *lens-fibres*. These are band-like, hexagonal in transverse section; their outline is beset with numerous fine ridges and furrows, which in neighbouring fibres, fitting the one into the other, form a firm connection between the fibres (Valentin, Henle, Kölliker, and others). The fibres of the peripheral portion are broader and thicker, and their substance less firm than those of the centre—*i.e.*, of the lens-nucleus. The substance of the lens-fibres is finely granular and delicately and longitudinally striated.

430. The lens fibres (Fig. 156) are arranged in concentric lamellæ, each consisting of a single layer of fibres joined by their broad surfaces. Each fibre is slightly enlarged at the extremities; and in each lamella the fibres extend from the anterior to the posterior surface. Their extremities are in contact with the ends of the fibres of the same lamella in the *sutures*, or the rays of the so-called *lens stars*. In the lens of the newborn child, the stars of both anterior and posterior lamellæ possess three such rays, while in the adult each of these rays has secondary rays. In these rays there is a homogeneous thin layer of an albuminous cement substance; a similar substance in minute quantity is also present between the lamellæ, and in it occur smaller or larger clefts and channels,

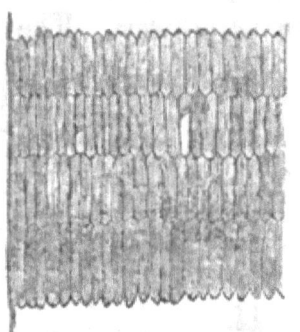

Fig. 156.—From a Section through the Lens of Dog.

Showing four lamellæ; in each the component lens-fibres are cut across; they appear as flattened hexagons. (Atlas.)

which evidently carry the nutritious fluid for the lens-fibres.

431. (2) The **vitreous body** is a fluid substance enclosed in a delicate hyaline capsule—the *membrana hyaloidea*. This membrane, at the margin of the fossa patellaris of the vitreous body—*i.e.*, the fossa in which the lens is lodged—but without covering it, passes as the zonula ciliaris, or zonula Zinnii, or *suspensory ligament* of the lens, to the margin of the latter, to which it is firmly adhering. So it adheres also to the surface of the ciliary processes. The zonula Zinnii is hyaline and firm, and is strengthened by numbers of bundles of minute stiff fibrils.

Between the suspensory ligament of the lens, the margin of the lens and of the fossa patellaris is a circular lymph space, called the **canalis Petiti**.

Beneath the membrana hyaloidea are found isolated nucleated granular-looking cells (the subhyaloid cells of Ciaccio), possessed of amœboid movement (Ivanoff).

432. The substance of the corpus vitreum appears differentiated by clefts, concentric in the peripheral, radiating in the central, part (Brücke, Hannover, Bowman, Ivanoff, Schwalbe). But these do not contain any distinct membranous structures (Stilling, Ivanoff, Schwalbe).

The canalis hyaloideus, or canal of Stilling, extends from the papilla nervi optici to the posterior capsule of the lens, and is lined with a continuation of the membrana hyaloidea.

433. In the substance of the corpus vitreum occur isolated nucleated cells; they have amœboid movements, and some contain vacuoles, from commencing degeneration. They are all identical with white blood-corpuscles (Lieberkühn, Schwalbe).

Fine bundles of fibrils are occasionally seen in the substance of the vitreous body.

CHAPTER XXXIX.

THE RETINA.

434. THE retina (Fig. 157) consists of the following layers, counting from inwards towards the choroid membrane:— (1) The membrana limitans interna, which is next to the membrana hyaloidea of the vitreous body; (2) the nerve-fibre layer; (3) the layer of ganglion cells; (4) the inner granular or inner molecular layer; (5) the layer of inner nuclei; (6) the outer granular, or outer molecular, or internuclear layer; (7) the layer of outer nuclei; (8) the membrana limitans externa; (9) the layer of rods and cones; and (10) the pigmented epithelium of the retina, or the tapetum nigrum mentioned above, which forms, at the same time, the inner lining epithelium of the choroid membrane.

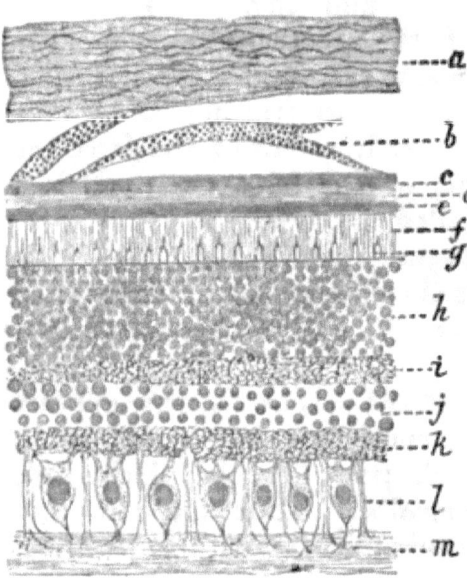

Fig. 157.—From a Transverse Section through the Eye of Sheep; Peripheral Portion of Retina.

a, The inner part of the sclerotic; b, the supra-choroidal (pigmented) lamellæ; i, d, the layers of the choroid coat; e, the pigmented epithelium of the retina; f, the layer of rods; g, the cones; h, the layer of outer nuclei; i, the outer molecular layer; j, the layer of inner nuclei; k, the inner molecular layer; l, the layer of ganglion cells, with the radial or Müller's fibres between; m, the layer of nerve-fibres. (Atlas.)

435. From this arrangement is excepted—(*a*) the papilla nervi optici, (*b*) the macula lutea and fovea centralis retinæ, and (*c*) the ora serrata of the retina.

(*a*) The papilla nervi optici, or the blind spot of

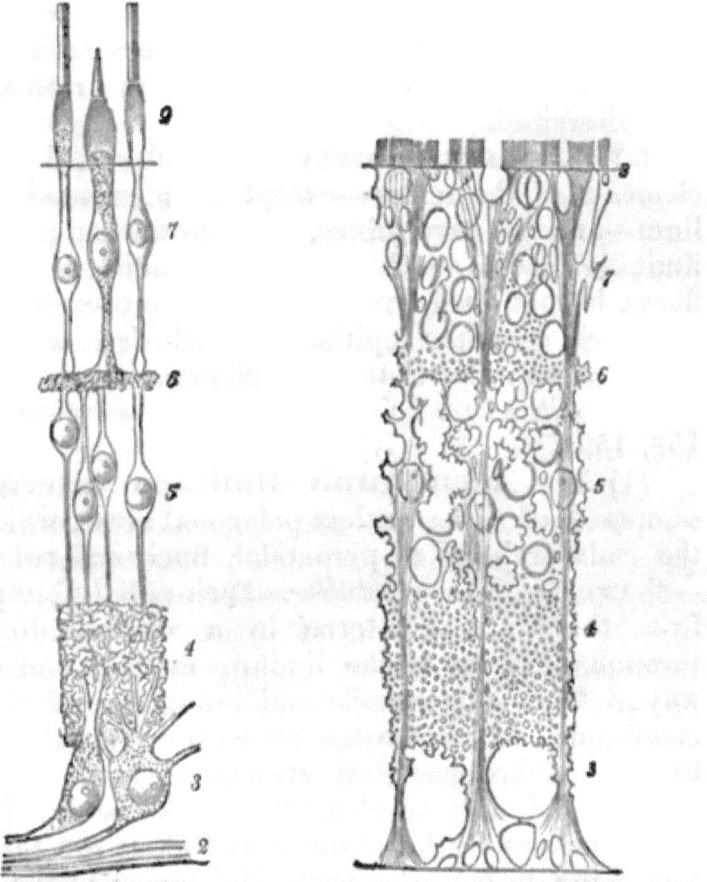

Fig. 158.—Diagram of the Nervous Elements of the Retina. Fig. 159.—Diagram of the Connective Tissue Substance of the same.

2, Nerve-fibres; 3, ganglion cells; 4, inner molecular layer; 5, inner nuclear layer; 6, outer molecular layer; 7, outer nuclear layer; 8, the membrana limitans externa; 9, the rods and cones. (Max Schultze.)

the retina, represents the entrance of the optic nerve-fibres into the retina; thence, as from a centre, they spread out in a radiating direction into the saucer-shaped retina, of which they form the internal layer. No other elements of the retina are present at the papilla, except a continuation of the limitans interna. At the papilla nervi optici the arteria and vena centralis nervi optici also enter, and spread out with their branches in the inner layers of the retina. A large lymph space is also found there.

(*b*) The **macula lutea** and **fovea** centralis will be considered after the various layers of the retina have been described.

(*c*) At the ora serrata all cellular and nuclear elements of the retina—except the pigmented epithelium—and the nerve-fibres, come to an end; but the limitans interna, with its peculiar radial or Müller's fibres, is continued over the ciliary processes in the shape of columnar epitheloid nucleated cells mentioned above: this is the pars ciliaris retinæ.

436. *Structure of the layers of the retina* (Figs. 158, 159).

(1) The **membrana limitans interna** is composed of more or less polygonal areas, which are the ends or bases of pyramidal, finely-striated fibres —the *radial fibres of Müller*. Each radial fibre passes from the limitans interna in a vertical direction through all layers to the limitans externa, and on its way gives off numerous lateral branchlets, fibrils, and membranes, which anastomose with one another so as to form a honeycombed stroma or matrix for all cellular and nuclear elements of the retinal layers. In the nerve-fibre layer the radial fibres are thickest, this being, in fact, the pyramidal basis; in the inner nuclear layer each possesses an oval nucleus.

437. (2) The **layer of nerve-fibres.**—The optic nerve-fibres at their entrance into the eyeball lose their medullary sheath, and only the transparent axis cylinder is prolonged into the retina. In man, medullated nerve-fibres in the retina are very exceptional; in the rabbit there are two bundles, whose

fibres retain their medullary sheath in the retina (Bowman). The nerve-fibres remain grouped in bundles in the retina, and even form plexuses. For obvious reasons, the number of nerve-fibres in the nerve-fibre layer diminishes towards the ora serrata.

438. (3) **The layer of ganglion cells.**—There is one stratum of these cells only, except in the macula lutea, where they form several strata. Each cell is multipolar, and possessed of a large nucleus. One process is directed inwards and becomes connected with a fibre of the nerve-fibre layer. Several processes pass from the opposite side of the cell, and enter the next outer layer, *i.e.*, the inner molecular layer.

According to Max Schultze and others, they break up there into a reticulum of fibrils which is part of this molecular layer; but according to Retzius, Mans, and Schwalbe, they simply pass through the inner molecular layer.

The ganglion cells are separated from one another by the radial fibres of Müller.

439. (4) **The inner molecular layer** is a fine and dense reticulum of fibrils, with a small amount of granular matter between. The fibrils are connected with lateral branchlets of the radial fibres of Müller. This layer is, on account of its thickness, a conspicuous part of the retina. In lower vertebrates it appears stratified.

440. (5) **The inner nuclear layer** contains in a honeycombed matrix of a hyaline stroma numerous nuclei, in two, three, or four layers. In the amphibian retina these form a larger number of layers. Some oblong nuclei of this layer belong, as has been mentioned above, to the radial fibres of Müller. Next to the molecular layer are small nuclei belonging to flattened branched cells (Vintschgau). But the great majority of the nuclei of this layer are slightly

oval, with a reticulum in their interior. Each belongs to a spindle-shaped cell, with a small amount of protoplasm around the nucleus; it is, in fact, a bipolar ganglion cell (Max Schultze), of which one process (the inner) passes as a fine varicose fibre into and through the inner molecular layer, to become connected with the outer processes of the ganglion cells (Retzius, Schwalbe), while the other or outer process passes into and through the next outer layer of the retina.

(6) **The outer molecular layer** is of exactly the same structure as the inner molecular layer—*i.e.*, a fine reticulum of fibrils—but is considerably thinner than the latter.

441. (7) **The outer nuclear layer** contains, in a honeycombed matrix, a large number of oval nuclei. In the retina of man and mammals these nuclei are always present in considerably greater numbers or layers than those of the inner nuclear layer, but in the amphibian animals the reverse is the case. They are smaller than the nuclei of the inner nuclear layer, and show often a peculiar transversely-ribbed differentiation of their contents (Henle, Krause). The honeycombed matrix of this layer is in connection with lateral branchlets of the radial fibres of Müller, with which it forms a sort of limiting delicate membrana propria at the outer surface of the layer; this is

442. (8) **The limitans externa.**—The nuclei of the outer nuclear layer next to this limitans externa are connected, in the retina of man and mammals, with the cones, while the nuclei farther inwards from the limitans externa are connected with the rods. In both instances the connection is established through holes in the limitans externa. Each nucleus of the outer nuclear layer is, in reality, that of a spindle-shaped cell with a minute amount of protoplasm; this is prolonged outwards, as the outer part

of the rod- or cone-fibre, to become connected with a rod or cone respectively, while inwards it passes into a longer, more conspicuous fibre, the inner part of the rod- or cone-fibre. This branches, and penetrating into the outer molecular layer, is lost with its branchlets among the fibrils of this layer.

443. (9) **The rods and cones.**— Each *rod* is of cylindrical shape, with rounded or conical outer extremity; it consists of an outer and inner member, joined by linear cement. Its substance is bright and glistening, and that of the outer member is composed of the neurokeratin of Kühne and Ewald. In the fresh state the outer member shows a more or less fine and longitudinal striation, due to longitudinal fine ridges and furrows (Hensen, Max Schultze). After certain reagents, such as serum, liquor potassæ, the outer rod-member disintegrates into numerous transverse, thin, homogeneous-looking discs (Hannover). The inner member in the human rods is slightly broader than the outer; it is pale or finely and longitudinally striated, and contains in many instances a peculiar lenticular structure; in the human and mammalian retina this is absent, but in its stead is a mass of longitudinal fibrils (Max Schultze). The inner member passes through a hole in the limitans externa, and becoming thinner, represents the outer part of the rod-fibre.

444. Each *cone* is composed of an outer, short, pointed, conical member, and an inner larger member with convex surface: this is the *body* of the cone. The outer member of the cone separates under certain conditions also into thin transverse discs. The body of the cone is longitudinally and finely striated. The outer extremity of the body of the cones in many birds, reptiles, and amphibia contains a spherical corpuscle of red, orange, yellow, green, or even blue colouration.

The cones are shorter than the **rods**, the pointed end of the former not reaching much farther than the junction between the outer and inner members of the rods.

In the macula lutea and fovea centralis of man and most mammals there are present cones only, and towards the peripheral portion of the retina they gradually decrease in numbers; in the peripheral part there are only rods. But in birds the cones exceed the rods everywhere.

In the bat and mole the macula lutea possesses no cones, and in the owl, rat, mouse, guinea-pig, and rabbit, they are few and small.

445. The outer members of the rods (only) contain in the fresh and living state a peculiar diffuse purplish colour (Leydig, Boll, Kühne): this is the visual purple or Rhodopsin of Kühne. When exposed to sunlight it passes through red, orange, and yellow, and finally disappears altogether — becomes bleached. There is no visual purple in the rods of Rhinolophus hipposideros, fowl and pigeon; in those retinæ in which the cones contain coloured globules (see

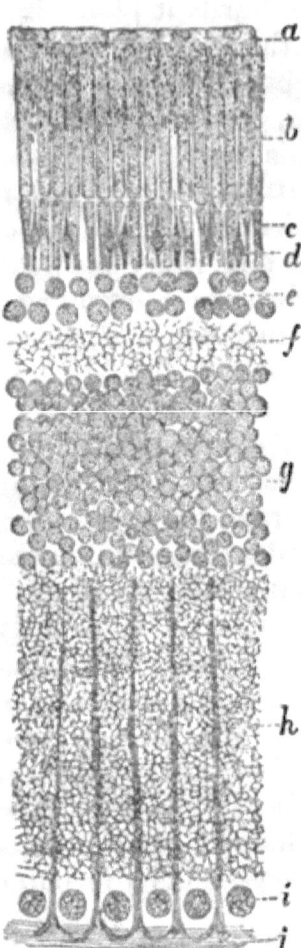

Fig. 160.—Vertical Section through the Frog's Retina.

a, The pigmented epithelium of the retina or tapetum nigrum; *b*, the outer members of the rods, those of cones between them; *c*, the inner members of the rods and cones; *d*, limitans externa; *e*, the outer nuclei; *f*, the outer molecular layer; *g*, the inner nuclei; *h*, the inner molecular layer; *i*, the nuclei of the ganglion cells; *j*, the nerve-fibres; the pyramidal extremities of the radial fibres are well shown. (Atlas.)

above) the surrounding rods are wanting in the visual purple.

The visual purple stands in an intimate relation to the pigmented epithelium of the retina, since a retina regains its visual purple after bleaching, when replaced on the pigmented epithelium (Kühne). This holds good, of course, only within certain limits.

446. (10) **The pigmented epithelium** (Fig. 160), or tapetum nigrum, is composed of polygonal protoplasmic cells, which, when viewed from the surface, appear as a mosaic, in which they are separated from one another by a thin layer of cement substance. Each cell shows an outer non-pigmented part, containing the slightly-flattened oval nucleus, and an inner part next to the rods and cones, which is full of pigmented crystalline rods (Frisch). This part is prolonged into numerous fine fibrils, each containing a row of the pigmented particles, and these fibrils pass between the outer members of the rods, to which they closely adhere, and which in reality become almost entirely ensheathed in them (M. Schultze). Each cell supplies a number of rods with these fibrils. Sunlight causes a protrusion of these fibrils from the cell body, whereas in the dark they are retracted (Kühne), in a manner similar to what takes place in pigmented connective tissue cells. (See par. 43.) The tint of this pigment is darker in dark than in light eyes. It is bleached by the light in the presence of oxygen (Kühne), but it persists in the absence of oxygen (Mays).

447. The **macula lutea** (Fig. 161) of man and ape contains a diffuse yellow pigment, between the elements of the retina (M. Schultze). In man and most mammals, as mentioned above, there are hardly any rods here, but cones only; these are longer than in other parts, and in the fovea centralis they are longest, and, at the same time, very thin. Since there are few rods here, the nuclei of the outer nuclear

layer are limited to a very few layers (generally about two) next to the membrana limitans externa. For this reason, the rest of the outer nuclear layer is occupied by the cone-fibres only, which in the fovea centralis pass in a slanting, or almost horizontal,

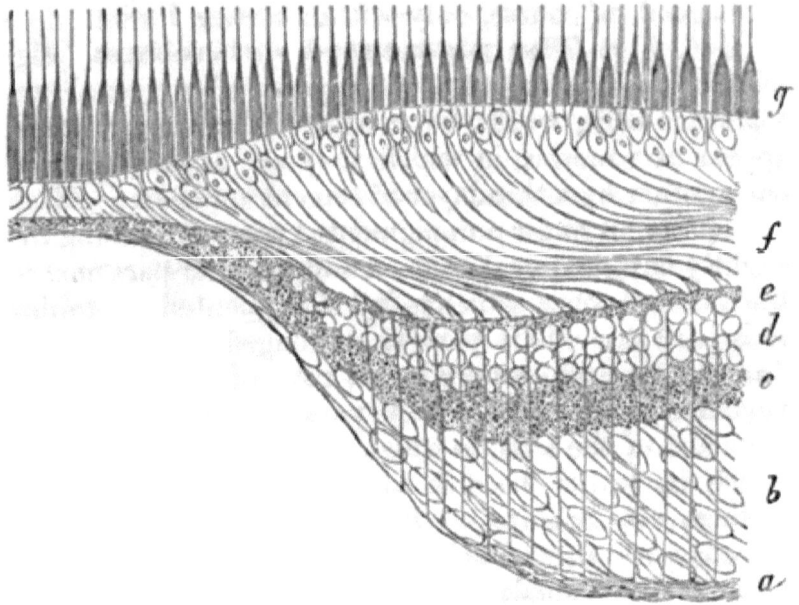

Fig. 161.—From a Vertical Section through the Macula Lutea and Fovea Centralis,

a, Nerve-fibres; b, ganglion cells; c, inner molecular layer; d, inner nuclei; e, outer molecular layer; f, cone-fibres; g, cones. (Diagram by Max Schultze.)

direction sideways into the outer molecular layer. The ganglion cells form several strata in the macula lutea. In the fovea centralis are present the cones (very long and thin), the limitans externa, the few nuclei representing the outer nuclear layer, a thin continuation of the inner molecular layer, and the limitans interna.

448. In the embryo, the primary optic vesicle becomes invaginated so as to form the optic cup, which consists of two layers—an outer, giving origin to the pigmented epithelium; and an inner, the retina

proper. In this the rods and cones, with their fibres and the nuclei of the outer nuclear layer, correspond to columnar epithelial cells (the *sensory epithelium*), while all the other layers—*i.e.*, the outer molecular, inner nuclear, inner molecular layer, ganglion cells, nerve-fibres, and limitans interna—represent Brücke's tunica nervea or Henle's stratum nerveum.

449. The **blood-vessels** of the retina. The branches of the arteria and vena centralis of the optic nerve can be traced into the retina in the layer of nerve-fibres and ganglion-cells, while the capillaries connecting the arteries with the veins extend through the layers up to the outer molecular layer.

The **lymphatics** of the retina exist as perivascular lymphatics of the retinal veins and capillaries (His). Lymph channels are present in the nerve-fibre layer.

450. The **lamina cribrosa** is the part of the sclerotic and choroid membrane through which the optic nerve-fibres have to pass in order to reach the papilla nervi optici. In the optic nerve the fibres are grouped in larger or smaller groups—not bundles, in the sense of those present in other nerves and surrounded by perineurium (see a former chapter)—but surrounded by septa of connective tissue, and these groups pass through corresponding holes of the sclerotic and choroid.

451. The **optic nerve** possesses three sheaths, composed of fibrous connective tissue—(*a*) an outer, or the dural ; (*b*) a middle, or arachnoidal ; and (*c*) an inner, or pial, sheath—which are continuations of the respective membranes of the brain. The pial sheath is, in reality, the perineurium, the whole optic nerve being comparable to a compound nerve-bundle as described in a former chapter. The dural sheath of the optic nerve, at its entrance into the lamina cribrosa, passes into the outer strata of the sclerotic, while the arachnoidal and pial sheaths pass into the inner strata of the sclerotic. Outside the dural sheath is a lymph

space—the supravaginal space; and also between these various sheaths are lymph spaces—the subdural or subvaginal space of Schwalbe, and the subarachnoidal space. The supravaginal and subvaginal spaces anastomose with one another (Michel).

452. Around the sclerotic is a lymph space limited by a fibrous membrane—the *Tenonian capsule*: the space is called the *Tenonian space*. The supravaginal space anastomoses with this Tenonian space, and into it pass also the lymph clefts in the suprachoroidal tissue (Schwalbe), by means of the lymph canalicular system of the sclerotic (Waldeyer). The suprachoroidal lymph spaces communicate also with the subarachnoidal space of the optic nerve.

CHAPTER XL.

THE OUTER AND MIDDLE EAR.

453. THE meatus auditorius externus is lined with a delicate skin, in structure identical with, but thinner than, the skin of other parts. The ceruminous glands have been mentioned and described before. The cartilage of the auricula and its continuation into the meatus auditorius externus is elastic cartilage.

454. The **membrana tympani** separating the outer from the middle ear has for its matrix a firm stratum of stiff trabeculæ of fibrous connective tissue, with numerous elastic fibrils and elastic membranes. This is the middle and chief stratum of the membrane: outwards it is covered with a delicate continuation of the skin of the meatus auditorius externus, and inwards with a continuation of the delicate mucous membrane lining the cavum tympani. In the middle stratum of the membrana tympani the trabeculæ radiate more or less from the junction of the manubrium mallei with the membrane; but towards the periphery many are

also arranged in a circular direction. The former belong to the outer, the latter to the inner, portion of the middle stratum.

The mucous membrane lining the tympanic surface of the membrane is delicate connective tissue, covered with a single layer of polyhedral epithelial cells.

The blood-vessels form capillary **networks for all three layers**—*i.e.*, a special network for the skin layer, a second for the middle stratum, and a third one for the mucous layer; the lymphatics are also arranged in this way. An intercommunicating system of lymphatic sinuses and clefts (Kessel) is left between the trabeculæ. The non-medullated nerve-fibres form plexuses for the skin and mucous layer; from these pass off fine fibrils, which form a sub-epithelial network, and from this the fibrils pass into the epithelium.

455. The **tuba Eustachii** is lined with a mucous membrane, which is a continuation of that lining the upper part of the pharynx, and therefore, like it, is covered on its inner or free surface with columnar ciliated epithelium. As in the pharynx, so also here, we find a good deal of adenoid tissue in the mucous membrane.

The cartilage of the tuba Eustachii in the adult approaches in structure the elastic cartilages of other parts.

456. The **cavum tympani**, including the cellulæ mastoideæ and the surface of the ossicula auditus, is lined with a delicate connective tissue membrane. Its free surface is covered with a single layer of polyhedral epithelial cells in the following regions: on the promontory of the inner wall of the cavity, on the ossicula auditus, on the roof of the cavity, and in the cellulæ mastoideæ; in all other parts it is columnar ciliated epithelium, like that lining the tuba Eustachii.

457. The three **ossicula auditus** are osseous substance covered with periosteum, which is covered with the delicate mucosa just described. The liga-

ments of the bones are, like other ligaments, made up of straight and parallel bundles of fibrous connective tissue. The articulation surface of the head of the malleus, of the incus, of the extremity of the long process of the incus, and of the stapes, are covered with hyaline (articular) cartilage.

CHAPTER XLI.

THE INTERNAL EAR.

458. THE osseous labyrinth consists of the vestibule, prolonged on one side into the cochlea, and on the other into the three semicircular canals, each of which possesses an ampulla at one extremity. The vestibule shows two divisions—the fovea hemispherica next to the cochlea, and the fovea hemielliptica next to the semicircular canals. The cochlea consists of two and half turnings twisted round a bony axis—the modiolus. From this a bony lamina extends towards the outer wall for each turn, but does not reach it: this is the lamina spiralis ossea. It extends through all turns, and it subdivides the cavity of each turn into an upper passage, or scala vestibuli, and a lower, or scala tympani. At the top of the cochlea the two scalæ pass into one another by the helicotrema. The scala vestibuli opens into the fovea hemispherica, while the scala tympani at its commencement—*i.e.*, at the proximal end of the first turn—would be in communication, by the fenestra rotunda, with the cavum tympani, were it not that this fenestra rotunda is closed by a membrane—the secondary membrane.

459. The semicircular canals start from, and return to, the fovea hemielliptica of the vestibule.

The fenestra ovalis leads from the cavum tympani into the vestibule—its hemispheric division; and this fenestra ovalis is, in the fresh condition, filled out by a membrane, in which the basis of the stapes is fixed, the circumference of this being nearly as great as that of the fenestra.

460. The osseous labyrinth in all parts consists of ordinary osseous substance, with the usual periosteum lining its outer surface and its inner cavities. These cavities contain the albuminous fluid called *perilymph*. But they are not filled out by this, since, in each of the two divisions of the vestibule, in each of the semicircular canals, and in the cochlea, is a membranous structure, analogous in shape to the corresponding division of the labyrinth. These membranous structures possess a cavity filled with the same albuminous fluid as above, called the *endolymph*. These structures are disposed thus—in the fovea hemispherica is a spherical sac, called the *saccule*; in the fovea hemielliptica is an elliptical sac, the *utricle*; in each of the three semicircular canals is a membranous semicircular tube, which possesses also an ampulla corresponding to the ampulla of the bony canal.

461. In the cochlea is a membranous canal, triangular in cross-section—the scala media or cochlear duct—which also twists two and half times from the basis to the apex of the cochlea, and is placed against the end of the lamina spiralis ossea so as to occupy a position between the peripheral part of the scala vestibuli and scala tympani.

462. The different divisions of the membranous labyrinth are connected with one another in this manner: the three semicircular (membranous) canals open into the utricle; this does not form a direct continuity with the saccule, but a narrow canal comes off both from the saccule and utricle; the two canals join into one minute membranous tube situated in the aqueductus vestibuli. At its distal end it enlarges into the saccus endolymphaticus, situated in a cleft of

v

the dura mater, covering the posterior surface of the petrous bone. The saccule is in communication with the cochlear canal or scala media, by a short narrow tube—the canalis reuniens of Reichert. Thus the cavity of the whole membranous labyrinth is in direct communication throughout all divisions, and it represents the inner lymphatic space of the labyrinth. There is no communication between the perilymph and endolymph, and the cavity of the membranous labyrinth stands in no direct relation to the cavum tympani, since the fenestra ovalis and fenestra rotunda both separate the perilymphatic space, or the cavity of the bony labyrinth, from the cavum tympani. The vibrations of the membrana tympani, transferred by the ossicula auditus to the fenestra ovalis, directly affect, therefore, only the perilymph. The fluctuations of this pass from the vestibule, on the one side, towards and into the perilymph of the semicircular canals; and on the other side, through the scala vestibuli, to the top of the cochlea, then by the helicotrema into the scala tympani, and find their conclusion on the membrana secundaria closing the fenestra rotunda. On their way they affect, of course, the membrane of Reissner (see below) separating the scala media from the scala vestibuli; and the vibrations of this membrane naturally affect the endolymph of the scala media and the terminations of the auditory nerve-fibres (see below).

463. Structure of **semicircular canals, utricle and saccule.**—The *membranous semicircular* canals are fixed by stiff bands of fibrous tissue to the inner periosteum of the one (convex) side of the osseous canal, so that towards the concave side there is left the space for the perilymph. A similar condition obtains with regard to the *saccule* and *utricle*, which are fixed by the inner periosteum to one side of the bony part.

The structure of the wall is the same in the semicircular canals, utricle and saccule. The above-mentioned fibrous ligaments of the periosteum form an *outer coat;* inside this is a glassy-looking *tunica propria.* At one side (the one away from the bone) this tunica propria forms numerous papillary projections. The internal surface of the membrane is covered with a single layer of polyhedral epithelial cells.

464. Each of the branches of the nervus vestibuli—*i.e.*, one for the saccule, one for the utricle, and three for the three ampullæ—possesses a ganglionic swelling. The nerve-branch, having passed through the membranous wall, enters special thickenings of the tunica propria, on that part of the membranous wall next to the bone; in the saccule and utricle the thickening is called *macula acustica,* in the ampullæ *crista acustica*

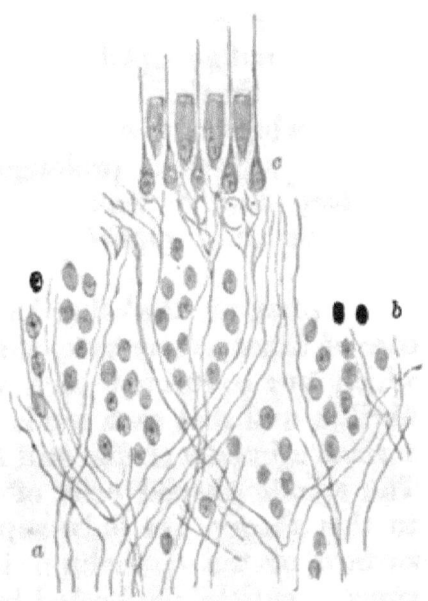

Fig. 162.—From a Transverse Section through the Macula Acustica of the Utricle of the Labyrinth of a Guinea-pig.

a, Medullated nerve fibres, forming plexuses; *b*, nuclei of the membrane; *c*, the sensory epithelium (diagrammatic); the spindle-shaped sensory-cells possess long auditory hairs projecting between the conical epithelial cells beyond the free surface. (Atlas.)

(Fig. 162) (M. Schultze). This thickening is a large villous or fold-like projection of the tunica propria, into which pass the nerve-fibres of the several branches. These fibres are all medullated nerve-fibres, and, ascending towards the internal or free surface of that projection, form a plexus. In this plexus are interspersed numerous nuclei. From the medullated fibres pass off minute bundles of primitive fibrillæ, which enter the epithelium that covers the free surface of the projection.

465. This epithelium is composed of a layer of columnar or conical cells, between which are wedged in spindle-shaped cells; both kinds possess an oval nucleus. According to Max Schultze and others, each of the spindle-shaped cells is connected by its inner process with the nerve-fibrillæ coming from underneath; whereas, towards and beyond the free surface, its outer process is prolonged into a long, thin, stiff, *auditory hair.* Max Schultze, therefore, calls the columnar cells *epithelial;* the spindle-shaped ones, *sensory.*

Retzius, on the other hand, maintains that, in the case of fishes at any rate, the epithelial cells are those which are connected each with a bundle of nerve-fibrillæ, and that each sends out over the internal free surface a bundle of fine stiff hairs—the auditory hairs. The spindle-shaped cells of Max Schultze, according to this theory, are only supporting cells. The free surface of the epithelium is covered with a homogeneous cuticle, perforated by holes which correspond to the epithelial cells and the auditory hairs.

On the internal surface of the macula and crista acustica are found the *otoliths*, rhombic crystals, and amorphous masses, chiefly of carbonate of lime, embedded in a gelatinous or granular-looking basis.

466. The **cochlea** (Fig. 163), as has been mentioned above, consists also of a bony shell and a membranous canal, the former surrounding the latter in the same way as the bony semicircular canal does the membranous—*i.e.*, the latter is fixed to the outer or convex side of the former. The difference between the cochlea and the semicircular canals is this, that in the cochlea there is a division of the perilymphatic

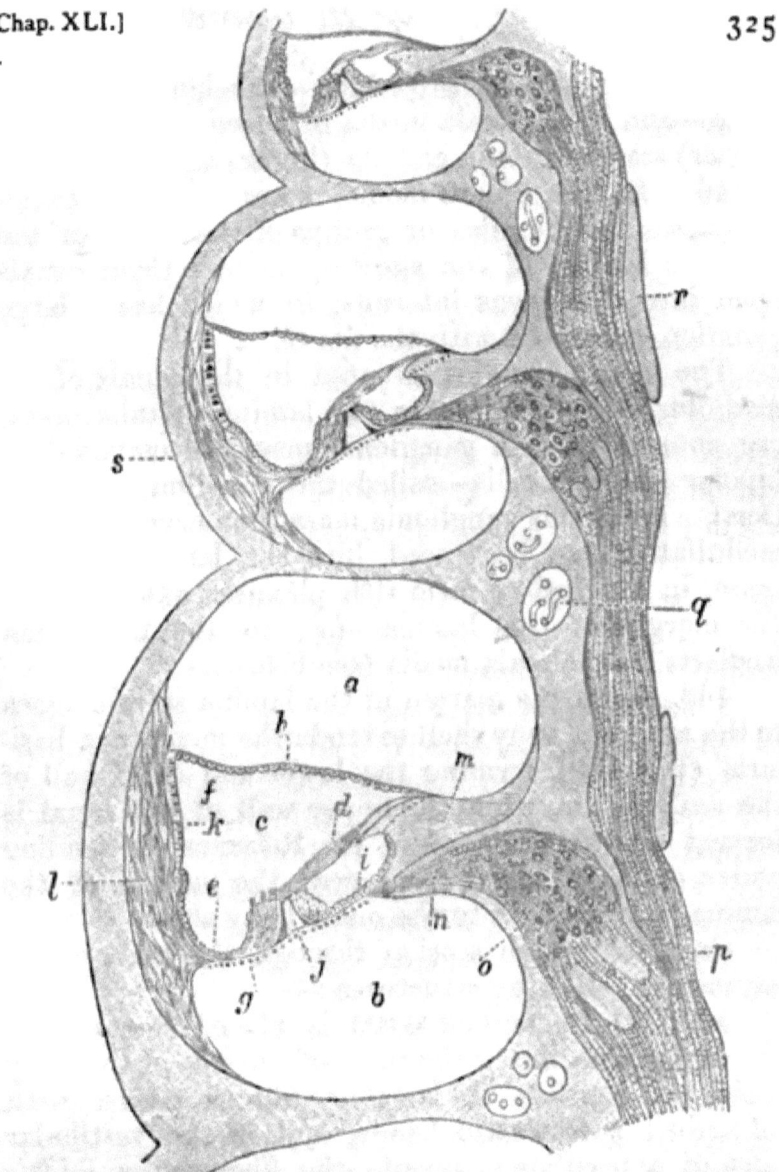

Fig. 163.—From a Vertical Section through the Cochlea of a Guinea-pig's Ear, seen in the long axis of the Modiolus.

a, The scala vestibuli ; *b*, the scala tympani ; *c*, the scala media ; *d*, the membrana tectoria ; *e*, the cells of Claudius ; *f*, the upper outer angle of the scala media ; *g*, the region of the outer hair cells on the membrana basilaris ; *h*, the membrane of Reissner ; *i*, the epithelium lining the sulcus spiralis (internus) ; *j*, the tunnel of Corti's arch ; *k* the stria vascularis ; *l*, the ligmentum spirale ; *m*, the crista spiralis ; *n*, the nerve-fibres in the lamina spiralis ossea ; *o*, the ganglion spirale ; *p*, the nerve-fibres in the modiolus ; *q*, channels in bone containing blood-vessels ; *r*, masses of bone in the modiolus ; *s*, the bony capsule. (Atlas.)

space by an osseous projection—the lamina spiralis ossea—and by the scala media into two scalæ, viz., the (upper) scala vestibuli, and the (lower) scala tympani.

467. In the osseous modiolus are numerous **parallel canals** for bundles or groups of the **fibres of the** cochlear branch of the auditory nerve; these canals open into the porus internus, in which lies a large ganglion connected with the nerve.

The nerve bundles situated in the canals of the modiolus, corresponding to the lamina spiralis ossea, are connected with ganglionic masses—composed of bipolar ganglion cells—called the ganglion spirale of Corti. From this ganglionic mass the nerve-fibres (all medullated) can be traced into the lamina spiralis ossea, in which they form rich plexuses extending to the margin of this lamina—*i.e.*, to the membrana basilaris of the scala media (see below).

468. From the margin of the lamina spiralis ossea to the external bony shell extends the membrana basilaris (Fig. 163), forming the lower and chief wall of the scala media, while the upper wall of the canal is formed by the membrane of Reissner, extending under an acute angle from near the margin of the lamina spiralis ossea to the outer bony shell.

On a transverse section through the scala media we see the following structures:—

469. (1) Its **outer wall** is placed close against the periosteum lining the internal surface of the bony shell; it consists of lamellar fibrous tissue, with numerous stiff elastic bands, and is the vestibular part of a peculiar ligament—the *ligamentum spirale* (Kölliker)—semi-lunar in cross section, and with its middle angular projection fixed to the outer end of the membrana basilaris.

470. (2) Its **inner wall** is represented by an exceedingly delicate membrane—the *membrane of Reissner;* this is also its upper wall, extending under

an acute angle from the upper outer angle of the scala media to the lamina spiralis ossea. But there it is not fixed on the osseous substance, but on a peculiar projection on this latter—the *crista spiralis* (Fig. 163, *m*), which is a sort of tissue intermediate between fibrous and osseous tissue, and added to the vestibular surface of the lamina spiralis ossea. This crista spiralis has on its inner surface—*i.e.*, that directed towards the scala media—a deep sulcus, called the *sulcus spiralis*, or sulcus spiralis internus; so that of the crista spiralis there are two labia to be distinguished—the labium vestibulare and the labium tympanicum; the former being the upper, the latter the lower, boundary of the sulcus spiralis.

471. (3) Between the labium tympanicum of the crista spiralis and the above-mentioned projection of the ligamentum spirale extends, in a straight direction, the *membrana basilaris*, forming the lower wall of the scala media. The scala media is lined on its whole internal surface with epithelium, this only being derived from the epithelium forming the wall of the auditory vesicle of the embryo, peculiarly modified in certain places. The scala tympani and scala vestibuli are likewise lined with a continuous layer of flattened cells—an endothelium, which only on the lower (tympanic) surface of the membrana basilaris is somewhat modified, being composed of granular looking irregular cells.

472. As regards the scala media, the epithelium lining its internal surface is of the folllowing aspect: Starting with the lower outer angle—*i.e.*, where the membrana basilaris is fixed to the ligamentum pirale—we find it a single layer of polyhedral or short columnar transparent cells, lining this outer angle—*the cells of Claudius*; ascending on the ligamentum spirale, the cells become shorter, more squamous; such are found over a slight projection on

the outer wall—*i.e.*, the ligamentum spirale accessorium—caused by a small blood-vessel, the vas prominens.

473. Then we come to the **stria vascularis**, lining nearly the upper two-thirds of the outer wall of the scala media. It consists of a layer of columnar and spindle-shaped epithelial cells, between which extend capillary blood-vessels from the ligamentum spirale, and in some animals (guinea-pig) clumps of pigment granules are found between them.

474. Then we pass from the upper angle of the scala on to the membrane of Reissner. This consists of a homogeneous thin membrana propria, covered on its outer vestibular surface with a layer of flattened endothelium, and on its inner surface—*i.e.*, that facing the scala media—with a layer of less flattened, smaller, polyhedral, epithelial cells.

475. We come next to the vestibular labium of the crista spiralis, on which peculiar cylindrical horizontal projections anastomose with one another: these are the *auditory teeth* (Huschke). The epithelium of Reissner's membrane is continued as small polyhedral cells into the grooves and pits between the auditory teeth, but over the teeth as large, flattened, squamous cells, which pass on, lining the sulcus spiralis, and, as such, cover also the tympanic labium of the crista spiralis. Now we arrive at the membrana basilaris, on which the epithelium becomes modified into the *organ of Corti*.

476. The **membrana basilaris** consists of a hyaline basement membrane, on which the organ of Corti is fixed; underneath this is the *tunica propria*, a continuation of the tissue of the ligamentum spirale, composed of fine parallel stiff fibrils (Hannover, Henle) stretched in a very regular and beautiful manner in the direction from the ligamentum spirale to the crista spiralis (Nuel). On the tympanic side

there is also a hyaline basement membrane. The endothelial cells covering this on the tympanic surface have been mentioned above.

477. The **organ of Corti** (Fig. 164).—Passing outwards from the epithelium lining the sulcus spiralis, we meet with small polyhedral epithelial cells in the region of the termination of the lamina spiralis ossea, next which are columnar-looking cells—the *inner supporting cells;* next to these is the *inner hair-cell—*

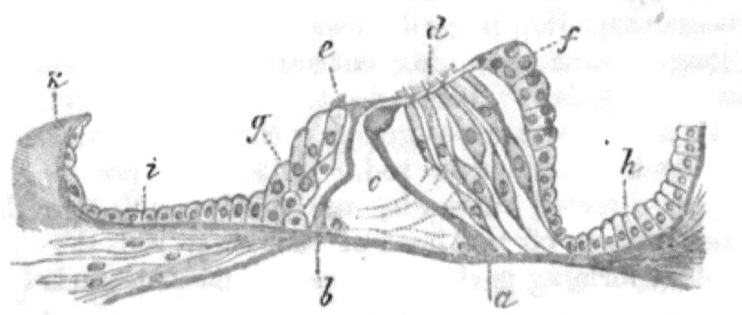

Fig. 164.—Organ of Corti of the Cochlea of a Guinea-pig.

a, Outer rod or pillar of Corti ; *b*, inner rod or pillar of Corti ; *c*, tunnel of Corti's arch; *d*, outer hair-cells; *e*, inner hair-cell ; *f*, outer supporting cells containing fat globules ; *g*, inner supporting cells; *h*, cells of Claudius ; *i*, epithelial cells lining the sulcus spiralis internus; *j*, nerve-fibres; *k*, part of crista spiralis. (Atlas.)

a columnar, or conical, epithelial cell, with a bundle of stiff hairs, or rods, extending beyond the surface. The inner hair-cells form a single file along the whole extent of the two and a half turns of the scala media.

478. Next to the inner hair-cell is the *inner rod, or inner pillar, of Corti,* and next to this the *outer rod, or outer pillar, of Corti.* Each forms a single file for the whole extent of the two and a half turns of the scala media. The two rods are inclined towards one another, and in contact with their upper extremity, or *head;* whereas their opposite extremity, the *foot,* rests under an acute angle on the membrana basilaris, on which it is firmly fixed. The rest of the rod is a slender, more or less cylindrical, piece—the *body.* The

outer rod is larger and longer than the inner, the latter being slightly bent in the middle. Owing to the position of the rods, the two files form an arch—the *arch of Corti*. Between it and the corresponding part of the basilar membrane is a space—the *tunnel* of the arch, triangular in cross section.

479. The substance of the rods, or pillars, **of Corti** is bright, highly refractive, and slightly and longitudinally striated.

The head of the inner rod is triangular, a short process extending inwards towards the inner hair-cell, a long process extending outwards over the head of the outer pillar. Outwards, the triangular head possesses a concave surface grasping the convex surface of the head of the outer rod. This latter possesses a process directed outwards, and firmly applied to the outer process of the head of the inner rod, the two together forming part of the membrana reticularis (*see* below).

The relation between the outer and inner rods is such that the head of one outer rod fits into those of about two inner rods.

480. At the foot, each rod has, on the side directed towards the tunnel, a granular, nucleated, lump of protoplasm, probably the remnant of the epithelial cell from which the lower half of the rod is derived; the upper part sometimes possesses a similar remnant, proving that this also has been formed by an epithelial cell, so that each rod is in reality derived from two epithelial cells (Waldeyer).

481. Next follow three or four rows of *outer hair-cells*, similar in size and structure to the inner hair-cells. Each of the outer hair-cells represents a file of hair-cells, extending on the membrana basilaris along the whole extent—*i.e.*, two and a-half turns—of the scala media. Each hair-cell possesses an oval nucleus and a number of stiff rods, or hairs, disposed

in the shape of a horseshoe in the outer part of the free surface of the cell.

Four, and even five, rows of hair-cells (Waldeyer), arranged in an alternating manner, are found in man.

The outer hair-cells are also called the cells of Corti; they are conical, and more or less firmly connected with a nucleated spindle-shaped cell—the cell of Deiters. The two cells are more or less fused together in their middle part (Nuel). The cell of Corti is fixed by a branched process to the membrana basilaris, while the cell of Deiters sends a process towards the surface, where it joins the membrana reticularis (*see* below).

482. Farther outwards from the last row of outer hair-cells are columnar epithelial cells, called the *outer supporting cells of Hensen*; they form the transition to the epithelium lining the outer angle of the scala media, *i.e.*, to the cells of Claudius.

In the guinea-pig, the outer supporting cells include fat globules.

483. The **medullated nerve-fibres**, which we traced in a former page to the margin of the lamina spiralis ossea, make rich plexuses in this, and pass through holes in it, in order to reach the organ of Corti on the membrana basilaris. Looking from the surface on this part, we notice a row of holes—the *habenula perforata* of Kölliker—a little to the inside of the region of the inner hair-cells. Numerous primitive fibrillæ pass there among small nucleated cells situated underneath the inner hair-cells: these are the granular cells. Some of these nerve-fibrillæ —the inner bundle of spiral nerve-fibres—become connected with the inner hair-cells; while others—the three outer bundles of spiral fibrils (Waldeyer)—pass, between the inner rods of Corti, right through the tunnel; and, further, penetrating between the outer

rods of Corti, they reach the outer hair-cells, with which they become connected (Gottstein, Waldeyer).

484. In connection with the outer process of the head of the inner and outer rods of Corti, mentioned above, is an elastic hyaline membrane—the *lamina* or *membrana reticularis*. It extends outwards over the organ of Corti to the supporting cells of Hensen, and possesses holes for the tops of the outer hair-cells and their hairs. The parts between the rods of Corti and between the outer hair-cells appear of the shape of phalanges—phalanges of Deiters. A short cuticular membrane extends from the head of the inner rod of Corti inwards to the inner supporting cells: it possesses holes for the tops of the inner hair-cells.

485. From the vestibular labium of the crista spiralis to the outer hair-cells of the organ of Corti extends a peculiar fibrillated membrane—the *membrana tectoria*. By means of it the sulcus spiralis internus is bridged over, and so converted into a canal.

486. As we ascend towards the top of the cochlea, all parts in the scala media decrease gradually in size. The organ of Corti, being of an epithelial nature, possesses no blood-vessels. From the anatomical relations of the organ of Corti, it appears most probable that the pillars, or rods, of Corti act as the supporting tissue, or framework, around which the other elements are grouped; and it seems likely that the hair-cells, with their rod-like hairs projecting freely into the endolymph, are the real sound-perceiving elements of the organ of Corti. Their connection with the terminal fibrillæ of the nerves points in the same direction.

CHAPTER XLII.

THE NASAL MUCOUS MEMBRANE.

487. THE lower part of the nasal cavity is lined with a mucous membrane, which has no relation to the olfactory nerve, and therefore is not connected with the organ of smell. It is covered with a stratified, columnar, ciliated epithelium of exactly the same nature as that of the respiratory passages—*e.g.*, the larynx and trachea. Large numbers of mucous secreting goblet-cells are met with in it. Below the epithelium is a thick hyaline basement membrane, and underneath this is a mucosa of fibrous tissue, with numerous lymph corpuscles in it. In many places this infiltration with lymph corpuscles amounts to diffuse adenoid tissue, or to perfect lymph follicles.

488. The mucosa contains in its most superficial layer the network of capillaries, but in the rest it includes a rich and conspicuous plexus of venous vessels.

In the deeper parts of the mucous membrane—*i.e.*, in the submucosa—are embedded smaller and larger glands, the ducts of which pass through the mucosa, and open on the free surface. Some of the glands are mucous; others are serous. In some cases (*e.g.*, guinea-pig) almost all glands are serous, and of exactly the same nature as those of the back of the tongue. In some places the mucous membrane is much thicker than in others, and then it contains larger glands, and between them bundles of non-striped muscular tissue.

489. In the upper or **olfactory region** (Fig. 165) of the nasal cavity, the mucous membrane is of a

different tint, being more of a brownish colour; it contains the ramifications of the olfactory nerve, and is the seat of the organ of smell.

490. The free surface is covered with a columnar

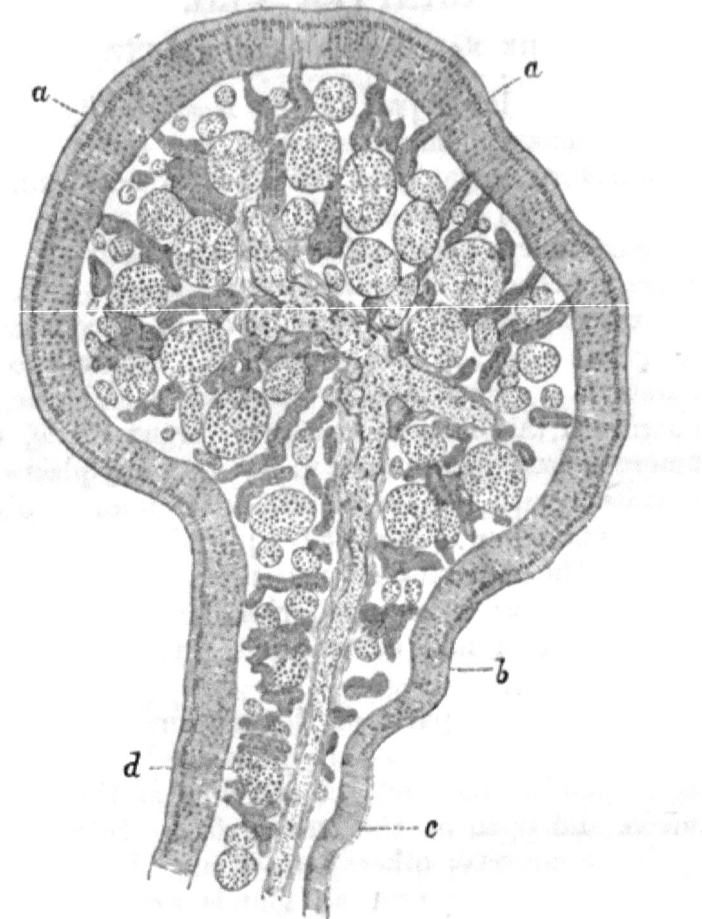

Fig. 165.—From a Section through the **Olfactory Region** of the Guinea-pig.

a, Thick olfactory epithelium; *b*, thin olfactory epithelium; *c*, ciliated non-olfactory epithelium; *d*, bone. The transverse sections of the olfactory nerve bundles and the tubular glands of **Bowman** are well seen. (Atlas.)

epithelium, composed of the following kinds of cells (Fig. 166):—

(*a*) A superficial layer of long columnar, or rather

conical, *epithelial cells*, each with an oval nucleus. In some places the free surface of these cells is covered with a bundle of cilia, similar to the superficial cells

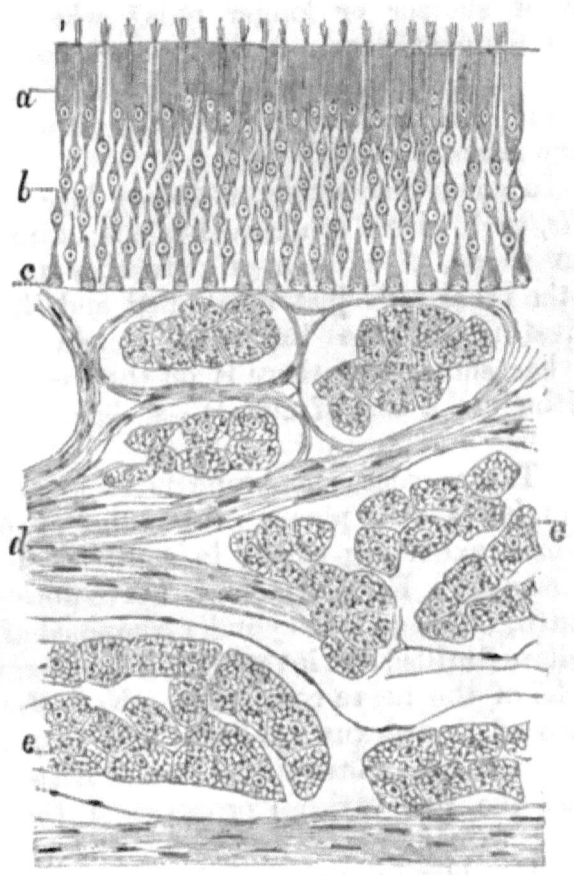

Fig. 166.—From a Vertical Section through the Olfactory Mucous Membrane of the Guinea-pig.

a, The epithelial cells; *b*, the sensory or olfactory cells; *c*, the deep epithelial cells; *d*, the bundles of olfactory nerve-fibres; *e*, the alveoli of serous (Bowman's) glands. (Atlas.)

of the respiratory part of the nasal cavity; in most places, however, the cilia are absent; the former condition obtains in those places which are in close proximity to the respiratory region.

(*b*) Between the epithelial cells extend spindle-shaped cells, each with a spherical, or very slightly oval, nucleus—the *sensory cells* (Max Schultze). Each cell sends one broad process towards the free surface, over which it projects in the shape of a small bundle of shorter or longer rods; whereas a fine varicose filament passes from the cell body towards the mucosa, and, as shown first by M. Schultze, becomes connected with a fibrilla of the network of the olfactory nerve-fibres.

(c) In some places there is a deep layer of *epithelial cells*, each with a spherical nucleus, of an inverted cone in shape, their pointed extremity passing between the other cells just mentioned and their broad basis resting on the basement membrane. Von Brunn has shown that there is on the free surface of the epithelium a sort of cuticle—a delicate limitans externa.

491. The mucous membrane is of loose texture, and contains a rich plexus of bundles of olfactory nerve-fibres, extending chiefly in a direction parallel to the surface. Each olfactory nerve-fibre is non-medullated, *i.e.*, is an axis cylinder composed of minute or primitive fibrillæ, and invested in a neurilemma with the nuclei of the nerve corpuscles. Near the surface the fibres of the plexus are thin, and they split up into the constituent fibrils which form a network; into this pass the fine varicose processes of the sensory cells above named.

492. The blood-vessels supply with capillary networks the superficial part of the mucous membrane and the numerous glands. These are the *glands of Bowman*, extending through the thickness of the mucous membrane. They are tubes, slightly branched, and gradually enlarging towards their distal end; in some parts they are more or less straight. In structure they are identical with serous glands, possessing a minute lumen, and being lined with a layer of columnar albuminous cells. The duct is a very fine

canal; it is that part of the gland that is situated in the epithelium of the free surface; it passes vertically through this, and consists of a fine limiting membrane, the continuation of the membrana propria of the gland-tube, and a layer of very flattened epithelial cells.

493. There is a definite relation between the size and number of the bundles of the olfactory nerve-fibres, the thickness of the olfactory epithelium, and the length of the gland-tubes. The size and number of the bundles of the nerve-fibres are determined by the thickness of the epithelium—*i.e.*, by the number of the sensory cells; the number and thickness of the olfactory nerve bundles determine the thickness of the mucous membrane, and the thicker it is, the longer are the glands of Bowman.

494. The **organ of Jacobson** is a minute tubular organ present in all mammals, and, as has been shown by Dursy and Kölliker, also in man. In mammals it is a bilateral tube, compressed from side to side, and situated in the anterior lower part of the nasal septum. Each tube is supported by a hyaline cartilage, in the shape of a more or less plough-shaped capsule—the *cartilage of Jacobson* —and opens in front directly into the nasal furrow (guinea-pig, rabbit, rat, &c.); or it leads into the canal of Stenson (dog), which passes through the canalis naso-palatinus, and opens immediately behind the incisor teeth on the palate. In all instances, however, it terminates posteriorly with a blind extremity.

495. The cavity of the tube is lined with stratified columnar epithelium, which on the lateral wall is ciliated in the guinea-pig and dog, and non-ciliated in the rabbit. The median wall—*i.e.*, the one next to the middle line—is lined with olfactory epithelium, identical with that of the olfactory region of the nasal

W

cavity. Branches of olfactory nerve-fibres also pass into the median wall, and behave in exactly the same manner as in the olfactory region. Numerous serous glands—belonging chiefly to the upper and lower wall—open into the cavity of the organ of Jacobson.

In the lateral wall there is in many instances a plexus of veins, extending in a longitudinal direction, and between the vessels are bundles of non-striped muscular tissue, thus constituting a sort of cavernous tissue.

CHAPTER XLIII.

THE DUCTLESS GLANDS.

496. I. The **hypophysis cerebri.**—The upper or smaller lobe belongs to the central nervous system. The lower or larger lobe is surrounded by a fibrous capsule, which sends numerous minute septa into the interior. These split up into numerous trabeculæ of fibrous tissue, which, by dividing and re-uniting, form a dense plexus, with smaller and larger, spherical or oblong, or even cylindrical spaces—the alveoli. In these lie spherical or oblong masses of epithelial cells. These epithelial cells are columnar, pyramidal, or polyhedral, each with an oval or spherical nucleus. Between the epithelial cells of the same group are found here and there small branched or spindle-shaped cells, with a small flattened nucleus. In some of the groups or alveoli of epithelial cells is a cavity, a sort of lumen, filled with a homogeneous gelatinous substance.

The interalveolar connective tissue contains a network of capillaries. Between the alveoli and the interalveolar tissue there are lymph sinuses, like those around the alveoli of other glands—*e.g.*, the salivary glands.

497. II. The **thyroid gland** (Fig. 167).—The framework of this gland is in many respects similar to that of other glands, there being an outer fibrous capsule, thicker and thinner septa, and finally the fine trabeculæ forming the septa between the gland alveoli. These are *closed vesicles* of a spherical or oval shape, and of various sizes. Each vesicle is lined with a single layer of polyhedral or columnar epithelial cells, each with a spherical or oval nucleus. There is a cavity, which differs in size according to the size of the vesicle. It contains, and is more or less filled with, a homogeneous, viscid, albuminous fluid—the so-called colloid. In this often occur degenerating nucleated lymph-corpuscles and coloured blood-corpuscles (Baber).

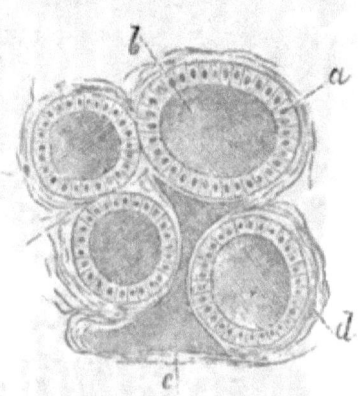

Fig. 167.—From a Section through the Thyroid Gland of Dog.

a, The epithelium lining the vesicles; *b*, the "colloid" contents of the vesicles; *c*, a lymphatic filled with the same material as the vesicles; *d*, the fibrous tissue between the gland vesicles.

498. The vesicles are surrounded by networks of blood capillaries. In the connective tissue framework lie networks of lymphatics; between the framework and the surface of the vesicles are lymph sinuses lined with endothelium (Baber). The large and small lymphatics are often filled with the same colloid material as the vesicles, and it is probable that this colloid material is produced in the vesicles, and carried away by the lymphatics, to be finally discharged into the circulating blood.

499. Its formation in the vesicles is probably due to an active secretion by the epithelial cells of the vesicles, and to a mixture with it, or maceration by it, of the effused blood mentioned above. In some

instances Baber found the amount of blood effused into the cavity of the vesicles very considerable, and hence it is justifiable to assume that the destruction of red blood-corpuscles forms one of the functions of the thyroid gland.

500. III. The **suprarenal bodies** (Fig. 168).—The suprarenal body is enveloped in a fibrous capsule; in connection with this are septa and trabeculæ passing inwards, and they are arranged differently in the cortex and in the medulla of the gland, as will be seen presently.

The *cortex* of the gland consists of an outer, middle, and inner zone, all three being directly continuous with one another. The outer one is the *zona glomerulosa;* it contains numerous spherical, or,

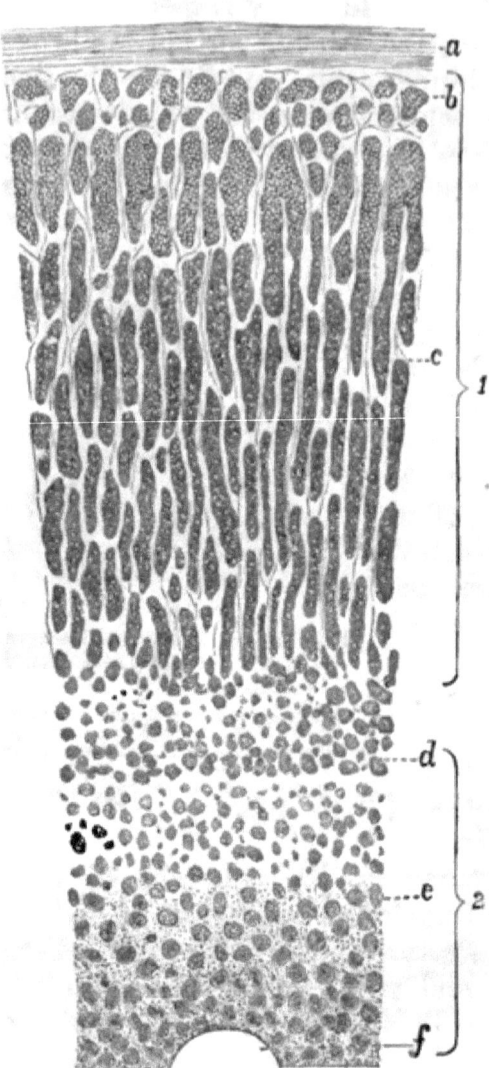

Fig. 168.—From a Vertical Section through the Suprarenal Body of Man.

1, Cortical substance; 2, medullary part; *a*, outer capsule; *b*, zona glomerulosa; *c*, zona fasciculata; *d*, zona reticularis; *e*, medulla; *f*, a large vein. (Elberth, in Stricker's Manual.)

more commonly, elongated, masses of epithelial cells. The cells are polyhedral or cylindrical, each with a spherical or oval nucleus. In some animals—as the dog, horse—the cells are thin and columnar, and arranged in a transverse manner. Occasionally a sort of lumen can be discerned in some of these cell masses.

501. Next follows the middle zone, or *zona fasciculata*. This is the most conspicuous and broadest part of the whole gland. It consists of vertical columns of polygonal epithelial cells, each with a spherical nucleus. The cell substance is transparent, and often contains an oil globule. The columns anastomose with their neighbours. Between the columns are fine septa of connective tissue carrying blood capillaries.

Between the cell columns and the connective septa are seen here and there lymph spaces, into which lead fine channels, grooved out between some of the cells of the columns.

502. Next follows the inner zona, or *zona reticularis*, composed of smaller or larger groups of polyhedral cells, with more or less rounded edges. These cell-groups anastomose with one another. The individual cells are slightly larger, and their substance is less transparent than those of the zona fasciculata. In the human subject they are slightly pigmented.

503. In the *medulla* we find cylindrical streaks of very transparent cells; the streaks are separated by vascular connective tissue. The cells are polyhedral, columnar, or branched. These cell-streaks anastomose with one another, and are directly continuous with the cell-groups of the zona reticularis of the cortex.

504. The cortex is richly supplied with dense networks of capillary blood-vessels; their meshes are polyhedral in the outer and inner zone, elongated in the middle zone, or zona fasciculata. In the medulla

numerous plexuses of veins are met with. In the centre of the suprarenal body lie the large efferent venous trunks. In the capsule (Kölliker, Arnold), and in the connective tissue around the central veins, are plexuses of lymphatic tubes with valves. The nerves are very numerous, and composed of non-medullated fibres; in the medulla they form rich plexuses. In connection with these and with those of the outer capsule are small ganglia (Holm, Eberth).

505. IV. The **glandula coccygea and intercarotica.**—The first of these is a minute corpuscle situated in front of the apex of the os coccygis, and was discovered by Luschka. The glandula carotica of Luschka (ganglion intercaroticum) is of exactly the same structure as the glandula coccygea.

506. Its framework is of about the same nature as that of other glands—a fibrous capsule and inner fibrous septa and trabeculæ. The septa and trabeculæ contain in some places bundles of non-striped muscular tissue (Sertoli).

507. The spaces of the **framework are** occupied by the parenchyma. This consists of spherical or cylindrical masses of cells connected into networks. The individual cells are polyhedral epithelial cells, each with a spherical nucleus. According to Luschka, in the new-born child they are ciliated. In the centre of each of the cell-masses lies a capillary blood-vessel, much convoluted and wavy.

Numerous non-medullated nerve-fibres forming a plexus are situated in the framework of the gland.

INDEX.

Absorption (see Lymphatics).
Achromatin, 9.
Acini of glands, 180, 182.
—— of liver, 211.
—— of pancreas, 209.
Adenoid reticulum, 44, 80, 92.
—— tissue, 92.
Admaxillary glands, 179.
Adventitia of arteries, 76, 78.
—— of capillaries, 80, 93.
—— of veins, 79.
—— lymphatic vessels, 102.
Agminated glands, 95.
Air cells, 220.
Alæ nasi, 46.
Albuginea of ovary, 258.
Albumin membrane of Ascherson, 273.
Albuminous cells, 183.
Alveolar cavity, 171, 178.
—— ducts, 65, 220.
Alveoli of glands, 180, 182.
—— of lung, 220.
—— of pancreas, 209.
Amœboid corpuscles, 9.
—— movement, 4, 5, 14, 41, 90, 93.
Ampulla, 320, 323.
Anterior commissure, 157.
—— nerve-roots from cord, 136.
Aorta, 65, 78.
Aponeurosis, 32.
Aqueductus cochleæ, 321
—— Sylvii, 150, 160.
—— vestibuli, 321.
Arachnoid membrane of spinal cord, 127.
—— of brain, 149.
Arachnoidal villi, 149.
Areolar tissue, 33.
Arrector pili, 64, 287.
Arteriæ ciliares breves, 305.
—— —— recurrentes, 305.
—— helicinæ, 256.
—— rectæ, 242.
Arteries, 65, 76.
Arterioles, Afferent, 40, 51.

Arterioles of ovary, 267.
Articular cartilage, 47, 60.
Articulation nerve-corpuscles, 121.
Arytenoid cartilage, 215.
Ascending loop-tube, 238.
Aster stage in nucleus, 8.
Auditory hairs, 324.
—— nerve, Origin of, 149.
—— ——, Division of, 148.
—— teeth, 328.
Auerbach's plexus, 170, 201, 207.
Auricle of heart, 75.
Auriculo-ventricular valves, 75.
Axis cylinder, 107.
—— —— process of sympathetic ganglion cell, 169.

Bartholin's glands, 270.
Basement membranes of skin, 277.
Basilar membrane of cochlea, 327, 328.
Bellini's ducts, 239.
Bile-duct, 213, 214.
Bile capillaries, 213.
Bilirubin, 13.
Bladder, 64, 243.
Blastoderm of chick, 2.
Blood, 10.
—— corpuscles, 10, 13.
—— ——, their origin, 15.
—— -glands of His, 92.
—— -plates of Bizzozero, 15.
—— -vessels in grey matter, 151.
Bone, 50.
——, Development of, 55.
—— cartilage, 50.
—— cells, 31, 53.
—— trabeculæ, 54.
—— corpuscles, 53.
Bowman's capsule, 234.
—— elastica anterior, 44.
—— glands, 336.
—— membrane, 295.
—— sarcous elements, 63.
Brain membranes, 149.
Brainsand, 163.

Bronchi, 65, 219.
Brownian molecular movement, 189.
Bruch, Glands of, 294.
Brücke's elementary organisms, 5.
—— oikoid and zooid, 12.
—— tunica nervea, 317.
Brunner's glands, 208, 209.
Buccal glands, 187.
Bulbus olfactorius, 159.
Bütschli's nuclear spindle, 9.

Calcification of bone, 55.
—— of cartilage, 47.
—— of dentine, 62.
Calices of kidney, 230.
Canal of Schlemm, 299.
—— of Stenson, 337.
—— of Stilling, 307.
Canalis hyaloideus, 307.
—— Petiti, 307.
—— reuniens, 322.
Canaliculi in bone, 52.
—— in cartilage, 48.
Capillaries of marrow-bone, 81.
—— of nerve system, 80.
Capillary bile-ducts, 213.
—— blood-vessels, 80, 201.
—— lymphatics, 86.
—— network in mucosa, 190.
—— sheaths, 228.
Capsule, External, of brain, 162.
——, Internal, of brain, 162, 163.
—— of Glisson, 210.
—— of kidney, 65.
—— of the spleen, 65, 225.
Cartilage, 45.
——, Articular, 47, 60.
——, cells, 31, 48.
——, Elastic, 49.
——, Fibrous, 48.
——, Hyaline, 46.
——, Lacunæ, 46.
—— of Jacobson, 337.
—— of Luschka, 215.
Cavernous tissues in genital organs, 255.
Cavities of tendon sheaths, 88.
Cavum tympani, 319.
Cells, 5.
——, Muscular, of blood-vessels, 65.
—— ——, of intestine, 65.
—— ——, of respiratory organs, 65.
—— ——, of stomach, 65.
—— ——, of urinary organs, 65.
—— in tadpole's tail, 35.
—— of Claudius, 327, 331.
Cellulæ mastoideæ, 319.

Cellular tissue, 33.
Cement of teeth, 174.
—— substance, 80, 82, 90.
—— —— of endothelium, 25.
—— —— of epithelium, 17.
—— —— of fibrous tissue, 32.
Central canal, 128.
—— grey nucleus, 134.
Centroacinous cells, 210.
Centrum ovale, 156, 157.
Cerebellum, 149, 152.
Cerebrum, 149, 156, 157.
Ceruminous glands, 278, 318.
Cervix of uterus, 266.
Chalice cells, 22.
Chondrin, 31, 45.
Chondroclasts, 62.
Choroidal portion of ciliary muscle, 64.
Choroid membrane, 303.
Chromatin, 9.
Chyle, 206.
Ciliary muscle, 65, 300.
—— nerve, 163.
—— processes, 303.
Ciliated cells, 21.
Circulus arteriosus in iris, 302.
Circumanal glands, 278.
Cisterna lymphatica magna, 89.
Clarke's columns, 137, 140.
Claustrum, 159, 162.
Cleavage of ovum, 2.
—— of white blood corpuscles, 15.
Clitoris, 270.
Cochlea, 320, 324.
Cohnheim's areas, 70.
Colloid, 339.
Colostrum corpuscles, 273.
Coloured blood corpuscles, 10.
Colourless blood corpuscles, 13.
Columnar epithelial cells, 17.
Commissure, Grey, of spinal cord, 128.
——, White, of spinal cord, 135.
Compound lymphatic glands, 98.
Conarium, 163.
Concentric bodies of Hassall, 97.
—— lamellæ, 53.
Cone fibre, 316.
Cones of retina, 313.
Coni vasculosi, 250.
Conjunctiva, 291.
—— blood-vessels, 293.
—— bulbi, 293.
—— lymphatics, 293.
—— nerves, 294.
—— palpebræ, 291, 293.
Connective tissue, 31.
Contraction wave, 67.

INDEX. 345

Contractility of corneal corpuscles, 37.
—— of pigment cells, 28.
Convolution in brain, 157.
—— in nucleus, 8.
Cordæ tendineæ, 75.
Cords of adenoid tissue, 94.
Corium, 274.
Cornea, 20, 295.
—— nerves in, 297.
Corneal cells, 36.
—— corpuscles, 34, 297.
Cornu Ammonis, 158, 159.
Cornua uteri, 267.
Corona radiata, 163.
Corpora albicantia, 161, 163.
—— cavernosa, 65, 255.
—— quadrigemina, 160.
—— striata, 161, 162.
Corpus callosum, 157.
—— Highmori, 244.
—— luteum, 262.
—— restiforme, 144, 142.
—— spongiosum, 255.
Corpuscles, Malphigian, 227, 233.
—— of blood, 10, 13.
—— of bone, 53.
—— of connective tissue, 36.
—— of Grandry, 120.
—— of Herbst, 115, 118.
—— of lymph, 90, 93.
—— of Meissner, 115, 118.
—— of muscle, 70.
—— of nerve, 108.
—— of Pacini, 115, 121, 291.
—— of Vater, 115.
——, Tactile, 115, 118.
Corti's arch, 330.
—— cells, 331.
—— ganglion, 326.
—— rods, 329.
Cortical layer of ovary, 258.
—— lymph-sinus, 101.
Costal cartilages, 46.
—— pleura, 89.
Cowper's glands, 254, 270.
Cremaster internus, 252.
Crescents of Gianuzzi, 183.
Cricoid cartilage, 46.
Crista acustica, 323.
—— spiralis, 327, 332.
Crus cerebri, 142, 161.
—— ——, Crusta of, 163.
Crusta petrosa, 171.
Crypts, 95.
—— of Lieberkühn, 203, 209.
Cuticle of Nasmyth, 172, 177.
Cutis anserina, 287.
—— vera, 274.

Cystic duct, 65.
Cytogenous tissue, 92.

Deiters' cells, 150, 331.
—— phalanges, 332.
—— processes, 141, 145.
Demilunes of Heidenhain, 183.
Dentinal canals, 173.
—— fibres, 62, 173.
—— sheaths, 173.
—— tubes, 62.
Dentine, 62, 172, 173.
Descemet's membrane, 44.
Diapedesis, 82.
Diaphragm, 73, 88.
Diaster stage in nucleus, 8.
Diffuse adenoid tissue, 93.
Dilatator, 65.
—— pupillæ, 302.
Direct division, 7, 9.
"Disc tactil," 121.
Discus proligerus, 261.
Disdiaclasts, 74.
Distal convoluted tubes, 238.
Division, Remak's mode of, 7.
Doyère's nerve-mount, 125.
Ductless glands, 338.
Ducts of pancreatic gland, 65.
—— of salivary gland, 65.
Ductus ejaculatorii, 253.
Dura mater, 127.
Dural sheath, 317.

Ear, External, 318.
——, Internal, 320.
Efferent lymphatics, 102.
—— medullated nerve-fibres, 116.
—— veins, 40.
Elastic fenestrated membrane of Henle, 44.
Elastin, 43.
Electric nerve, 111.
Eleidin, 20.
Elementary fibrillæ, 107.
—— fibrils, 32.
—— organisms, 4.
Enamel, 171.
—— cap, 175.
—— cells, 175, 177.
—— organ, 175, 177.
—— prisms, 171.
End-bulbs of Krause, 115, 119, 294.
Endocardium, 74, 76.
Endochondral formation of bone, 55.
Endolymph, 321.
Endomysium, 66.

Endoneurium, 105.
Endothelial cells, structure, 25.
—— membrane, 25.
Endothelium, 25.
Endotheloid cell-plates, 92.
Engelmann's lateral disc, 68.
Epicerebral space, 150.
Epidermis, 19, 274.
Epididymis, 65, 250.
Epiglottis, 49, 93.
Epineurium, 104.
Epithelial cells, 16.
—— division of, 24.
—— regeneration of, 24.
Epithelium, 19.
Epitheloid layer, 56.
Eustachian tube, 319.
External arcuate fibres, 144, 147.
—— capsule of brain, 162.
Eye, 290.
—— -lashes, 292.
—— -lids, 291.

Fasciæ, 87.
Fascicles, 66.
Fasciculus cuneatus, **144**.
—— gracilis, 144.
—— of Goll, 132, 144.
—— of Türk, 132, 142, 145.
—— pyramidal direct, 132.
Fat cells, 39.
—— —— and starvation, 41.
Femur, **60**.
Fenestra ovalis, 320.
—— **rotunda**, 320.
Fenestrated membrane, 30, 33, 127.
—— —— of Henle, 77.
Fertilisation of ovum, 2.
Fibræ arcuatæ, 296.
Fibres, Connective tissue, 31.
——, Elastic tissue, 50.
—— of muscle, 63, 66, 72.
—— of nerves, 103.
—— of Purkinje, 75.
Fibrillæ of connective tissue, 32.
—— of muscle, 64.
—— of nerve, 107.
Fibro-cartilage, 48.
Fibrous tissue development, 42.
Fillet, 160.
Fissura orbitalis, 65.
Fissures of spinal cord, 130.
Fœtal tooth papilla, 176.
Follicles, Lieberkühn's, 208.
——, Lymph, 92, 94, 100.
——, Sebaceous, 286.
——, Thymus, 96.
Fornix conjunctivæ, 293.

Fornix vaginal, 269.
Fossa glenoidalis, 48.
—— navicularis, 254.
—— patillaris, 307.
—— Sylvii, 159.
Fovea centralis, 310, 314, **315**
—— hemielliptica, 320.
—— hemispherica, 320.
Fundus, 266.
Funiculus of Rolando, 144.

Gall-bladder, 65.
Ganglia, Cerebro-spinal, 138, 141.
——, Sympathetic, 168.
Ganglion cells, 138, 164.
Gasserian ganglion, 163.
Gelatinous tissue, 45.
Geniculate ganglion, 163.
Genital corpuscles of Krause, 120.
—— end-corpuscles, 115.
—— organs (male), 244.
—— —— (female), 257.
Germ reticulum of von Ebner, 247.
Germinal endothelial cells, 28.
—— epithelium, 258.
—— spots, 1, 259.
—— vesicle, 1, 259.
Germinating cells, 29.
Giraldé's organ, 251.
Gland, Prostate, 253.
Glands, Bartholini, 270.
——, Brunner, 208, 209.
——, **Buccal**, 187.
——, **Carotic**, 342.
——, Ceruminous, 278, **318**.
——, Coccygeal, 342.
——, Harder, 295.
——, Krause, 293.
——, Lachrymal, 294.
——, Lieberkühn's, **17**.
——, Littré, 254.
——, Lymphatic, 80, 84, **100**.
——, Meibomian, 292.
——, Mohl, 292.
——, Mucous, 179, **191**.
——, Peptic, 197.
——, Peyer, 95, 205.
——, Pyloric, 199.
——, Salivary, 178.
——, Sebaceous, 287.
——, Solitary, 95.
——, Submaxillary, **179**.
——, Sweat, 276.
Glandulæ agminatæ, 200.
—— lenticulares, 200.
—— Pacchioni, 149.
—— uterinæ, 266.
Glans clitoridis, 270.
—— penis, 255.

Glassy membrane, 231.
Glisson's capsule, 210.
Globulin, 12.
Glomerule, 159.
Glutin, 31, 33.
Glycogen, 14.
Goblet cells, 18, 22.
Graafian follicles, 258, 263.
Grandry's corpuscles, 120.
Granular formation of Meynert, 159.
—— layer of Purkinje, 174.
Granules in blood, 15.
Granulosa membrana, 259.
Grey commissure, 128.
Ground lamellæ, 53.
—— plexus of Arnold, 121.
—— substance, 117.
Growing capillaries, 82.

Habenula perforata, 331.
Hæmatin, 13.
Hæmatoidin, 13.
Hæmatoplasts, 15.
Hæmin crystals, 13.
Hæmoglobin, 12.
—— crystals, 13.
Hair, 278.
—— bulb, 282, 285.
——, Development of, 285.
—— fibres, 282.
—— follicles, 278, 280.
—— knob, 285.
——, Marrow of, 282.
——, New formation of, 284.
—— papilla, 285.
——, Root of, 282.
—— —— -sheath of, 281.
—— sac, 280.
——, Shaft of, 279, 283.
Harder's gland, 295.
Haversian canals, 53.
—— lamellæ, 53.
—— spaces, 54.
Heart and blood-vessels, 74.
Helicotrema, 320.
Hemisphere of brain, 152.
Henle, Fenestrated membranes of, 44, 77.
——, Fibres of, 44.
——, Sheath of, 105.
——, Stratum nerveum of, 317.
——, Tubes of, 237.
Hensen's cells, 331.
Hepatic cells, 212.
—— duct, 65.
—— lobules, 211.
—— veins, 211.
Herbst, Corpuscles of, 115, 118.

Hilum of glands, 98.
—— of salivary glands, 180.
—— of spleen, 226.
Hippocampus, 159.
Homogeneous elastic membranes, 44.
Howship's lacunæ, 62.
Hyaline cartilage, 46.
Hyaloid membrane, 307.
Hypophysis, 161, 338.

Ileum, 205.
Incremental lines of Salter, 174.
Incus, 320.
Indirect division, 7.
Infundibula, 65, 161, 163.
—— of bronchiole, 220.
—— of gland, 188.
Inner molecular layer, 311.
—— nuclear layer, 311.
Interarticular cartilages, 48.
Interfascicular spaces, 34, 36.
Interglobular spaces of Czermak, 174.
Interlobar ducts, 180.
Interlobular bile-ducts, 213, 214.
—— connective tissue of liver, 210.
—— ducts, 180.
Intermediary cartilage, 48.
—— plexus, 121.
—— zone, 200.
Intermembranous formation of bone, 55, 61.
Intermuscular fibrils, 122.
Internal capsule of brain, 163.
Interstitial lamellæ, 53.
Intervertebral discs, 48.
Intestine, Large, 201.
——, Small, 201.
Intima of arteries, 76, 77.
Intralobular bile-vessels, 213.
—— ducts, 180.
Intranuclear network,
Iris, 300.
——, Blood-vessels of, 302.
——, Lymph-clefts of, 302.
——, Lymph-sinuses of, 302.
——, Nerve-fibres of, 302.
Island of Reil, 162.

Jacobson's organ, 337, 338.

Karyokinesis, 7, 43, 247.
Kidney, 229.
——, Afferent arterioles of, 240.
—— blood-vessels, 240.
—— glomerulus, 234.

Kidney lymphatics, 214, 215, 243.
—— parenchyma, 231.
—— vessels, 241.
Kölliker's osteoclasts, 62.

Labia pudendi majora, 287.
Labium tympanicum, 327.
—— vestibulare, 327.
Labyrinth, Osseous, 320.
Lachrymal glands, 294.
Lacunæ Morgagni, 254.
—— of bone, 52, 53.
—— of cartilage, 46.
—— of lymphatics, 86.
Lamellæ of bone, 54.
—— of cornea, 296.
—— of lens, 306.
Lamina cribosa, 317.
—— elastica of cornea, 295.
—— fusca, 299.
—— reticularis, 332.
—— spiralis ossea, 320.
—— vitrea, 301, 305.
Langerhans' granular layer, 20.
Larynx, 215.
Lateral basilar process, 158.
—— nucleus, 145.
—— tract, 143.
Lens, 305.
—— fibres, 306.
—— stars, 306.
Lenticular glands, 95.
Ligamentum denticulatum, 128.
—— latum, 65.
—— pectinatum, 299.
—— pulmonis, 221.
—— spirale accessorium, 326, 328.
—— suspensory of lens, 307.
Limitans externa, 312.
—— interna, 310.
Lines of Schreger, 174.
Liquor folliculi, 260.
—— sanguinis, 10.
Littré's glands, 254.
Liver, 88, 210.
—— vessels of, 211, 212.
Lobes of pancreas, 209.
—— of salivary gland, 180.
—— of thymus gland, 96.
—— of lung, 220.
Lobules of liver, 211.
—— of lung, 220.
—— of salivary glands, 180.
—— of thymus gland, 96.
Lung, 88, 219.
—— blood-vessels, 223.
—— lymphatics, 223.
Lymph, 90.
Lymphatic capillaries, 86.

Lymphatic clefts, 79, 86.
—— glands, 80, 84, 100.
—— rootlets, 86.
—— sinuses, 88, 101.
—— tissue, 85.
—— vessels, 79, 84, 92.
Lymphatics, 65, 201.
—— in mucosa, 190.
Lymph-canal system in cornea, 36.
—— -canalicular system, 86, 90, 297.
—— cavities, 88.
—— corpuscles, 90, 93.
—— follicles, 92, 94.
—— hearts, 90.
Lymphoid cells, 29.

Macula acustica, 323.
—— lutea, 310, 311, 314, 315.
Malleus, 320.
Malpighian corpuscles of kidney, 233.
—— —— of spleen, 227.
—— pyramids of kidney, 231.
—— stratum of skin, 19.
Mammary gland, 270.
Manubrium mallei, 318.
Marrow of bone, 50.
Matrix of osseous substance, 52.
Meatus auditorius externus, 318.
Meckel's ganglion, 163.
Media of arteries, 76.
Median lateral fissure, 132.
Mediastinum **testis, 214.**
Medulla oblongata, **142.**
—— of gland, 99.
Medullary cylinders, **100.**
—— lymph-sinus, 101.
—— ray, 231.
—— sheath of nerve-fibres, 107, 108.
Medullated nerve-fibres, 106, 114, 137.
Meibomian glands, 292.
Meissner's corpuscles, 191, 291.
—— plexus, 170, 201, 207.
Membrana basilaris, 327, 328.
—— chorio-capillaris, 304.
—— Descemeti, 297.
—— granulosa, 259.
—— hyaloidea, **307.**
—— **secundaria, 322.**
—— supra-choroidea, **303.**
—— tectoria, 332.
—— tympani, 318.
Membranes of Krause, 67, 74.
Mesencephalon, 159.
Mesentery, 88.
Mesogastrium, 30.
Migratory cells, 41.

Milk, 273.
—— globules, 272.
—— tooth, 177.
Modiolus, 320, 326.
Motor ganglion cells, 141.
Movement of cilia, 22.
Mucin, 23.
Mucosa, 189.
——, Lymph follicles of, 200.
Mucous cells, 183.
—— glands, 179, 191.
—— membrane, 189.
Muco-salivary glands, 179.
Mucus, formation of, 23.
Müller's fibres, 310.
—— muscle, 300.
Muscle bundles, 64.
—— cells, 64.
—— corpuscles, **70.**
—— fibres, 63, 72.
—— tissue, striped, 66, 189.
—— ——, Non-striped, 63, 65.
Muscular compartments, 67.
Muscularis externa, 196.
—— mucosæ, 64, 195.
Musculus ciliaris **Biolani, 292.**
Myeloplax, 6.
Myeloplaxes of Robin, 51, 62.

Nail, 288.
—— cells, 289.
—— groove, 288.
—— substance, 289.
Nasal mucous membrane, 333.
—— septum, 46.
Nerve bundles, 104.
—— corpuscles, 108.
—— end plate, 124.
—— endings, 112, **115.**
—— **fibres,** 103.
—— **plexus,** 106, **111,** 114.
Network of fibrillæ, 111.
Neurilemma, 108.
Neuroglia, **133.**
—— cells, **133.**
—— fibrils, 133, **150.**
—— of Virchow, 45.
—— tissue, 130.
Neurokeratin, 107, 109, 135, 313.
Nipple, 65.
Non-medullated nerve-fibres, 111, 114.
Norris's blood corpuscles, 13.
Nuclear layer in bulbus olfactorius, 159.
—— zone, 306.
Nuclein, 6.
Nucleoli, 6.
Nucleus, Structure of, **6.**

Nucleus caudatus, 162.
—— cuneatus, 147.
—— dentatus, 147.
—— gracilis, 147.
—— lenticularis, 162.
Nuclei, Inner, of retina, 311.
——, Outer, of retina, 312.
Nymphæ, 270.

Odontoblasts, 173, 174, 177.
Œsophagus, 194.
Oil-globule, 39, 40.
Olfactory cells, 336.
—— nerves, 111, 159.
Olivary bodies, 143.
—— nucleus, 147.
Omentum of cat, 29.
—— of frog, 89.
—— of guinea-pig, 40.
—— of rat, 32.
——, Structure of, 33.
Optic nerve, 317.
—— nerve-fibres, 310.
—— tract, 162.
—— vesicle, 316.
Ora serrata, 310.
Organ of Corti, 328, 329, 332.
—— of Giraldé, 251.
—— of Jacobson, 337, 338.
Ossein, 52.
Osseous labyrinth of ear, 321.
—— lamellæ, 52.
—— substance from osteoblasts, 61.
Ossicula auditus, 319.
Ossifying cartilage, 48.
Osteoblasts forming bone, 50, 57, 61.
Osteoclasts, 62.
Osteogenetic layer, **49.**
Otoliths, 324.
Oval nucleus, 39.
Ovary, 65, 257.
——, Development of, 263.
——, Lymphatics of, 268.
——, Nerves of, 268.
Oviduct, 65, 265.
Ovum, 1, 259.

Pacinian corpuscles, 115, 121, 291.
Palate, 187.
Palmæ plicatæ, 266.
Palpebræ, 293.
Pancreas, 208, 209.
Papilla circumvallata, **192.**
—— filiformis, 191.
—— foliata, 193.
—— fungiformis, 191.
—— nervi optici, 309.

Papillary hair of Unna, 280.
—— muscle, 75.
Paraglobulin, 12.
Parenchyma of testis, 253.
Parenchymatous cartilage, 48.
Parietal cells, 199.
Pars ciliaris retinæ, 303, 305, 310.
—— membranacea, 254.
—— prostatica, 254.
Pedunculated hydatid of Morgagni, 251.
Pedunculus cerebelli, 144, 152.
Penis, 255.
—— corpora cavernosa, 255.
—— nerve endings, 120.
Peptic glands, 197.
Peribronchial lymphatics, 220, 224.
Pericardial cavities, 88.
Pericellular space, 141.
Perichondrium, 45, 46.
Perilymph, 321.
Perimysium, 66.
Perineurium, 104, 121.
Periosteal bone, 59, 61.
—— formation, 55.
—— processes of Virchow, 55.
Periosteum, 50.
Peripheral nerve-endings, 112, 115.
Peritoneal cavities, 88.
Peritoneum, 65.
Perivascular lymphatics, 89, 224.
—— lymph-spaces, 151.
—— spaces of His, 141.
Pes, 161.
Peyer's glands, 95, 205.
—— patch, 95.
Pharynx, 190.
—— tonsil, 94, 190.
Pia mater, 80, 127.
Pial sheath, 317.
Pigment cells, 23, 37.
Plasma, 10.
—— cells, 41, 100.
Pleura pulmonalis, 65, 220.
Pleural cavities, 88.
Plexus choroideus, 150.
—— myentericus, 173, 201, 217.
—— of Meissner, 201, 207.
—— venosus vaginal's, 269.
Plicæ villosæ, 197, 204.
Pons Varolii, 142, 152, 154, 159.
Porta hepatis, 210.
Portio Mülleri, 300.
—— vaginalis uteri, 266.
Posterior nerve roots from spinal cord, 136.
Prickle cells, 23.
Primitive dental groove, 175.
—— fibrillæ, 107, 111, 138.

Primitive fibrils, 69, 113.
—— ora, 263.
Prostate, 253.
Protoplasm, 1.
——, Structure of, 6.
Protoplasmic membrane, 39.
Proximal convoluted tubule, 235.
Pulp tissue, 227.
Pulvinar, 162.
Purkinje's ganglion cells, 152, 154.
Pyloric glands, 196, 199.
Pyramid of Ferrein, 233.
Pyramidal decussation, 142.
—— tracts, 142.

Rami capsulares, 214.
Ranvier's constrictions, 108.
—— nodes, 108.
Raphe, 147, 156.
Red blood corpuscles, 10, 12.
Reissner's membrane, 326, 328.
Remak's fibrous layer, 77.
—— nerve-fibre, 111.
Rete Malpighii, 19.
—— mucosum, 19.
—— testis, 250.
Reticular cartilage, 49.
—— formation, 156.
Retina, 308.
——, Blind spot of, 309.
——, Blood-vessels of, 317.
——, Ganglion cells of, 311.
——, Lymphatics of, 317.
Rhodopsin of Kuhne, 314.
Rods and cones, 313.
Rollet's secondary substance, 68.
Rosette stage in nucleus, 8.
Rugæ, 268.

Saccules, 321, 322.
Saccus endolymphaticus, 321.
Saliva, 189.
Salivary cells, 182.
—— glands, 178.
—— ——, Blood-vessels of, 186.
—— ——, Ducts of, 180.
—— ——, Lobes of, 180.
—— ——, Lobules of, 180.
—— ——, Lymphatics of, 186.
—— ——, Nerves of, 186.
Sarcode of Dujardin, 5.
Sarcolemma, 67, 68, 72.
Scala tympani, 320.
—— vestibuli, 320.
Schultze's protoplasm, 5.
Schwann's cells, 5.
Sclerotic, 298.
Scrotum, 64.
Sebaceous follicles, 236.

INDEX.

Semicircular canals, 320, 322.
Semilunar valves, 75, 86.
Seminal cells, 247.
—— tubules, 246, 252.
Sensory ganglion cells. 141.
Septum cisternæ lymphaticæ, 30.
Serous glands, 178, 188.
—— membranes, 89.
Sesamoid cartilages, 48.
Sharpey's perforating fibres, 54.
Sheath of Henle, 116, 119.
—— of Schwann, 108.
Simple axis cylinders, 111, 114.
—— lymphatic glands, 92.
Skin, 274.
—— blood-vessels of, 289.
—— lymphatics of, 290.
—— nerves of, 290.
Solitary glands, 95.
—— lymph follicles, 205.
Spaces of Fontana, 299.
Spermatoblasts, 247.
Spermatozoa, 249.
Sphincter papillæ, 301.
Sphincters, 65.
Spinal cord, 127.
—— —— grey matter, 128, 136.
—— —— white matter, 128, 130.
Spiral tubule, 237.
Spleen, 225.
——, Capsule of, 225.
——, Lymphatics of, 229.
——, Nerve-fibres of, 229.
——, Parenchyma of, 226.
——, Pulp of, 226.
——, Red blood corpuscles of, 16.
——, Trabeculæ of, 226.
Spongy bone substance, 54.
Squamous epithelial cells, 18.
Sternal cartilage, 47.
Stigmata, 82.
Stomach, 196.
Stomata, 29, 82, 89, 224.
Stratified columnar epithelium, 21.
—— pavement epithelium, 20.
Stratum adiposum, 276.
—— cinereum, 160.
—— corneum, 19, 274.
—— gelatinosum, 159.
—— glomerulosum, 159.
—— lucidum, 19.
—— Malpighii, 274.
—— opticum, 160.
Stria vascularis, 328.
Stroma, 12.
Subarachnoidal spaces, 88, 127, 150, 318.
—— tissue, 128.

Subcutaneous lymphatics, 88.
Subcutaneous tissue, 275.
Subdural spaces, 88, 127, 150.
Subendocardial tissue, 74, 75.
Subepithelial endothelium of Debove, 35.
Subhyaloid cells, 307.
Submaxillary ganglion, 163.
Submucosa, 189.
Submucous lymphatics, 88.
Subpericardial nerve branches, 76.
—— tissue, 74.
Substantia gelatinosa, 134.
—— nigra, 161.
Subvaginal space, 318.
Sudoriferous canal, 276, 277.
Sulcus hippocampi, 159.
—— spiralis, 327, 328.
Superior pedunculus cerebelli, 162.
Suprachoroidal tissue, 300.
Suprarenal bodies, 340.
Supravaginal space, 318.
Sweat glands, 65, 276, 290.
Sympathetic system, 166.
Synovial cavities, 88.

Tactile corpuscles, 115, 118.
—— hairs, 286.
Tapetum nigrum, 301, 303, 305, 315.
Tarsal plate, 292, 293.
Taste buds, 193.
—— cells, 193.
—— goblets, 193.
Teeth, 171.
—— cement, 174.
—— development, 175.
—— pulp, 174.
Tegmental cells, 193.
Tegmentum, 160, 161.
Teichmann's crystals, 13.
Tendon cells, 33.
Tendons, 87.
Tenonian capsule, 318.
—— space, 318.
Tensor choroideæ, 300.
Terminal bronchi, 221.
Testis, 244.
Thalamencephalon, 161.
Thalamus opticus, 157, 161, 162.
Thoracic duct, 84.
Thymus follicles, 96.
—— gland, 96.
Thyroid cartilage, 46.
—— gland, 339, 340.
Tongue, 190.
—— serous glands of, 191.

Tonsils, 95.
Touch-cells of Merkel, 115, 119.
—— -corpuscles of Merkel, 120.
Trabeculæ carneæ, 75.
—— of lymphatics, 65.
—— of spleen, 65.
Trachea, 46, 65, 217.
Tractus olfactorius, 159.
—— opticus, 162.
Transitional epithelium, 21.
Tuba Eustachii, 319.
Tuber cinereum, 161, 163.
Tubercle of Rolando, 145.
Tubes of epididymis, 252.
Tunica adnata, 244.
—— albuginea, 244.
—— dartos, 65.
—— fibrosa, 261.
—— propria, 328.
—— vaginalis, 244.
Tyson's glands, 255.

Ureter, 230, 243.
Urethra, Female, 269.
——, Male, 254.
Urinary tubules, 231.
Uterus, 64, 265.
Utricle, 321.
Uvea, 301.

Vagina, 65, 268.
Varicose nerve-fibres, 110.
Vas deferens, 65, 252.
—— rectum, 250.
Vasa efferentia, 250.
Vascularisation of cartilage, 55.
Vater's corpuscles, 115.
Veins, 65.
——, Intima of, 79.
——, Media of, 79.
—— of the bones, 78.

Veins of the brain, cord, gravid uterus, membranes, and retina, 80.
——, Valves of, 79.
Vena axillaris, azygos, cava, cruralis, hepatica, intima, iliaca, mesenterica, poplitea, renalis, spermatica, and umbilicalis, 80.
—— jugularis, and subclavia, 79.
Venæ rectæ, 242.
—— stellatæ, 242.
—— vorticosæ, 305.
Venous radicles, 227.
—— sinuses, 227.
Ventricle, Fourth, 148.
Ventricles, 76.
Vesiculæ seminales, 65, 252.
Vestibulum, 270.
Virchow's crystals, 13.
Visceral pericardium, 74.
—— peritoneum, 194.
Vitreous body, 307.

White blood corpuscles, 13, 16.
—— commissure, 135.
—— fibrous tissue, 30, 31.
—— substance of brain, 151.
—— —— of cord, 110.
—— —— of Schwann, 107, 108.
Wolffian body, 258.
Wreath arrangement of nucleus, 8.

Yellow elastic cartilage, 49.
—— —— tissue, 43.

Zona fasciculata, 341.
—— glomerulosa, 340.
—— pellucida, 259.
—— reticularis, 341.
—— vasculosa, 257.
Zonula ciliaris, 307.

MANUALS
FOR
Students of Medicine.

THIS Series has been projected to meet the demand of Medical Students and Practitioners for compact and authoritative Manuals which shall embody the most recent discoveries and present them to the reader in a cheaper and more portable form than has till now been customary in Medical Works.

Each Manual will contain all the information required for the Medical Examinations of the various Colleges, Halls, and Universities in the United Kingdom and the Colonies.

The Authors will be found to be either Examiners or the leading Teachers in well-known Medical Schools. This will ensure the practical utility of the Series, while the introduction of the results of the latest scientific researches, British and Foreign, will recommend them also to Practitioners who desire to keep pace with the swift strides that are being made in Medicine and Surgery.

In the rapid advance in modern Medical knowledge, new subjects have come to the front which have not as yet been systematically handled, nor the facts connected with them properly collected. The treatment of such subjects will form an important feature of this Series.

New and valuable Illustrations will be freely introduced. The Manuals will be printed in clear type, upon good paper. They will be of a size convenient for the pocket, and bound in red cloth limp, with red edges. They will contain from 300 to 540 pages, and will be published at prices varying from 4s. 6d. to 7s. 6d.

(*For List of Manuals see over.*)

Manuals for Students of Medicine.

I.—**Elements of Histology.** By E. KLEIN, M.D., F.R.S., Joint-Lecturer on General Anatomy and Physiology in the Medical School of St. Bartholomew's Hospital, London. *Second Edition.* 6s.

II.—**Surgical Pathology.** By A. J. PEPPER, M.B., M.S., F.R.C.S., Surgeon and Teacher of Practical Surgery at St. Mary's Hospital. 7s. 6d.

III.—**Surgical Applied Anatomy.** By FREDERICK TREVES, F.R.C.S., Senior Demonstrator of Anatomy and Assistant Surgeon at the London Hospital. 7s. 6d.

IV.—**Clinical Chemistry.** By CHARLES H. RALFE, M.D., F.R.C.P., Assistant Physician at the London Hospital. 5s.

V.—**The Dissector's Manual.** By W. BRUCE CLARKE, M.B., F.R.C.S., and C. B. LOCKWOOD, F.R.C.S., Demonstrators of Anatomy St. Bartholomew's Hospital Medical School. Price 6s.

VI.—**Human Physiology.** By HENRY POWER, M.B., F.R.C.S., Examiner in Physiology, Royal College of Surgeons of England. About 7s. 6d.

VII.—**Physical Physiology.** By J. McGREGOR-ROBERTSON, M.A., M.B., Muirhead Demonstrator of Physiology, University of Glasgow.

VIII.—**Materia Medica and Therapeutics: An Introduction to Rational Treatment.** By J. MITCHELL BRUCE, M.D., F.R.C.P., Lecturer on Materia Medica at Charing Cross Medical School and Physician to the Hospital.

IX.—**Comparative Physiology and Anatomy.** By F. JEFFREY BELL, M.A., Professor of Comparative Anatomy at King's College.

X.—**Operative Surgery.** By EDWARD BELLAMY, F.R.C.S., Surgeon and Lecturer on Anatomy at Charing Cross Hospital; Examiner in Anatomy, Royal College of Surgeons.

Other volumes will follow in due course.

Cassell & Company, Limited, London; and all Booksellers.

Authoritative Work on Health by **Eminent Physicians and Surgeons.**

The Book of Health.

A Systematic Treatise for the Professional and General Reader upon the Science and the Preservation of Health . **21s.**

CHAP.   CONTENTS.

1. Introductory.
   By W. S. SAVORY, F.R.S.
2. Food and its Use in Health.
   By SIR RISDON BENNETT, M.D., F.R.S.
3. The Influence of Stimulants and Narcotics on Health.
   By T. LAUDER BRUNTON, M.D., F.R.S.
4. Education and the Nervous System.
   By J. CRICHTON-BROWNE, LL.D., M.D.
5. The Influence of Exercise on Health.
   By JAMES CANTLIE, F.R.C.S.
6. The Influence of Dress on Health.
   By FREDERICK TREVES, F.R.C.S.
7. The Influence of our Surroundings on Health.
   By J. E. POLLOCK, M.D.
8. The Influence of Travelling on Health.
   By J. RUSSELL REYNOLDS, M.D., F.R.S.
9. Health at Home.
   By SHIRLEY MURPHY, M.R.C.S.
10. Health in Infancy and Childhood.
    By W. B. CHEADLE, M.D.
11. Health at School.
    By CLEMENT DUKES, M.D.
12. The Eye and Sight.
    By HENRY POWER, F.R.C.S.
13. The Ear and Hearing.
    By G. P. FIELD, M.R.C.S.
14. The Throat and Voice.
    By J. S. BRISTOWE, M.D., F.R.S.
15. The Teeth.
    By CHARLES S. TOMES, F.R.S.
16. The Skin and Hair.
    By MALCOLM MORRIS.
17. Health in India.
    By SIR JOSEPH FAYRER, K.C.S.I., F.R.S., and J. EWART, M.D.
18. Climate and Health Resorts.
    By HERMANN WEBER, M.D.

Edited by MALCOLM MORRIS.

Cassell & Company, Limited, London; and all Booksellers.

Important Work on Sanitation.

Our Homes, and How to Make them Healthy.

With numerous Practical Illustrations. Edited by SHIRLEY FORSTER MURPHY, *Medical Officer of Health to the Parish of St. Pancras; Hon. Secretary to the Epidemiological Society, and to the Society of Medical Officers of Health.* 960 pages. Royal 8vo, cloth . . . . **15s.**

CONTENTS.

Health in the Home. By W. B. RICHARDSON, M.D., LL.D., F.R.S.
Architecture. By P. GORDON SMITH, F.R.I.B.A., and KEITH DOWNES YOUNG, A.R.I.B.A.
Internal Decoration. By ROBERT W. EDIS, F.S.A., and MALCOLM MORRIS, F.R.C.S. Ed.
Lighting. By R. BRUDENELL CARTER, F.R.C.S.
Warming and Ventilation. By DOUGLAS GALTON, C.B., D.C.L., F.R.S.
House Drainage. By WILLIAM EASSIE, C.E., F.L.S., F.G.S.
Defective Sanitary Appliances and Arrangements. By PROF. W. H. CORFIELD, M.A., M.D.
Water. By PROF. F. S. B. FRANÇOIS DE CHAUMONT, M.D., F.R.S.; ROGERS FIELD, B.A., M.I.C.E.; and J. WALLACE PEGGS, C.E.
Disposal of Refuse by Dry Methods. By THE EDITOR.
The Nursery. By WILLIAM SQUIRE, M.D., F.R.C.P.
House Cleaning. By PHILLIS BROWNE.
Sickness in the House. By THE EDITOR.
Legal Responsibilities. By THOS. ECCLESTON GIBB.
&c., &c.

Fourth and Cheap Edition. Price 1s. 6d.; cloth, 2s.

A Handbook of Nursing

For the Home and for the Hospital. By CATHERINE J. WOOD, *Lady Superintendent of the Hospital for Sick Children, Great Ormond Street.*

"A book which every mother of a family ought to have, as well as every nurse under training."—*Guardian.*
"The best book of its kind."—*Nonconformist.*

CASSELL & COMPANY'S COMPLETE CATALOGUE, *containing particulars of several Hundred* Volumes, *including* Bibles *and* Religious Works, *Illustrated and* Fine-Art *Volumes, Children's Books, Dictionaries, Educational Works, History, Natural History, Household and Domestic Treatises, Science, Travels, &c., together with a Synopsis* of their *numerous Illustrated Serial Publications, sent post free on application.*

Cassell & Company, Limited, London; and all Booksellers.

CLINICAL MANUALS
For Practitioners and Students of Medicine.

Complete Monographs on Special Subjects.

THE object of this series is to present to the Practitioner and Student of Medicine original, concise, and complete monographs on all the principal subjects of Medicine and Surgery, both general and special.

It is hoped that the series will enable the Practitioner to keep abreast of the rapid advances at present made in medical knowledge, and that it will supplement for the Student the comparatively scanty information contained in the general text-books.

The Series will form a complete Encyclopædia of Medical and Surgical Science in separate volumes.

The Manuals will be written by leading Hospital Physicians and Surgeons, whose work on each special subject may be considered to be authoritative.

[P.T.O.

The following are among the Subjects that will be dealt with in the earlier Volumes:—

THE PULSE — HEART DISEASES — INSANITY — SYPHILIS — LUNG DISEASES — KIDNEY DISEASES — GOUT AND RHEUMATISM — AFFECTIONS OF THE TONGUE AND TESTICLE — INTESTINAL OBSTRUCTION — DISEASES OF JOINTS — FRACTURES AND DISLOCATIONS, SURGICAL AND MEDICAL DISEASES OF CHILDREN — CONTINUED FEVERS — AFFECTIONS OF THE UTERUS AND OVARIES — MIDWIFERY — SURGICAL DISEASES OF THE KIDNEY — ORTHOPÆDIC SURGERY — TREATMENT OF WOUNDS — DISEASES OF THE EYE, EAR, SKIN, AND THROAT.

Each Manual will be printed in clear type upon good paper. They will be of a size convenient for the pocket, substantially bound in blue cloth limp, with blue edges. They will contain from 270 to 540 pages, and will be freely Illustrated by Original Chromo-Lithographs and Woodcuts, when required, and will be published at prices varying from 5s. to 8s. 6d.

IN ACTIVE PREPARATION.

I.—*SYPHILIS.* By JONATHAN HUTCHINSON, F.R.S., Consulting Surgeon to the London Hospital, &c.

II.—*INSANITY, including HYSTERIA.* By G. H. SAVAGE, M.D., Medical Superintendent, Bethlem Royal Hospital, and Lecturer on Mental Diseases at Guy's Hospital.

III.—*THE PULSE.* By W. H. BROADBENT, M.D., F.R.C.P., Physician to St. Mary's Hospital.

Other Volumes will be announced in due course.

CASSELL & COMPANY, Limited, Ludgate Hill, London.

www.ingramcontent.com/pod-product-compliance
Lightning Source LLC
Chambersburg PA
CBHW030744250426
43672CB00028B/389